AF876980

Advances in Seed Priming

Amitava Rakshit • Harikesh Bahadur Singh
Editors

Advances in Seed Priming

Springer

Editors
Amitava Rakshit
Department of Soil Science and Agricultural Chemistry
Institute of Agricultural Sciences, BHU
Varanasi, Uttar Pradesh, India

Harikesh Bahadur Singh
Department of Mycology and Plant Pathology
Institute of Agricultural Sciences, BHU
Varanasi, Uttar Pradesh, India

ISBN 978-981-13-0031-8 ISBN 978-981-13-0032-5 (eBook)
https://doi.org/10.1007/978-981-13-0032-5

Library of Congress Control Number: 2018944402

© Springer Nature Singapore Pte Ltd. 2018
This work is subject to copyright. All rights are reserved by the Publisher, whether the whole or part of the material is concerned, specifically the rights of translation, reprinting, reuse of illustrations, recitation, broadcasting, reproduction on microfilms or in any other physical way, and transmission or information storage and retrieval, electronic adaptation, computer software, or by similar or dissimilar methodology now known or hereafter developed.
The use of general descriptive names, registered names, trademarks, service marks, etc. in this publication does not imply, even in the absence of a specific statement, that such names are exempt from the relevant protective laws and regulations and therefore free for general use.
The publisher, the authors and the editors are safe to assume that the advice and information in this book are believed to be true and accurate at the date of publication. Neither the publisher nor the authors or the editors give a warranty, express or implied, with respect to the material contained herein or for any errors or omissions that may have been made. The publisher remains neutral with regard to jurisdictional claims in published maps and institutional affiliations.

Printed on acid-free paper

This Springer imprint is published by the registered company Springer Nature Singapore Pte Ltd.
The registered company address is: 152 Beach Road, #21-01/04 Gateway East, Singapore 189721, Singapore

We give credit to God for granting us health, elegance, perception, strength, and act of kindness to bring about this task.

Preface

Seed is an important component of agriculture, contributing significantly to the booming production of food and feed crops across the different agro-ecological regions of the world with constant challenges with reference to production, storage, and quality control. In addition to these challenges, in the era of climate change, different stressors may lead to reduction in seed germination with poor and unsynchronized seedling emergence, poor establishment of crop stand, destruction of the root cell structure, and thus resulting in a significant decrease in the yield of agricultural crops. To conquer these confrontations, a number of seed technologies that augment germination and synchronization of seedling emergence under difficult environmental conditions have been developed. In order to encourage environmental friendly sustainable agriculture across different parts of the universe, priming technology, using inorganic chemicals, plant extract, and beneficial microorganisms, increases the yields of crops while reducing the environmental burden of disproportionate use of chemical fertilizers. As of now, it is a significant tool for hastening seed germination rate, guaranteeing consistent and homogeneous seedling emergence, and improving stand establishment and seedling vigor. The various priming options available are hydropriming, halopriming, osmopriming, thermopriming, solid matrix priming, and biopriming. Priming seeds with these divergent agents provide an innovative, cost-effective, and environmentally sound solution for improving seed quality and crop health and attaining better yields. Further, these collections will provide a much needed platform to discuss the emerging issues and problems in seed priming and will come out with a well-defined strategy to overcome the different stresses within the basic framework of sustainable development goals keeping in mind the challenges of food, environment, and livelihood security. This edited book is a comprehensive one and offers an authoritatively sound and lucid documentation of issues pertaining to priming, molecular biology, and agronomic incentives. The book is intended for use by the students, scientists, extension workers, and policy makers with an in-depth view.

Varanasi, Uttar Pradesh, India

Amitava Rakshit
Harikesh Bahadur Singh

Acknowledgment

An edited book of this expense does not become possible without contributions of several willing souls. Our sincere gratitude to the colleagues and students who helped us in our endeavor to bring this book to light. We are grateful to Prof. Rattan Lal, Prof. B.D. Singh, and Prof. Panjab Singh for their mammoth impact, unwavering encouragement, and support.

Finally the production team members deserve special appreciation for guiding us through the process of publishing a new work. Last but not the least, we should thank our family, immediate and extended, who always encouraged us to continue the massive task. In spite of the best efforts, it is possible that some errors may have crept into the compilation. Each of the chapter has been the primary responsibility of the invited author/group of authors: We have also read and critiqued all the chapters with extraordinary case. We shall be highly obligated to receive constructive comments and suggestions from the readers for further improvement in the future editions.

Varanasi, Uttar Pradesh, India

Amitava Rakshit
Harikesh Bahadur Singh

Contents

Part I Concepts

1 **Seed Priming: New Vistas and Contemporary Perspectives** 3
Puspendu Dutta

2 **Impact of Seed Priming on the Modulation of Physico-chemical and Molecular Processes During Germination, Growth, and Development of Crops** 23
Bandana Bose, Mahesh Kumar, Rajesh K. Singhal, and Sananda Mondal

3 **Seed Priming: An Emerging Technology to Impart Abiotic Stress Tolerance in Crop Plants** 41
Shambhu Krishan Lal, Sudhir Kumar, Vijay Sheri, Sahil Mehta, Panditi Varakumar, Babu Ram, Bhabesh Borphukan, Donald James, Dhirendra Fartyal, and Malireddy K. Reddy

4 **Recent Advances in Abiotic Stress Tolerance of Plants Through Chemical Priming: An Overview** 51
Muhammad Arslan Ashraf, Ali Akbar, Sajjad Hassan Askari, Muhammad Iqbal, Rizwan Rasheed, and Iqbal Hussain

5 **Seed Priming Technology in the Amelioration of Salinity Stress in Plants** 81
Aditya Banerjee and Aryadeep Roychoudhury

6 **Seed Priming with Plant Growth Regulators to Improve Crop Abiotic Stress Tolerance** 95
Elouaer Mohamed Aymen

7 **Addressing Stresses in Agriculture Through Bio-priming Intervention** 107
Deepranjan Sarkar, Sumita Pal, Ms. Mehjabeen, Vivek Singh, Sonam Singh, Subhadip Pul, Jancy Garg, Amitava Rakshit, and H. B. Singh

8 Role of Microbial Seed Priming and Microbial Phytohormone in Modulating Growth Promotion and Defense Responses in Plants 115
Vivek Singh, Anupam Maharshi, Dhananjaya P. Singh, Ram Sanmukh Upadhyay, Birinchi Kumar Sarma, and Harikesh Bahadur Singh

9 Potential of Biopriming in Enhancing Crop Productivity and Stress Tolerance 127
Ahmad Mahmood and Ryota Kataoka

10 Stimulating Plant Tolerance Against Abiotic Stress Through Seed Priming 147
Mona Gergis Dawood

Part II Case Studies on Priming

11 Seed Priming: A Low-Cost Technology for Resource-Poor Farmers in Improving Pulse Productivity 187
Malay K. Bhowmick

12 Studies on Seed Priming in Pepper (*Capsicum annuum* L.) 209
Nusret Ozbay

13 Effect of Different Seed Priming Treatments on Germination and Seedling Establishment of Two Threatened Endangered Medicinal Plants of Darjeeling Himalaya 241
Dhiman Mukherjee

14 Seed Priming on Germination, Growth and Flowering in Flowers and Ornamental Trees 263
Anjana Sisodia, Minakshi Padhi, A. K. Pal, Kalyan Barman, and Anil K. Singh

15 Role of SNP-Mediated Nitric Oxide Priming in Conferring Low Temperature Tolerance in Wheat Genotype (*Triticum aestivum* L.): A Case Study in Indian Northern Plains 289
Md. Afjal Ahmad, Pravin Prakash, and H. B. Singh

16 Seedling Bio-priming with *Trichoderma* spp. Enhances Nitrogen Use Efficiency in Rice 297
Preeti Priya, Kartikay Bisen, Amitava Rakshit, and H. B. Singh

Contributors

Md. Afjal Ahmad Department of Plant Physiology, Institute of Agricultrural Sciences, BHU, Varanasi, Uttar Pradesh, India

Ali Akbar Department of Botany, Government College University Faisalabad, Faisalabad, Pakistan

Muhammad Arslan Ashraf Department of Botany, Government College University Faisalabad, Faisalabad, Pakistan

Sajjad Hassan Askari Department of Botany, Government College University Faisalabad, Faisalabad, Pakistan

Elouaer Mohamed Aymen High Institute of Agronomy, Chott Mariem, Sousse, Tunisia

Aditya Banerjee Post Graduate Department of Biotechnology, St. Xavier's College (Autonomous), Kolkata, West Bengal, India

Kalyan Barman Department of Horticulture, Institute of Agricultural Sciences, Banaras Hindu University, Varanasi, Uttar Pradesh, India

Malay K. Bhowmick Rice Research Station (Government of West Bengal), Chinsurah, West Bengal, India

Kartikay Bisen Department of Mycology and Plant Pathology, Institute of Agricultural Sciences, BHU, Varanasi, Uttar Pradesh, India

Bhabesh Borphukan International Centre for Genetic Engineering and Biotechnology, New Delhi, India

Bandana Bose Department of Plant Physiology, Institute of Agricultural Sciences, Banaras Hindu University, Varanasi, India

Mona Gergis Dawood Botany Department, National Research Centre, Cairo, Egypt

Puspendu Dutta Department of Seed Science and Technology, Uttar Banga Krishi Viswavidyalaya, Pundibari, West Bengal, India

Dhirendra Fartyal International Centre for Genetic Engineering and Biotechnology, New Delhi, India

Jancy Garg Department of Soil Science and Agricultural Chemistry, Institute of Agricultural Sciences, BHU, Varanasi, India

Iqbal Hussain Department of Botany, Government College University Faisalabad, Faisalabad, Pakistan

Muhammad Iqbal Department of Botany, Government College University Faisalabad, Faisalabad, Pakistan

Donald James International Centre for Genetic Engineering and Biotechnology, New Delhi, India

Ryota Kataoka Faculty of Life and Environmental Sciences, University of Yamanashi, Yamanashi, Japan

Mahesh Kumar Department of Plant Physiology, Institute of Agricultural Sciences, Banaras Hindu University, Varanasi, India

Sudhir Kumar ICAR-Indian Institute of Agricultural Biotechnology, Ranchi, Jharkhand, India

Shambhu Krishan Lal International Centre for Genetic Engineering and Biotechnology, New Delhi, India

ICAR-Indian Institute of Agricultural Biotechnology, Ranchi, Jharkhand, India

Anupam Maharshi Department of Mycology and Plant Pathology, Institute of Agricultural Sciences, BHU, Varanasi, Uttar Pradesh, India

Ahmad Mahmood Faculty of Life and Environmental Sciences, University of Yamanashi, Yamanashi, Japan

Ms. Mehjabeen Department of Soil Science and Agricultural Chemistry, Institute of Agricultural Sciences, BHU, Varanasi, India

Sahil Mehta International Centre for Genetic Engineering and Biotechnology, New Delhi, India

Sananda Mondal Plant Physiology Section, Department of ASEPAN, Institute of Agriculture, Visva-Bharati, Sriniketan, India

Dhiman Mukherjee Department of Agronomy, Bidhan Chandra Krishi Viswavidyalaya, Kalyani, West Bengal, India

Nusret Ozbay Department of Horticulture, Agricultural Faculty, Bingol University, Bingol, Turkey

Minakshi Padhi Department of Horticulture, Institute of Agricultural Sciences, Banaras Hindu University, Varanasi, Uttar Pradesh, India

A. K. Pal Department of Horticulture, Institute of Agricultural Sciences, Banaras Hindu University, Varanasi, Uttar Pradesh, India

Sumita Pal Department of Soil Science and Agricultural Chemistry, Institute of Agricultural Sciences, BHU, Varanasi, India

Pravin Prakash Department of Plant Physiology, Institute of Agricultrural Sciences, BHU, Varanasi, Uttar Pradesh, India

Preeti Priya Department of Soil Science and Agricultural Chemistry, Institute of Agricultural Sciences, BHU, Varanasi, Uttar Pradesh, India

Subhadip Pul Department of Soil Science and Agricultural Chemistry, Institute of Agricultural Sciences, BHU, Varanasi, India

Amitava Rakshit Department of Soil Science and Agricultural Chemistry, Institute of Agricultural Sciences, BHU, Varanasi, Uttar Pradesh, India

Babu Ram International Centre for Genetic Engineering and Biotechnology, New Delhi, India

Rizwan Rasheed Department of Botany, Government College University Faisalabad, Faisalabad, Pakistan

Malireddy K. Reddy International Centre for Genetic Engineering and Biotechnology, New Delhi, India

Aryadeep Roychoudhury Post Graduate Department of Biotechnology, St. Xavier's College (Autonomous), Kolkata, West Bengal, India

Deepranjan Sarkar Department of Soil Science and Agricultural Chemistry, Institute of Agricultural Sciences, BHU, Varanasi, India

Birinchi Kumar Sarma Department of Mycology and Plant Pathology, Institute of Agricultural Sciences, Banaras Hindu University, Varanasi, Uttar Pradesh, India

Vijay Sheri International Centre for Genetic Engineering and Biotechnology, New Delhi, India

Anil K. Singh Department of Horticulture, Institute of Agricultural Sciences, Banaras Hindu University, Varanasi, Uttar Pradesh, India

Dhananjaya P. Singh ICAR-National Bureau of Agriculturally Important Microorganisms, Maunath Bhanjan, India

Harikesh Bahadur Singh Department of Mycology and Plant Pathology, Institute of Agricultural Sciences, BHU, Varanasi, Uttar Pradesh, India

Sonam Singh Department of Soil Science and Agricultural Chemistry, Institute of Agricultural Sciences, BHU, Varanasi, India

Vivek Singh ICAR-National Bureau of Agriculturally Important Microorganisms, Maunath Bhanjan, India

Department of Botany, Institute of Science, Banaras Hindu University, Varanasi, India

Rajesh K. Singhal Department of Plant Physiology, Institute of Agricultural Sciences, Banaras Hindu University, Varanasi, India

Anjana Sisodia Department of Horticulture, Institute of Agricultural Sciences, Banaras Hindu University, Varanasi, Uttar Pradesh, India

Ram Sanmukh Upadhyay Department of Botany, Institute of Science, Banaras Hindu University, Varanasi, India

Panditi Varakumar International Centre for Genetic Engineering and Biotechnology, New Delhi, India

About the Editors

Amitava Rakshit an IIT Kharagpur alumnus, is presently the faculty member in the Department of Soil Science and Agricultural Chemistry at the Institute of Agricultural Sciences, Banaras Hindu University (IAS, BHU). Dr. Rakshit had worked in the Department of Agriculture, Government of West Bengal, in administration and extension roles. He visited Scandinavia, Europe, Africa, and UAE pertaining to his research work and presentation. He was awarded with TWASNxt Fellow (Italy), BioVision.Nxt Fellow (France), Darwin Now Bursary (British Council), Young Achiever Award, and Best Teacher's Award at UG and PG level. He is serving as review college member of British Ecological Society, London, since 2011. He is the fellow of Society of Earth Scientists. He has published 100 research papers, 25 book chapters, 28 popular articles, 1 manual, and coauthored 12 books.

Harikesh Bahadur Singh is presently a Professor of Mycology and Plant Pathology at IAS, BHU. He served state agriculture university, central university, and CSIR institute in teaching, research, and extension roles. In recognition of Prof. Singh's scientific contributions and leadership in the field of plant pathology, he was honored with several prestigious awards, notable being, CSIR Prize for Biological Sciences, Vigyan Bharti Award, Prof. V.P. Bhide Memorial Award, BRSI Industrial Medal Award, Bioved Fellowship Award, Prof. Panchanan Maheshwari Award, IPS Plant Pathology Leader Award, CSIR–CAIRD Team Award, Environment Conservation Award, CST Vigyan Ratna Award, and many more. Prof. Singh has been the fellow of National Academy of Agricultural Sciences. Prof. Singh has written 2 books, several training modules and manuals, and more than 160 research publications, and has more than 18 US patents and 3 PCTs to his credit.

Part I

Concepts

1 Seed Priming: New Vistas and Contemporary Perspectives

Puspendu Dutta

Abstract

Seed germination and uniform plant stand in field are of most critical stages of crop growth that determine the final yield. Crop production is very often hampered under suboptimal conditions, and such effect is principally attributed to poor or uneven germination and unsynchronized seedling emergence. Seed priming is an age-old and simple but effective technique to enhance germination percentage and speed and to achieve uniform plant stand and better yield in a wide range of environmental conditions. However, various priming protocols differ in their effectiveness depending on a complex interaction of factors including plant species or genotypes, water potential of priming agents, duration of treatment and environmental features. Basically priming is physiological advancement of seeds, and it involves the initiation of pre-germinative metabolisms through soaking of seeds in water or solution of other conventional priming agents under controlled condition. But, adjusting of the priming protocols by accurate timing to stop the treatment followed by rapid drying is of major importance to overcome the problem of seed storability due to prolonged treatment. The use of hydrotime analysis or digital image technology has been specified to be useful for optimization of priming protocols. As an alternative to the conventional methods and under the context of contemporary issues of agricultural pollution, the use of physical methods or nanoparticles for seed priming has been evidenced to be advantageous in several aspects. Though these latest methods of priming are receiving great attention by researchers in recent times, the detailed physiological and molecular mechanisms of seed priming with physical methods and/or nanoparticles and their impact on crop plants and environment as well as on human health still remain to be fully explored.

P. Dutta (✉)
Department of Seed Science and Technology, Uttar Banga Krishi Viswavidyalaya, Pundibari, West Bengal, India

© Springer Nature Singapore Pte Ltd. 2018

A. Rakshit, H. B. Singh (eds.), *Advances in Seed Priming*,
https://doi.org/10.1007/978-981-13-0032-5_1

Keywords
Seed priming · Conventional methods · Physiological basis · Promising techniques · Future needs

1.1 Introduction

Seed germination and thereby emergence of seedling are thought to be the most critical stages in the plant life cycle that ultimately determine field production of any crop plant through optimum plant stand (Gelmond 1978). The performance of a crop largely depends on rapid and synchronous seedling emergence as slow germination often exposes crop plants to adverse environment conditions (Osburn and Schroth 1989). Thus, researchers have been engaged in developing strategies and techniques for improving seed germination, plant establishment and growth of crop species for many years. A simple way to improve seed germination and seedling establishment and consequently field performance of plants is through seed priming (McDonald 2000). Seed priming is an age-old technique for improving germination and achieving high vigour, leading to better stand establishment and yield. The Greek Theophrastus (371–287 BC) first observed quick seed germination when cucumber (*Cucumis sativus* L.) seeds were soaked in water prior to sowing (Theophrastus Enquiry in Plants Book VII, I.6).

Seed priming is a value-added technique performed on a given seed lot, and this method is useful for quick germination and uniform emergence (Taylor et al. 1998). It has been proved to improve seed germination and seedling establishment of many field and horticultural crops (Basra et al. 2003; Sadeghian and Yavari 2004). Several benefits of seed priming have been reported including rapid and synchronized germination, increased nutrient uptake, relieved phytochrome-induced photo- and thermodormancy, increased range of germination temperature, improved water use efficiency and synchronous maturity of crops (Hill et al. 2008). 'Priming' literally means the triggering of stress tolerance particularly to moderate and intermittent stress. Seed priming also has long been known as a potential way to promote crop performance by enhancing tolerance of plant to biotic and abiotic stresses (Bruce et al. 2007). It actually causes to improve germination and also seed vigour which is a complex agronomic trait controlled by multiple genetic and environmental factors (Rajjou et al. 2012; Jisha et al. 2013).

In spite of delivering many benefits and having successful commercial use of seed priming, its popular application is often hindered due to available contradictory reports on deleterious effects of priming on seed longevity during storage (Tarquis and Bradford 1992). Difficulties also arise in standardizing priming protocol due to occurrence of great variations in effects among species, cultivars and even seed lots. Thus, the detailed understanding of physiological basis of priming effect can ease the optimization of this technique. The conventional priming protocols are relatively time-consuming with high labour costs and sometimes may have adverse impact on environment. So it necessitates the development of fast, effective and more environment-friendly priming protocols. With this background, this chapter describes the current understandings on physiological basis of seed priming and recent advances in priming methods as a tool to combat with present-day challenges.

1.2 Physiological Basis of Seed Priming

Basically priming is a water-based technique that allows controlled seed hydration to trigger 'pre-germinative metabolism' but does not allow the seed for transition towards full germination. Though the interest of seed priming has been demonstrated since long, a comprehensive physiological basis of this fascinating technique remains poorly understood until recently. But a better understanding of pre-germinative metabolism during priming treatment and the subsequent germination will help to use this simple and cheap but unique technique in a more efficient way.

When a dry seed is kept in water, it passes through three distinct phases before it germinates (Bewley 1997). Phase I is imbibition when there is rapid initial water uptake due to low water potential of seed than outside. During this phase, initially, there is water movement in apoplastic spaces, proteins are synthesized from existing mRNAs and DNA and mitochondria are repaired. In phase II, there is activation of metabolic activities and repairing processes along with synthesis of proteins by translation of new mRNAs and synthesis of new mitochondria, whereas phase III is associated with regaining capacity of rapid water uptake and initiation of growing processes linked with cell elongation that leads to radicle protrusion. Priming allows a seed to hydrate up to a seed moisture content involving the entire phase I and before the end of phase II when the germination still remains a reversible process and just short of radicle protrusion (Bray 1995). Thus, priming actually activates 'pre-germinative metabolism' that includes a broad range of physiological functioning. This activates DNA repair pathways and ROS scavenging systems (that impart for seed repair response) and also helps in preserving genome integrity (Paparella et al. 2015). Taylor et al. (1998) has reported that the advanced germination status of primed seeds actually contributes for increased germination under stressful conditions. Besides it also facilitates the initiation of many germination-related activities such as enhanced energy metabolism, early reserve mobilization, embryo expansion and endosperm weakening (Pandita et al. 2007). Priming also enhances the specific stress-responsive systems which include induced accumulation of LEA and heat shock proteins (Catusse et al. 2011), activation of catalase and other antioxidant scavenging enzymes (Chen 2011) and upregulation of genes encoding peroxiredoxin (Li et al. 2005). Figure 1.1 describes a generalized physiology of priming effects in seed.

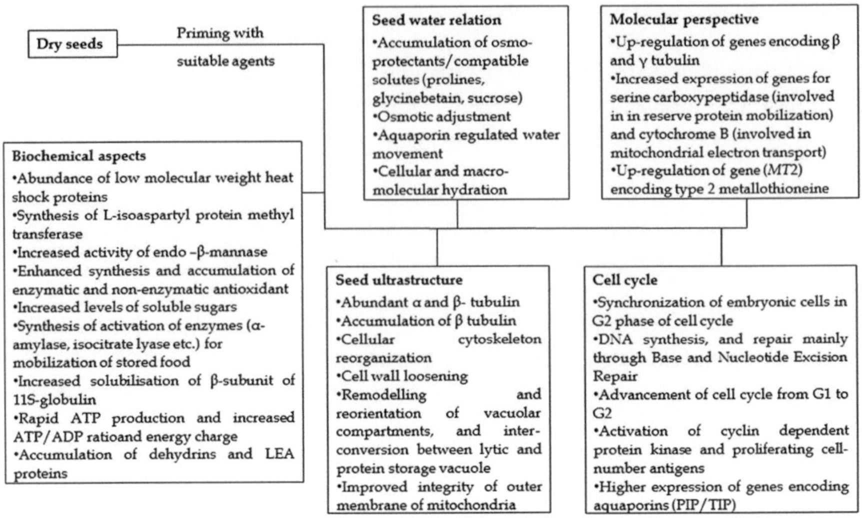

Fig. 1.1 Generalized physiology of priming effects in seed (Adapted from Benamar et al. 2003; Chen et al. 2012; Capron et al. 2000; Zhang et al. 2015; Corbineau et al. 2000; Varier et al. 2010; Paparella et al. 2015)

1.3 Conventional Methods of Seed Priming

Seed priming techniques commonly include several categories based on the choice of priming substances, i.e. hydro-priming, osmo-priming, chemo-priming, hormo-priming, solid matrix-priming, bio-priming, nutrio-priming and thermo-priming (Farooq et al. 2007; Chen 2011). Nonetheless, the efficacy of seed priming technique largely depends on plant species/genotypes, seed morphological and physiological parameters as well as the technique of seed priming. Priming is routinely used to treat many horticultural crops mainly vegetables and flowers, but large-scale use of priming cereals or other field crops is limited (Murungu et al. 2004). Same priming protocol may often have varied effect among species, cultivars and even seed lots (Pill et al. 1994). A summary on effects of different priming agents in crops is given in Table 1.1.

1.4 Latest Approaches of Seed Priming

The use of eco-friendly and cheap methods for enhancing production to cope with the needs of a growing population is a common interest of agricultural scientists. Several proprietary priming agents like (a) EasyPrime/EasyDormex (targeted crops are brinjal, tomato, *Brassica*, melon and lettuce), (b) Advantage®/Xbut®/Emergis® (targeted crops are sugar beet, vegetables and flowers) and (c) Thermocare™/Promoter™ (targeted crops are lettuce, onion and carrot) are available for commercial uses in many crops (Paparella et al. 2015). But researchers are continuously

Table 1.1 Effect of different conventional priming treatments on seed germination and crop performance

Method	Agent	Description	Crop	References
Hydro	Water (hydration/ hydration-dehydration)	Improved growth of nursery seedling, and subsequently crop growth increased panicle number m^{-2}, filled grain $panicle^{-1}$, 1000 grain weight and yield	Rice	Farooq et al. (2007) and Mahajan et al. (2011)
		Improved field stand and plant growth both at vegetative maturity stage and yield	Maize	Nagar et al. (1998) and Murungu et al. (2004)
		Improved endosperm weakening, germination and seedling development	Brinjal	Anese et al. (2011)
		Improved germination under salinity, increased tillers and grain yield	Wheat	Meena et al. (2013)
Osmo	Polyethylene glycol (PEG)	Improved seedling establishment, improved germination, growth and yield	Wheat	Salehzade et al. (2009)
		Improved germination and lowered mean germination time	Rice	Ruan et al. (2002)
		Increased germination, radicle and shoot length	Marigold	Ganji Arjenaki et al. (2011)
		Improved germination under temperature and water stress	Spinach	Chen et al. (2010)
		Increased germination and strengthened antioxidant systems	Sorghum	Zhang et al. (2015)
	Glycine-betaine	Increased seedling emergence and reduced mean emergence time	Rice	Chen et al. (2005)
	Chitosan (large polysaccharides)	Increased germination, lipase activity, GA and IAA levels	Peanut	Zhou et al. (2002)
		Increased resistance to diseases and improved seed quality	Wheat	Reddy et al. (1999)
		Improved germination and seedling growth under low temperature	Maize	Guan et al. (2009)
	Polyamine	Synchronized and enhanced germination and improved seedling length, fresh and dry weight and leaf score	Rice	Farooq et al. (2009)

(continued)

Table 1.1 (continued)

Method	Agent	Description	Crop	References
Halo	KNO_3	Improved germination, early seedling growth, increased plant height, leaf area and yield attributing parameters	Rice	Srivastava and Bose (2012) and Basra et al. (2005)
		Increased germination and emergence percentage, radicle and plumule length, plant height and dry weight	Soybean	Ahmadvand et al. (2012)
Halo	$CaCl_2$	Increased germination percentage, seedling length and weight	Fennel	Hoseini et al. (2013)
		Increased germination percentage, biological yield, harvest index and tassel weight	Maize	Vazirimehr et al. (2014)
		Increased activity of total amylase and protease in germinating seed under salt stress	Sorghum	Kadiri and Hussaini (1999)
		Increased germination speed and seedling vigour	Maize	Kulkarni and Eshanna (1988)
		Improved growth of seedling and stand establishment	Rice	Farooq et al. (2007)
	NaCl	Best germination percentage	Coriander	Fredj et al. (2013)
		Improved germination and germination index	Safflower	Elouaer and Hannachi (2012)
Solid matrix	Calcium-aluminium silicate	Increased seedling vigour, germination percentage and fruit yield	Okra	Sharma et al. (2013)
	Sand	Improved seed emergence and seedling density, increased root number and length	Rice	Hu et al. (2005)
Nutrio	Zn	Improved seed emergence, field performance and grain yield	Hybrid maize	Afzal et al. (2006)
	Mo	Increased plant height, number of pods $plant^{-1}$, test weight, seed yield, straw yield, water use efficiency	Chickpea	Singh et al. (2014)
	Urea	Increased germination, enhanced antioxidative enzyme activity, soluble sugar and proline	Chinese cabbage	Yan (2015)

(continued)

Table 1.1 (continued)

Method	Agent	Description	Crop	References
Bio	*Trichoderma*	Improved seedling establishment and yield	Rice	Rahman et al. (2015)
		Synchronized seed germination and plant growth	Soybean	Entesari et al. (2013)
		Increased activity of superoxide dismutase, peroxidize, glutathione reductase and glutathione s-transferase	Tomato	Mastouri et al. (2012)
	Azotobacter/*Azospirillum*	Increased crop growth, dry matter accumulation and yield	Maize	Sharifi (2011)
		Increased yield, 1000 grain weight and harvest index	Barley	Mirshekari et al. (2012)
Hormo	Salicylic acid	Improved germination rate, seed stamina index and seedling fresh and dry weight	Fennel	Farahbakhsh (2012)
		Improved germination and antioxidant capacity	Rice	Kata et al. (2014)
	Brassino-steroids	Increased antioxidative response of seedling under salt stress	Alfalfa	Zhang et al. (2007)
	GA_s	Enhanced emergence, germination percentage and speed of germination, stimulation in activity of enzymes related to germination	Soybean	Maske et al. (1997) and Bassi et al. (2011)
		Improved speed of germination, seedling dry weight and vigour index	French bean	Sarika et al. (2013)
		Enhanced germination and emergence of seedling, increased vegetative growth and photosynthetic activity at vegetative stage	Wheat	Bassi (2005) and Sharma et al. (2010)
		Synchronized and enhanced germination, improved root length and leaf score	Rice	Farooq et al. (2007)
		Increased germination and seed vigour	Maize	Kumari et al. (2017)

engaged in finding out some different easy methods for seed priming which can be used as alternative to conventional protocols, and at the same time the latest techniques should be rapid, cost-effective and environmentally safe. Among the latest techniques, use of physical agents and nanoparticles for seed priming is rapidly developing.

1.4.1 Priming with Physical Agents

Physical methods of seed priming have immense potential and advantages over the conventional methods based on chemical substances (Aladjadjiyan 2012). As such, seed priming with physical treatment can be an alternative approach by considering the contemporary issues of agricultural pollution due to injudicious use of chemical compounds (Araujo et al. 2016). These physical methods also are affordable, clean and environmentally safe, and there is every prospect of using these methods to increase seed germination and seedling vigour in a high-throughput scale (Haq et al. 2012). Thus, physical methods for seed priming are being preferred than conventional methods in particular chemical priming since present-day agriculture relies on idea of sustainability with little or no adverse effects on environment (Pretty et al. 2006). Though physical treatments are good alternatives to raise agricultural production along with improvement in plant protection and post-treatment storability (Aladjadjiyan 2012), it is obvious that most of the physical and/or extra-terrestrial environmental factors have genotoxic effects particularly at high levels. But interestingly, stimulatory effects of physical treatments at low doses have been reported by many in recent times (De Micco et al. 2014), and the phenomenon of boosting up of plant characteristic effects by toxic agents at lower doses is known as hormesis. As such, hormesis actually represents an adaptive response of living organisms to moderate or intermittent stress, due to stimulation of cellular defence and repair mechanisms which will otherwise be missed in absence of that stress agents (Mattson 2008). Various physical treatments have been reported for their stimulatory effects when used for seed priming (Araujo et al. 2016) (Fig. 1.2).

Among the physical methods, electromagnetic priming has been described as eco-friendly, cheap and non-invasive technique (Bilalis et al. 2012). Recent researches have shown that seed priming using magnetic field can improve germination rate, vigour and seedling biomass along with tolerance to biotic and abiotic stresses which is possibly attributed to reduced production of superoxide radicles along with increasing activities of antioxidant enzymes (Bhardwaj et al. 2012; Araujo et al. 2016). Another physical agent, ionizing radiation like gamma (γ)-rays is a powerful mutagenic tool in the agricultural sciences, and it can directly interact with cellular components. However, the biological effects of γ-rays are strongly depended on the intensity, dose rate and exposure time. Seed irradiation with γ-rays at doses lower than 100 Gy can remarkably enhance germination percentage and seedling establishment. This γ-ray-induced enhancement particularly at low-dose irradiation can be attributed to favourable changes in the hormonal signalling network in plant cells and trigger the antioxidative capacity which acts as a boost in

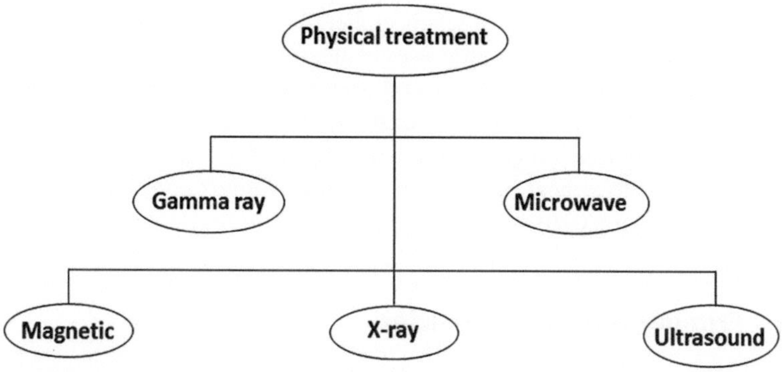

Fig. 1.2 Different types of physical treatments used for seed priming

pre-germination metabolism leading to early dormancy breaking, improved germination and overcome of daily stress factors such as fluctuations of light intensity and temperature in growth conditions (Qi et al. 2015).

Ultraviolet rays are physical agents which fall under the group of nonionizing radiation and are extremely harmful to organisms. It is because UV radiation can penetrate into living tissues and is potential to cause DNA damage and membrane injury. However, few works in recent times have showed that UV radiation can elicit some positive responses like inducing capability to mitigate environmental stresses through increased free radical scavenging activities, increasing total soluble phenols and enhancing the activities of L-phenylalanine ammonia lyase and tyrosine ammonia lyase (Araujo et al. 2016). Microwaves are also known as components of nonionizing electromagnetic spectrum which is able to induce thermal as well as nonthermal effects in biological systems (Banik et al. 2003). The detailed studies to evaluate the impact of microwaves in seed technology are really inadequate to establish it as promising approach for seed priming. But few researches have clear indication of beneficial effects of this physical agent on seedling growth and biomass accumulation (Jakubowski 2010, Talei et al. 2013).

Ultrasound is a novel physical method that involves the application of mechanical waves having frequency in the range between 20 and 100 kHz for seed pretreatment. During the recent years, ultrasonic waves are being immensely applied as an efficient technique for breaking seed dormancy and improving the germination characteristics. This technique is unique among existing seed pretreatment methods in that it is fast, energy-saving and environmentally friendly (Goussous et al. 2010). The mechanisms of action of the ultrasound for seed invigoration are probably clearer than other physical priming methods. Ultrasound priming process enforces the uptake of water and activates enzymatic and other biological reactions in the seed resulting in a faster and more uniform germination. Ultrasound priming processes actually impose a mechanical pressure on the seed coat, and this pressure increases the seed's porosity which is called as acoustic cavitation. This increase in porosity leads to improve water uptake and oxygen availability. Thereby, ultrasonic priming enhances mass transfer of the extra-absorbed water and allows it to react freely with the cell embryo. As a result gibberellic acid is released and causes an increase in the rate of metabolic processes in aleurone cells due to increasing activities of hydrolyzing enzymes. Then the endosperm nutrients ultimately mobilize due to cell membrane disruption (Miano et al. 2015). All these series of responses finally lead to enhanced germination percentage and speed. Stimulation of seed germination by ultrasound treatment has been reported in various crops (Yaldagard et al. 2008; Goussous et al. 2010).

1.4.2 Priming with Nanoparticles

Application of nanotechnology is almost every field of science owing to the extensively research that being undertaken. Among the latest line of technological innovations, nanotechnology is a versatile field and is a rapidly developing discipline. It is substantially influencing almost all existing fields of science, and introduction of nanomaterials in agriculture has immense possibility of using this unique technology as a solution to several agricultural and environmental challenges. Nanotechnology certainly occupies a prominent position in transforming agriculture and food production (Fraceto et al. 2016) and holds the potential to rejuvenate agriculture. The unique properties of materials at nanoscale make them suitable for the design and development of novel tools in support of sustainable agriculture. Introduction of nanomaterials in agriculture can potentially contribute to address the issue of sustainability under the future challenges of ever-increasing global population and climate change (Parisi et al. 2015). Nanoparticles have been shown to be an attractive alternative for the manufacture of nanofertilizers, which are more effective and efficient than traditional fertilizers. Thus the exploitation of nanotechnology and nanomaterials in agriculture can reduce the application of excess chemical fertilizers to crops and contribute to precision farming as it reduces fertilizer wastage and in turn environmental pollution due to agricultural malpractices (Upadhyaya et al. 2017). Because of their impact on crop nutritional quality and stress tolerance in plants, the application of nanoparticles is increasing. Nanopriming is a new technique of using nanoparticles of less than 100 nm size, and it enhances the germination percentage as well as seedling dry weight and vigour in most of the crops. Seed priming with nanoparticles of micronutrients has been reported as a potential new method for increasing germination percentage and seedling development and vigour (Ghafari and Razmjoo 2013). The increased germination seedling growth rate under nanopriming has been attributed to enhancing water and nutrient uptake due to more penetration through seed coat. Table 1.2 summarizes the effects of different physical and nanopriming methods on seed germination and other plant responses.

Table 1.2 Summary of effect of physical and nanopriming methods on seed germination and plant characteristic

Method	Agent	Description	Crop	References
Physical	Magnetic field	Increased antioxidant potential under soil flooding	Wheat	Balakhnina et al. (2015)
		Increased germination parameters and seedling biomass, increased light harvesting capacity, reduced production of superoxide radicals	Soybean	Baby et al. (2011) and Shine et al. (2011)
		Improved seedling growth, leaf water status and photosynthesis under water stress	Maize	Anand et al. (2012)
		Increased plant height, root length, root biomass and yield	Okra	Haq et al. (2012)
		Increased activities of superoxide dismutase, catalase and glutathione reductase	Cucumber	Bhardwaj et al. (2012)
	UV radiation	Improved tolerance to salinity	Lettuce	Ouhibi et al. (2014)
		Increased germination speed and seedling vigour and reduced susceptibility to root interacting fungi	Mungbean Groundnut	Siddiqui et al. (2011)
	Gamma-radiation	Improved morphological traits like plant height, shoot number, panicle length and seed number panicle^{-1}	Rice	Maity et al. (2005)
		Improved germination, plant growth and pigment contents	Maize	Marcu et al. (2013)
		Improved seed germination, photosynthetic capacity and seed yield	Okra	Hegazi and Hamideldin (2010)
	X-rays	Enhanced leaf growth	Date palm	Al-Enezi et al. (2012)
		No toxic effect in terms of germination and leaf development under increasing irradiation dose	Tomato	De Micco et al. (2014)
	Microwaves	Increased germination percentage and rate, primary shoot and root length	Rice	Talei et al. (2013)
		Enhanced germination and vigour index after exposure during 20s	Barley	Iuliana et al. (2013)
		Highest biomass growth in seed potato germs	Potato	Jakubowski (2010)

(continued)

Table 1.2 (continued)

Method	Agent	Description	Crop	References
Physical	Electro	Increased growth and yield	Rice	Bera and Maity (2003)
		Increased female/male flower ratio, increased length of seedling	Bottle gourd	Rahman and Yasmin (1993)
	Ultrasonic	Improved germination percentage and speed of germination	Chickpea, wheat, watermelon	Goussous et al. (2010)
		Improved germination percentage and increased activities of superoxide dismutase and catalase	Sesame	Shekari et al. (2015)
		Improved water uptake and germination percentage and reduced time for germination	Barley	Yaldagard et al. (2008)
Nano	Calcium-phosphate nanoparticles	Improves seedling growth and stimulates both enzymes and metabolites related with antioxidative responses	Rice	Upadhyaya et al. (2017)
	ZnO nanoparticles	Increased seed germination, growth and yield	Peanut	Prasad et al. (2012)
		Improved germination, reduced time to flowering, increased seed weight per umbel and 100 seed weight	Onion	Laware and Raskar (2014)
	SiO_2 Nanoparticles	Enhanced root length, root volume and seedling dry weight	Rice	Adhikari et al. (2013)
		Increased germination, nitrate reductase activity and nutrient uptake	Soybean	Lu et al. (2002)
	Silver nanoparticles	Increased germination percentage, plumule and radicle length	Wheat	Salehi and Tamaskani (2008)
	Iron nanoparticles	Increased pod number, 100 seed weight and total yield	Chickpea	Valadkhan et al. (2015)

1.5 Seed Priming as a Tool to Combat Contemporary Challenges

Global climate change is inevitable with time, but kind of changes which occurred in the last several decades due to combination of natural and human-made causes are really a matter of great apprehension. As a consequence of rapid climate change, there is a rise in mean temperatures together with more frequent episodes of water deficit and heat waves. Thus, agricultural production is currently under intense challenges of various potentially adverse environmental conditions such as water deficit, high salinity, extreme temperature, submergence, etc. (Jisha et al. 2013) apart from various biotic stress factors. These stresses adversely affect the plant growth leading to substantial and unpredictable loss in crop production. Nevertheless, one of the biggest challenges of present day's agricultural system is to improve production sustainably to feed up the ever-increasing global population but with minimum impact on environment (Edmondson et al. 2014). It has been reported that one of the major obstacles to high yield and production of crop plants is the lack of synchronized crop establishment due to poor weather and soil conditions (Mwale et al. 2003).

Seed germination and subsequent seedling growth are critically influenced by environmental causes, especially moisture, temperature and salinity (Zahedifar 2013). The poor or uneven germination and seedling growth lead to nonsynchronous maturity and thereby render the possibilities of mechanization and ultimately lead to great financial losses (Ghiyasi et al. 2008). However, it is evident that a previous exposure to moderate stress can affect the subsequent responses and eventually prepare the plants to respond to future stresses in a better way (Li et al. 2014). Priming makes the plants able to respond more rapidly and more effectively for stress protection through developing a higher level of fitness on a given stress. Thus, seed priming is the simple and best solution for higher germination, seedling establishment and yield especially when the crops are grown under stressful and/or unfavourable condition (Singh et al. 2015). Seed priming can effectively attenuate the detrimental effects of stresses during germination and radicle emergence and elicit rapid positive response by triggering efficient plant hormones and antioxidant systems. The improvement in seed quality by priming has been attributed principally to reduced membrane damage or reduced lipid peroxidation through effective scavenging of reactive oxygen species generated during stressful conditions (Chiu et al. 2006). Priming actually leads to promote germination and seed vigour of different plant species through favourable changes in cellular, subcellular and molecular levels under diverse environmental conditions such as salinity, water stress and high- or low-temperature stresses (McDonald 2000). As a consequence, primed seeds result in increased levels of biotic and abiotic stress tolerance and crop yield.

1.6 Limitation and Ways to Overcome

Nonetheless, in spite of exerting many beneficial effects, contradictory reports are available regarding adverse consequences of priming of seed quality. The main drawback of conventional priming techniques is reported as increased deterioration during storage which has been ascribed to loss of seed desiccation tolerance due to prolonged treatments and thereby reduced seed longevity (Tarquis and Bradford 1992) that largely depends on the conditions (temperature, humidity, air composition) in which seeds are kept immediately after priming (Schwember and Bradford 2005). However, several post-priming protocols have been developed for amelioration of negative effect of conventional priming on seed longevity (Gurusinghe and Bradford 2001). Another challenge arises from the internal variability of priming itself, which is that not all priming protocols will lead to an improved seed germination performance and inappropriate priming conditions may render seeds vulnerable to stresses by inducing the degradation of protective proteins (Capron et al.

2000). Thus, it is always essential to first determine an optimal priming protocol specific to plants with regard to the germination performance and its effect on seed stress tolerance. The priming protocols can be optimized by using 'hydrotime concept' which can also be useful in minimizing this setback of reduced longevity due to prolonged treatment. It is because the 'hydrotime concept' describes the relations of seed water potential to time required for radicle emergence (Bradford and Still 2004). Hydrotime is measured in MPa h^{-1} or MPa day^{-1}, and this analysis can allow prediction of the pattern of germination in terms of speed, uniformity and stress tolerance which can vary with different cultivars, seed lots and treatments (Paparella et al. 2015). Thus, hydrotime analysis is useful for evaluating the physiological status of a seed lot since all the parameters used for hydrotime analysis can reveal any possible abnormal response in terms of germination and seedling establishment under stressful condition (Bradford and Still 2004). Additionally, a non-invasive method, namely, digital image technology (DIT), can extensively be exploited for accurate analysis of vigour as well as seed purity since the said technology has immense potential for precise and automated assessment of seed morphological and physiological features (Dell'Aquila 2009). Thus, the effectiveness of different priming methods on germination and seedling establishment can be analysed by using DIT (Mahajan et al. 2011) which allows high-resolution and accurate analysis of germination parameters on a single-seed basis.

Further the detailed impacts of physical treatments on morpho-structural aspects, pre-germinative metabolism, molecular mechanisms and/or gene expressions are also yet to be fully explored, and this gap of knowledge renders hindrance in their successful application (Araujo et al. 2016). And the effectiveness of physical treatments largely depends on several factors such as type, dose and dose rate of radiation and magnetic field as well as plant characteristics such as species, cultivars, age, ploidy level and complexity of target organ or tissue (De Micco et al. 2014). Despite of considerable advances in identifying possible application nanotechnology in agriculture, many issues like characterization and quantification of nanomaterials in different surrounding substances as well as their impact on environment and on human health remain to be fully resolved till date because there are virtually no studies on the potential environmental impact of nanoparticles when used in agriculture.

Thus, the application of physical methods or nanoparticles for seed priming still requires an extensive research for determining detailed impact of these methods on plants and also to define the optimal treatments in terms of dose, dose rate and exposure time which certainly varies with crop species, genotypes and environmental conditions. Further understanding the physiological and molecular mechanisms underlying the improved performance of physical and nano-primed seeds and thereby implicating the combined application of these methods with environmental factors are very important for developing improved priming protocols.

1.7 Conclusion

In the era of modern and precision agricultural technologies, there is a demand that each and every seed should be healthy and readily germinate to produce a vigorous seedling to ensure better crop yield. Seed priming, a very simple and unique technique, has long been identified as a potential tool to improve seed germination, reduce seedling emergence time, produce vigorous plant, reduce time to flower, improve crop tolerance to abiotic stresses and ultimately improve grain yield. But, optimizing the priming protocol by hydrotime analysis of individual species/cultivar would alleviate the several difficulties that arise during conventional priming methods. Recent developments of seed priming particularly through physical methods and/or by using nanomaterials have been found to be very promising ways over the conventional seed priming techniques for sustainability in agricultural production. The application of physical methods or nanoparticles for seed priming still requires an extensive research for determining the detailed impact of these methods on plants and environment as well as on human health. This is further required to define the optimal treatment in terms of dose, dose-rate and exposure time and their interactive effects on particular species/genotypes under specific environmental conditions before these technologies may make significant contribution to agriculture.

References

Adhikari T, Kundu S, Subba Rao A (2013) Impact of SiO_2 and Mo nano particles on seed germination of rice (*Oryza sativa* L.). Int J Agric Food Sci Technol 4(8):809–816

Afzal S, Nadeem A, Zahoor A, Qaiser M (2006) Role of seed priming with zinc in improving the hybrid maize (*Zea mays*) yield. Am-Eurasian J Agric Environ Sci 13:301–306

Ahmadvand G, Soleimani F, Saadatian B, Pouya M (2012) Effect of seed priming on germination and emergence traits of two soybean cultivars under salinity stress. Int Res J Appl Basic Sci 3:234–241

Aladjadjiyan A (2012) Physical factors for plant growth stimulation improve food quality. In: Aladjadjiyan A (ed) Food production-approaches, Challenges and tasks. InTech, Rijeka, pp 145–168

Al-Enezi N, Al-Bahrany A, Al-Khayri J (2012) Effect of X-irradiation on date palm seed germination and seedling growth. Emirates J Food Agric 24:415–424

Anand A, Nagarajan S, Verma APS, Joshi DK, Pathak PC, Bhardwaj J (2012) Pre-treatment of seeds with static magnetic field ameliorates soil water stress in seedlings of maize (*Zea mays* L.). Indian J Biochem Biophys 49:63–70

Anese S, Da Silva EAA, Davide AC, Rocha Faria JM, Soares GCM, Matos ACB, Toorop PE (2011) Seed priming improves endosperm weakening, germination, and subsequent seedling development of *Solanum lycopersicum* St. Hil. Seed Sci Technol 39:125–139

Araujo SS, Paparella S, Dondi D, Bentivoglio A, Carbonera D, Balestrazzi A (2016) Physical methods for seed invigoration: advantages and challenges in seed technology. Front Plant Sci 7:646. https://doi.org/10.3389/fpls.2016.00646

Baby SM, Narayanaswamy GK, Anand A (2011) Superoxide radical production and performance index of Photosystem II in leaves from magneto primed soybean seeds. Plant Signal Behav 6:1635–1637

Balakhnina T, Bulak P, Nosalewicz M, Pietruszewski S, Włodarczyk T (2015) The influence of wheat *Triticum aestivum* L. seed pre-sowing treatment with magnetic fields on germination, seedling growth, and antioxidant potential under optimal soil watering and flooding. Acta Physiol Plant 37:59. https://doi.org/10.1007/s11738-015-1802-2
Banik S, Bandyopadhyay S, Ganguly S (2003) Bio effects of microwave – a brief review. Bioresour Technol 87:155–159
Basra SMA, Farooq M, Khaliq A (2003) Comparative Study of Pre-Sowing Seed Enhancement Treatments in Fine Rice (Oryza sativa L.). Pakistan J Life Soc Sci 1:21–25
Basra SMA, Farooq M, Tabassum R (2005) Physiological and biochemical aspects of seed vigour enhancement treatments in fine rice (*Oryza sativa* L.). Seed Sci Technol 33:25–29
Bassi G (2005) Seed priming for in-vigourating late sown wheat (*Triticum aestivum*). Crop Improv 32:121–123
Bassi G, Sharma S, Gill BS (2011) Pre-sowing seed treatment and quality in-vigouration in soybean (*Glycine max* (L) Merrill). Seed Res 31:81–84
Benamar A, Tallon C, Macherel D (2003) Membrane integrity and oxidative properties of mitochondria isolated from imbibing pea seeds after priming or accelerated ageing. Seed Sci Technol 13:35–45
Bera AK, Maity U (2003) Enhancing yield potential in rice by pre-sowing electric current treatment of rice gain. Plant Arch 4(1):17–24
Bewley JD (1997) Seed germination and dormancy. Plant Cell 9(7):1055
Bhardwaj J, Anand A, Nagarajan S (2012) Biochemical and biophysical changes associated with magneto priming in germinating cucumber seeds. Plant Physiol Biochem 57:67–73. https://doi.org/10.1016/Jplaphy.2012.05.008
Bilalis DJ, Katsenios N, Efthimiadou A, Karkanis A, Efthimiadis P (2012) Investigation of pulsed electromagnetic field as a novel organic pre-sowing method on germination and initial growth stages of cotton. Electromagn Biol Med 31:143–150. https://doi.org/10.3109/15368378.2011.624660
Bradford KJ, Still DW (2004) Application of hydrotime analysis in seed testing. Seed Tech 26:75–85
Bray CM (1995) Biochemical processes during the osmopriming of seeds. In: Kigel J, Galili G (eds) Seed development and germination. Marcel Dekker, New York, pp 767–789
Bruce TJA, Matthes MC, Napier JA, Pickett JA (2007) Stressful "memories" of plants: evidence and possible mechanisms. Plant Sci 173:603–608
Capron I, Corbineau F, Dacher F, Job C, Côme D, Job D (2000) Sugar beet seed priming: effects of priming on germination, solubilization of 11-S globulin and accumulation of LEA proteins. Seed Sci Res 10:243–254
Catusse J, Meinhard J, Job C, Strub JM, Fischer U, Pestova E, Westohoff P, Van Dorselaer A, Job D (2011) Proteomics reveals potential biomarkers of seed vigor in sugar beet. Proteomics 11:1569–1580
Chen K (2011) Antioxidants and dehydrin metabolism associated with osmopriming-enhanced stress tolerance of germinating spinach (*Spinacia oleracea* L. cv. Bloomsdale) seeds. Graduate theses and dissertations, Paper 10471, Iowa State University
Chen D, Gunawardena TA, Naidu BP, Fukai S, Basnayake J (2005) Seed treatment with gibberellic acid and glycine-betaine improves seedling emergence and seedling vigour of rice under low temperature. Seed Sci Technol 33:471–479
Chen K, Arora R, Arora U (2010) Osmopriming of spinach (*Spinacia oleracea* L. cv. Bloomsdale) seeds and germination performance under temperature and water stress. Seed Sci Technol 38:45–57
Chen H, Chu P, Zhou Y, Li Y, Liu J, Ding Y, Tsang EW, Jiang L, Wu K, Huang S (2012) Overexpression of AtOGG1, a DNA glycosylase/AP lyase, enhances seed longevity and abiotic stress tolerance in Arabidopsis. J Exp Bot 63:4107–4121
Chiu KY, Chuang SJ, Sung JM (2006) Both anti-oxidant and lipid-carbohydrate conversion enhancements are involved in priming-improved emergence of *Echinacea purpurea* seeds that differ in size. Sci Hortic 108:220–226

Corbineau F, Özbingol N, Vineland D, Côme D (2000) Improvement of tomato seed germination by osmopriming as related to energy metabolism. In: Black M, Bradford KJ, Vasquez-Ramos J (eds) Seed biology advances and applications: proceedings of the sixth international worshop on seeds. Merida, Mexico. CABI, Cambridge.; 2000, pp 467–474

De Micco V, Paradiso R, Aronne G, De Pascale S, Quarto M, Arena C (2014) Leaf anatomy and photochemical behaviour of *Solanum lycopersicum* L. plants from seeds irradiated with low-LET ionising radiation. Sci World J 2014:428141. https://doi.org/10.1155/2014/428141

Dell'Aquila A (2009) New perspectives for seed germination testing through digital imaging technology. Open Agric J 3:37–42

Edmondson JL, Davies ZG, Gaston KJ, Leake JR (2014) Urban cultivation in allotments maintains soil qualities adversely affected by conventional agriculture. J Appl Ecol 51:880–889. https://doi.org/10.1111/1365-2664.12254

Elouaer MA, Hannachi C (2012) Seed priming to improve germination and seedling growth of safflower (*Carthamus tinctorius*) under salt stress. Eur Asian J Bio Sci 6:76–84

Entesari M, Sharifzadeh F, Ahmadzadesh M, Farhangfar M (2013) Seed biopriming with *Trichoderma* species and *Pseudomonas fluorescent* on growth parameters, enzyme activity and nutritional status of soybean. Int J Agron Plant Prod 4:610–619

Farahbakhsh H (2012) Germination and seedling growth in primed and primed seeds of fennel as affected by reduced water potential induced by NaCl. Int Res J Appl Basic Sci 3:737–744

Farooq M, Basra SMA, Khan MB (2007) Seed priming improves growth of nursery seedlings and yield of transplanted rice. Arch Agron Soil Sci 53:311–322

Farooq M, Wahid A, Lee DJ (2009) Exogenously applied polyamines increase drought tolerance of rice by improving leaf water status, photosynthesis and membrane properties. Acta Physiol Plant 31:937–945. https://doi.org/10.1007/s11738-009-0307-2

Fraceto LF, Grillo R, de Medeiros GA, Scognamiglio V, Rea G, Bartolucci C (2016) Nanotechnology in agriculture: which innovation potential does it have? Front Environ Sci 4:20

Fredj MB, Zhani K, Hannachi C, Mehwachi T (2013) Effect of NaCl priming on seed germination of four coriander cultivars (*Coriandrum sativum*). Eur Asian J Bio Sci 7:21–29

Ganji-Arjenaki F, Amini-Dehghani M, Jabbari R (2011) Effects of priming on seed germination of marigold (*Calendula officinalis*). Adv Environ Biol 5:276–280

Gelmond H (1978) In: Gupta VS (ed) Problems in crop seed germination in crop physiology. Oxford and IBH publishing Co., New Delhi, pp 1–78

Ghafari H, Razmjoo J (2013) Effect of foliar application of nano-iron oxidase, iron chelate and iron sulphate rates on yield and quality of wheat. Int J Agric Plant Prod 4(11):2997–3003

Ghiyasi M, Seyahjani AA, Tajbakhsh M, Amirnia R, Salehzade H (2008) Effect of osmopriming with Polyethylene glycol (8000) on germination and seedling growth of wheat (Triticum aestivum L.) seeds under salt stress. Res J Bio Sci 3(10):1249–1251

Goussous SJ, Samarah NH, Alqudah AM, Othman MO (2010) Enhancing seed germination of four crop species using an ultrasonic technique. Exp Agric 46(2):231–242

Guan YJ, Hu J, Wang XJ, Shao CX (2009) Seed priming with chitosan improves maize germination and seedling growth in relation to physiological changes under low temperature stress. J Zhejiang Univ Sci B 10(6):427–433

Gurusinghe S, Bradford K (2001) Galactosyl-sucrose oligosaccharides and potential longevity of primed seeds. Seed Sci Res 11:121–133

Haq Z, Jamil Y, Irum S, Randhawa MA, Iqbal M, Amin N (2012) Enhancement in germination, seedling growth and yield of radish (*Raphanus sativus*) using seed pre-sowing magnetic field treatment. Pol J Environ Stud 21(2):369–374

Hegazi AZ, Hamideldin N (2010) The effect of gamma irradiation on enhancement of growth and seed yield of okra [*Abelmoschus esculentus* (L.) Monech] and associated molecular changes. J Hortic For 2:38–51

Hill H, Bradford KJ, Cunningham J, Taylor AG (2008) Primed lettuce seeds exhibit increased sensitivity to moisture during aging. Acta Hortic 782:135–141

Hoseini M, Baser S, Jahandideh E (2013) Response of fennel to priming techniques. Ann Rev Res Biol 3:124–130

Hu J, Zhu ZY, Song WJ, Wang JC, Hu WM (2005) Effects of sand priming on germination and field performance in direct-sown rice (*Oryza sativa* L.). Seed Sci Technol 33:243–248
Iuliana C, Caprita R, Giancarla V, Sorina R (2013) Response of barley seedlings to microwaves at 2.45GHz. Sci Pap Anim Sci Biotechnol 46:185–191
Jakubowski T (2010) The impact of microwave radiations at different frequencies on the weight of seed potato germs and crop of potato tubers. Agric Eng 6:57–64
Jisha KC, Vijayakumari K, Puthur JT (2013) Seed priming for abiotic stress tolerance: an overview. Acta Physiol Plant 35:1381–1396
Kadiri M, Hussaini MA (1999) Effect of hardening pre-treatment on vegetative growth, enzyme activities and yield of *Pennisetum americanum* and *sorghum bicolour* L. Glob J Pure Appl Sci 5:179–183
Kata LP, Bhaskaran M, Umarani R (2014) Influence of priming treatments on stress tolerance during seed germination of rice. In J Agric Environ Biotechnol 7:225–232
Kulkarni GN, Eshanna MR (1988) Effect of pre-soaking of corn seed on seed quality. Seed Res 16:37–40
Kumari N, Rai PK, Bara BM, Singh I (2017) Effect of halo priming and hormonal priming on seed germination and seedling vigour in maize (*Zea mays* L.) seeds. J Pharm Phytochem 6(4):27–30
Laware SL, Raskar S (2014) Influence of zinc oxide nanoparticles on growth, flowering and seed productivity in onion. Int J Curr Microbiol App Sci 3(7):874–881
Li F, Wu X, Tsang E, Cutler AJ (2005) Transcriptional profiling of imbibed *Brassica napus* seed. Genomics 86:718–730. https://doi.org/10.1016/Jygeno.2005.07.006
Li X, Cai J, Liu F, Dai T, Cao W, Jiang D (2014) Cold priming drives the sub-cellular antioxidant systems to protect photosynthetic electron transport against subsequent low temperature stress in winter wheat. Plant Physiol Biochem 82:34–43
Lu CM, Zhang CY, Wen JQ, Wu GR, Tao MX (2002) Research on the effect of nanometer materials on germination and growth enhancement of *Glycine max* and its mechanism. Soybean Sci 21:68–172
Mahajan G, Sarlach RS, Japinder S, Gill MS (2011) Seed priming effects on germination, growth and yield of dry direct-seeded rice. J Crop Improv 25:409–417
Maity JP, Mishra D, Chakraborty A, Saha A, Santra SC, Chanda S (2005) Modulation of some quantitative and qualitative characteristics in rice (*Oryza sativa* L.) and mung (*Phaseolus mungo* L.) by ionizing radiation. Radiat Phys Chem 74:391–394. https://doi.org/10.1016/Jradphyschem.2004.08.005
Marcu D, Damian G, Cosma C, Cristea V (2013) Gamma radiation effects on seed germination, growth and pigment content, and ESR study of induced free radicals in maize (*Zea mays*). J Biol Phys 39:625–634. https://doi.org/10.1007/s10867-013-9322-z
Maske VG, Dotale RD, Sorte PN, Chore CN (1997) Germination, root and shoot studies in soybean as influenced by GA3 and NAA. J Soils Crops 7:147–149
Mastouri F, Bjorkman T, Harman GE (2012) *Trichoderma harzianum* enhances antioxidant defence of tomato seedlings and resistance to water deficit. Mol Plant – Microbe Interact 25:1264–1271
Mattson MP (2008) Hormesis defined. Ageing Res Rev 7:1–7
McDonald MB (2000) Seed priming. In: Black M, Bewley JD (eds) Seed technology and its biological basis. Sheffield Academic Press, Sheffield, pp 287–325
Meena RP, Sendhil R, Tripathi SC, Chander S, Chhokar RS, Sharma RK (2013) Hydro-priming of seed improves the water use efficiency, grain yield and net economic returns of wheat under different moisture regimes. SAARC J Agric 11:149–159
Miano AC et al (2015) Effect of ultrasound technology on barley seed germination and vigour. Seed Sci Technol 43:297–302
Mirshekari B, Hokmalipour S, Sharifi RS et al (2012) Effect of seed biopriming with plant growth promoting rhizobacteria (PGPR) on yield and dry matter accumulation of spring barley (*Hordeum vulgare* L.) at various levels of nitrogen and phosphorus fertilizers. J Food Agric Environ 10:314–320

Murungu FS, Chiduza C, Nyamuga P, Clark LJ, Whalley WR, Finch WE (2004) Effects of on-farm seed priming on consecutive daily sowing occasion on the emergence and growth of maize in semi-arid Zimbabwe. Field Crop Res 89:49–57

Mwale SS, Hamusimbi C, Mwansa K (2003) Germination, emergence and growth of sunflower (*Helianthus annuus* L.) in response to osmotic seed priming. Seed Sci Technol 31:199–206

Nagar R, Dadlani PM, Sharma SP (1998) Effect of hydro-priming on field emergence and crop growth of maize genotypes. Seed Sci Res 26:1–5

Osburn RM, Schroth MN (1989) Effect of osmopriming sugar beet deed on exudation and subsequent damping-off caused by *Pythium ultimum*. Phytopathology 78:1246–1250

Ouhibi C, Attia H, Rebah F, Msilin N, Chebbi M, Aarrouf J et al (2014) Salt stress mitigation by seed priming with UV-C in lettuce plants, growth, antioxidant activity and phenolic compounds. Plant Physiol Biochem 83:126–133

Pandita VK, Anand A, Nagarajan S (2007) Enhancement of seed germination in hot pepper following presowing treatments. Seed Sci Technol 35:282–290

Paparella S, Araújo SS, Rossi G, Wijayasinghe M, Carbonera D, Balestrazzi A (2015) Seed priming: state of the art and new perspectives. Plant Cell Rep 34:1281–1293

Parisi C, Vigani M, Rodriguez-Cerezo E (2015) Agricultural nano-technologies: what are the current possibilities? Nano Today 10:124–127

Pill WG, Crossan CK, Frett JJ, Smith WG (1994) Matric and osmotic priming of *Echinacea purpurea* (L.) Moench seeds. Sci Hortic 59:37–44

Prasad TNVKV, Sudhakara P, Sreenivasulua Y, Lathaa P, Munaswamya V, Raja Reddya K, Sreeprasad TS, Sajanlalb PR, Pradeepb T (2012) Effect of nanoscale zinc oxide particles on the germination, growth and yield of peanut. J Plant Nutr 35(6):905–927

Pretty JN, Noble AD, Bossio D, Dixon J, Hine RE, Penning De Vries FWT et al (2006) Resource-conserving agriculture increases yields in developing countries. Environ Sci Technol 40:1114–1119

Qi W, Zhang L, Wang L, Xu H, Jin Q, Jiao Z (2015) Pre-treatment with low-dose gamma irradiation enhances tolerance to the stress of cadmium and lead in *Arabidopsis thaliana* seedlings. Ecotoxicol Environ Saf 115:243–249

Rahman MA, Yasmin F (1993) Effects of electric current treatments to seeds on bottle gourd: seed NPK and seedling leaf pigments. Chittagong Univ Stud Part II: Sci 17(2):37–41

Rahman M, Ali J, Masood M (2015) Seed priming and *Trichoderma* application: a method for improving seedling establishment and yield of dry direct seeded boro (winter) rice in Bangladesh. Univ J Agric Res 3(2):59–67

Rajjou L, Duval M, Gallardo K, Catusse J, Bally J, Job C, Job D (2012) Seed germination and vigor. Annu Rev Plant Biol 63:507–533

Reddy MVB, Arul J, Angers P, Couture L (1999) Chitosan treatment of wheat seeds induces resistance to *Fusarium graminearum* and improves seed quality. J Agric Food Chem 47:67–72

Ruan S, Xue Q, Tylkowska K (2002) Effects of seed priming on germination and health of rice (Oryza sativa L.) seeds. Seed Sci Technol 30:451–458

Sadeghian SY, Yavari N (2004) Effect of water deficient stress on germination and early seedling growth in sugar beet. J Agron Crop Sci 190:138–144

Salehi M, Tamaskani F (2008) Pretreatment effect of nanosilver on germination and seedling growth of wheat under salt stress. Proceeding of 11th Iranian Congress in Seed Sciences and Echnology. Gorgan, Iran

Salehzade H, Shishvan MI, Ghiyasi M, Forouzin F, Siyahjani AA (2009) Effect of seed priming on germination and seedling growth of wheat [*Triticum aestivum* (L).]. Res J Biol Sci 4:629–631

Sarika G, Basavaraju GV, Bhanuprakash K, Chaanakeshava V, Paramesh R, Radha BN (2013) Investigation on seed viability and vigour of aged seed by priming in French bean. Veg Sci 40:169–173

Schwember AR, Bradford KJ (2005) Drying rates following priming affect temperature sensitivity of germination and longevity of lettuce seeds. Hortic Sci 40:778–781

Sharifi RS (2011) Study of grain yield and some of physiological growth indices in maize (*Zea mays* L.) hybrids under seed biopriming with plant growth promoting rhizobacteria (PGPR). J Food Agric Environ 189:3–4
Sharma S, Gupta A, Shekhar J (2010) Growth parameters in late sown rainfed wheat as influenced by seed priming and foliar applied growth regulators. J Res SKUAST 1:63–69
Sharma AD, Rathore SVS, Srinivasan K, Tayagi R (2013) Comparison of various seed priming method for seed germination, seedling vigour and fruit yield in okra (*Abelmoschus esculentus* L. Moench). Sci Hortic 165:75–81
Shekari F, Mustafavi SH, Abbasi A (2015) Sonication of seeds increase germination performance of sesame under low temperature stress. Acta Agric Slov 105:203–212
Shine MB, Guruprasad KN, Anand A (2011) Enhancement of germination, growth, and photosynthesis in soybean by pre-treatment of seeds with magnetic field. Bioelectromagnetics 32:474–484
Siddiqui A, Dawar S, Javed Zaki M, Hamid N (2011) Role of ultraviolet (UV-C) radiation in the control of root infecting fungi on groundnut and mung bean. Pak J Bot 43:2221–2224
Singh S, Bawa S, Singh S, Sharma SC, Kumar V (2014) Effect of seed priming with molybdenum on performance of rainfed chickpea (*Cicer arietinum* L.). J Res Punjab Agric Univ 51:124–127
Singh H, Jassal RK, Keng JS, Sandhu SS, Kang H, Grewal K (2015) Seed priming techniques in field crops – a review. Agric Rev 36(4):251–264
Srivastava AK, Bose B (2012) Effect of nitrate seed priming on phenology, growth rate and yield attributes in rice (*Oryza sativa* L.). Vegetos 25:174–181
Talei D, Valdiani A, Maziah M, Mohsenkhah M (2013) Germination response of MR219 rice variety to different exposure times and periods of 2450 MHz microwave frequency. Sci World J 2013:408026. https://doi.org/10.1155/2013/408026
Tarquis AM, Bradford KJ (1992) Prehydration and priming treatments that advance germination also increase the rate of deterioration of lettuce seed. J Exp Bot 43:307–317
Taylor AG, Allen PS, Bonnett MA, Bradford KJ, Burris JS, Misra MK (1998) Seed enhancements. Seed Sci Res 8(2):245–256. https://doi: 10.1017/S0960258500004141
Theophrastus. Enquiry into plants, Book VII, I.6
Upadhyaya H, Begum L, Dey B, Nath PK, Panda SK (2017) Impact of calcium phosphate nanoparticles on rice plant. J Plant Sci Phytopathol 1:001–0010
Valadkhan M, Mohammadi K, Nezha MTK (2015) Effect of priming and foliar application of nanoparticles on agronomic traits of chickpea. Biol Forum – Int J 7(2):599–602
Varier A, Vari AK, Dadlani M (2010) The subcellular basis of seed priming. Curr Sci 99:450–456
Vazirimehr MR, Ganjali HR, Rigi K, Keshtehgar A (2014) Effect of seed priming on quantitative traits corn. Int J Plant Sci 4:134–140
Yaldagard M, Mortazavi SA, Tabatabaie F (2008) Influence of ultrasonic stimulation on the germination of barley seed and its alpha-amylase activity. Afr J Biotech 7:2465–2471
Yan M (2015) Seed priming stimulate germination and early seedling growth of Chinese cabbage under drought stress. S Afr J Bot 99:88–92
Zahedifar M (2013) Corn germination and seedling characteristics as influenced by seed- priming with potassium nano-chelate and potassium sulfate fertilizers under salinity stress conditions. Thai J Agric Sci 46(4):219–229
Zhang S, Hu J, Zhang Y, Xie XJ, Knapp A (2007) Seed priming with brassinolide improves lucerne (*Medicago sativa* L.) seed germination and seedling growth in relation to physiological changes under salinity stress. Aust J Agric Res 58:811–815
Zhang F, Yu J, Johnston CR, Wang Y, Zhu K, Lu F, Zhang Z, Zou J (2015) Seed priming with polyethylene glycol induces physiological changes in sorghum (*Sorghum bicolour* L. Moench) seedlings under suboptimal soil moisture environments. PLoS One 10(10):e0140620. https://doi.org/10.1371/journal.pone.0140620
Zhou YG, Yang YD, Qi YG, Zhang ZM, Wang XJ, Hu XJ (2002) Effects of chitosan on some physiological activity in germinating seed of peanut. J Peanut Sci 31:22–25

Impact of Seed Priming on the Modulation of Physico-chemical and Molecular Processes During Germination, Growth, and Development of Crops

2

Bandana Bose, Mahesh Kumar, Rajesh K. Singhal, and Sananda Mondal

Abstract

Seed is the prime input in agriculture sector, and production of quality seed is the immense challenge in front of agriculturist to achieve the goal of food security. Present sceario emphasis that the world population is increasing day by day resulting to quick exhaustion of natural resources leading to climate change which accelerates the issue of abiotic (heat, cold, drought, and salt) and biotic stress in plants. These abiotic and biotic stresses are often interrelated and cause undesirable physiological, morphological, biochemical, and molecular change that affect plant growth and development and ultimately yield. Time to time various plant breeding and molecular techniques developed to solve the problem of abiotic and biotic stresses. However, alternatively, some simple and economical techniques are also in race to address this problem. Seed priming is one of them, approved by many agriculturists for better crop stand establishment and growth, even under adverse environmental conditions. The present chapter deals with the different types of seed priming methods and their scope in mitigating abiotic and biotic stresses. Further, mechanisms of seed "priming-induced" physiological, biochemical, and molecular changes in regulation to stress tolerance were extensively explained in the light of the latest research work carried in this direction.

Keywords

Seed priming · Coating · Seed hardening · Abiotic stress · Biotic stress

B. Bose (✉) · M. Kumar · R. K. Singhal
Department of Plant Physiology, Institute of Agricultural Sciences, Banaras Hindu University, Varanasi, India

S. Mondal
Plant Physiology Section, Department of ASEPAN, Institute of Agriculture, Visva-Bharati, Sriniketan, India

© Springer Nature Singapore Pte Ltd. 2018

A. Rakshit, H. B. Singh (eds.), *Advances in Seed Priming*,
https://doi.org/10.1007/978-981-13-0032-5_2

Abbreviations

AQP	Aquaporin
CAT	Catalase
DEPs	Differentially expressed proteins
DHY	Dehydrin
LEA	Late embryogenic abundance
MDA	Malondialdehyde
POD	Peroxidase
ROS	Reactive oxygen species
SMP	Solid matrix priming
SOD	Superoxide dismutase

2.1 Introduction

Biologically, seed is the mature ovule that consists of an embryo and stores food materials for germination and contains a protective covering (seed coat), or it can also be defined as a small embryonic plant enclosed within a protecting covering called seed coat along with endosperm (Dieter and Bouman 1995). Agriculturally, the term seed represents any living material that can be sown and gives rise to a fully functional plant, e.g., seeds of potato is a part of a tuber and the setts of sugarcane are the parts of the stem. Seeds play very important role in input technology in agriculture sector due to its easy handling and transporting from one place to another. Therefore, in agriculture, demand and supply of quality seeds hold the center position. Further the quality of seeds is a determinant factor for the yield of particular crop. Hence, the seeds of high quality are always in demand. Basically, high-quality seeds show synchronized germination and the formation of potential seedlings, those which are able to attend the optimum level of genetic potential. Always the good-quality seed has attracted the attention of researchers to get proper production potential of a particular crop. The quality seed in present day can be achieved by various means where the basic and applied knowledge of plant physiology, genetics, and seed technology all are integrated to improve the criteria of quality of seeds. It is the ultimate goal of successful companies that breed crop plants for seed production (Bishaw et al. 2007).

The crop growers as well as the kitchen garden growers always have faced some problems, associated with the seeds, like more time taken for germination and less germination percentage leading to yield loss. Seeds especially of some flowers and herbs are often found quite difficult to germinate and using certain techniques to increase the rate in which the sprout has been the focus of a large amount of scientific research. From last two decades, several seed enhancement/invigoration technologies are implemented to enrich the seed quality.

Seed enhancement/seed invigoration is a range of treatments of seeds that improves their performance after harvesting and conditioned but before they are sown. They include priming, steeping, hardening, pregermination, pelleting, encrusting, film-coating, and tagging but exclude treatments for control of seed-borne pathogens (Halmer 2006). They are used to improve seed sowing, germination, and seedling growth by altering seed vigor and/or the physiological state of the seed. The alteration may improve vigor or the physiological state of the seed and finally improve yield potential by enhancing uniformity of germination, early seedling vigor, and healthy seedling.

Seed priming is most commonly used at farmer's field. It improves germination, germination speed, seedling vigor, root length, seedling dry weight, dry matter production, photosynthetic efficiency, and many other plant growth traits. Other than this it also improves biochemical status of plant by improving α-amylase activity and soluble sugar contents during seed germination even in low temperature (Anaytullah and Bose 2007) and nitrate reductase activity and nitrogen content in growing seedlings in normal growing condition in wheat crop in respect to non-primed seeds (Sharma and Bose 2006).

It is well established that seed priming treatment found to ameliorate the adverse effects of biotic and abiotic (drought, salinity, flooding, heat, cold, heavy metal) stress responses in affected plants via altering the antioxidant metabolism (Kausar and Ashraf 2003; Basra et al. 2005; Guan et al. 2009; Nayaka et al. 2010; Kumar et al. 2016). It improves the stress memory and boosts antioxidant system by improving activity of SOD, catalase, MDA, glutathione reductase, ascorbic acid, and stress protein like late embryogenesis abundant (LEA), dehydrin and aquaporin (AQP) proteins (Mittal and Dubey 1995; Bohnert and Shen 1999; Vander et al. 2006; Anaytullah et al. 2012). In this chapter we are summarizing the types of seed enhancement technique specifically seed priming in respect to their roles in modulation of physiological and molecular mechanism during germination and post-germination phases as well as how it helps in the amelioration of abiotic and biotic stress responses in crops/plants during their developmental process.

2.2 Seed Invigoration Techniques and Their Use in Agriculture

Seed invigoration technology includes *priming* [A pre-sowing hydration treatments include noncontrolled water uptake systems (here water is freely available and not restricted by the environment) and controlled systems (it regulates seed moisture content preventing in the completion of germination)] (Taylor et al. 1998), *pelleting* (it adds thicker artificial coverings to seeds, which can be used to cover irregular seed shapes and add chemicals to the pellet matrix, e.g., of sugar beet or vegetable seeds; the pellet matrix consists of filling materials and glue; it is also used to increase the size of very small horticultural seeds), and *coating* (*film-coating* methods allow the chemicals to be applied in a form of synthetic polymer that is sprayed onto the seeds and provides a solid, thin coat covering on them; the advantage of the polymers is that they adhere tightly to the seed and prevent loss of active materials like fungicides, nutrients, colorants, or plant hormones); further in *seed hardening* (kind of seed priming) technology, the seeds are allowed to be hydrated either in presence of water or in presence of various organic/inorganic solutions just before the emergence of radical and then dehydrated(hardened) under the forced air to get its initial starting weight (Sharma and Bose 2006). *Physical seed invigoration*, comparatively new one, offers an approach where the conventional method of seed treatment by using chemicals is ignored, and that has been replaced by using irradiation with microwaves and ionizing radiations, found to be a promising pre-sowing seed treatment; magneto priming is also one of them (Araújo et al. 2016).

Generally basic and applied seed research projects focus on embryo growth and on the different seed-covering layers (e.g., testa, endosperm, pericarp), which are the determinants of seed quality and exhibit the biodiversity of seed structures. Seed germination is controlled by various external environmental factors (light, temperature, water) and also by internal factors like plant hormones (gibberellins, abscisic acid, ethylene, auxin, cytokinins, and brassinosteroids) as endogenous regulators. The utilization of plant hormones and inhibitors as well as their biosynthesis and action in seed treatment technologies affects seed germination and seedling emergence. The genes, enzymes, signaling components, and downstream targets of some plant hormones provide molecular marker for seed quality and seedling performance (The Seed Biology Place).

By considering all these points, in this chapter it will be discussed how seed priming (a plant physiological technology) improves the germination, growth and development, and yield potential as well as quality of the produce of various agriculturally important crop plants by protecting them from various hazardous environmental conditions.

2.3 Types of Seed Priming

The term "seed priming" was coined by Malnassy (1971) and deals with a practice which promotes rapid and uniform seedling emergence consequently beneficial for better establishment of crops in field condition. Seed priming is of many types depending upon the priming material, which includes hydro-priming (continuous or successive addition of a limited amount of water to the seeds), osmo-conditioning or osmo-priming (exposing seeds to relatively low external water potential), halo-priming (pre-sowing soaking of seeds in salt solution), hormonal priming (priming solutions containing the limited amount of plant growth regulators or hormones), nutri-priming (seeds are soaked in solutions containing the plant growth-limiting nutrients instead of being soaked just in water), bio-priming (coating of seeds with biocontrol agents), redox priming (it represents the redox state of cell and regulates the key processes in growth and development as well as stress tolerance in response to any external stimuli; plants modify their redox state, and the extent of change is dependent on the nature of the stimulus itself, the dose and the time to which the tissue is exposed as stated by Miller et al. (2009)), solid matrix priming (mixing seeds with a solid or semisolid material and measured amount of water) (term was coined by Taylor et al. 1988), and pre-sowing soaking (soaking of seeds either in water or in any solution of low water potential before sowing) (Bose et al. 1982a, b, c; Ashraf and Foolad 2005) (Fig. 2.1).

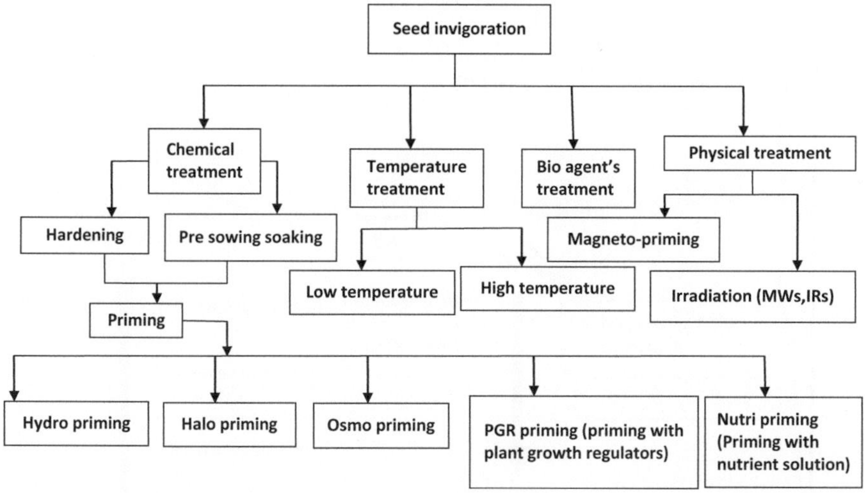

Fig. 2.1 Represent the type of seed invigoration technique

2.4 Seed Priming and Physiological Changes During the Process

Nowadays, seed treatment technology is an important link between seed producers and crop production industry. Its aim is to allow the seed treatment product to be used in such a form that represents best quality in the market. One of them is seed priming, which is an innovative concept of treating seeds using various solvents including water which activates physiological processes of seeds. Generally osmo-regulators like PEG, mannitol, glycerol, etc. are being extensively used in seed treatment for various purposes. If graph is plotted between seed water content (imbibition/osmosis) and time, then non-primed seed and primed seed represent three subsequent phases in a stepwise manner. First phase (phase I) represents the entry of water in the seed by the process of adsorption called imbibition, which is similar in both cases. Second phase (phase II) represents hydration process in non-primed seeds. In case of primed seed hydration treatment allows controlled imbibition and induction of the pre-germinative metabolism ("activation"), but radicle emergence is prevented, represented by extended second phase. Last phase (phase III) represents the germination and post-germination phase which is again similar in case of primed and non-primed seeds (Rajjou et al. 2012) (Fig. 2.2).

In seed priming the hydration treatment is to be stopped before desiccation tolerance is lost. An important problem is to stop the priming process at the right moment; this time it depends on the species, genotype, and the types of seed. Priming solutions can be supplemented with plant hormones or beneficial microorganisms. The seeds can be dried back for storage, distribution, and planting. Priming can induce the germination by improving speed and synchronization of seed germination (Bose and Tandon 1991); it can improve seed vigor which requires very short or no activation time during germination. It may introduce a wider range of temperature for

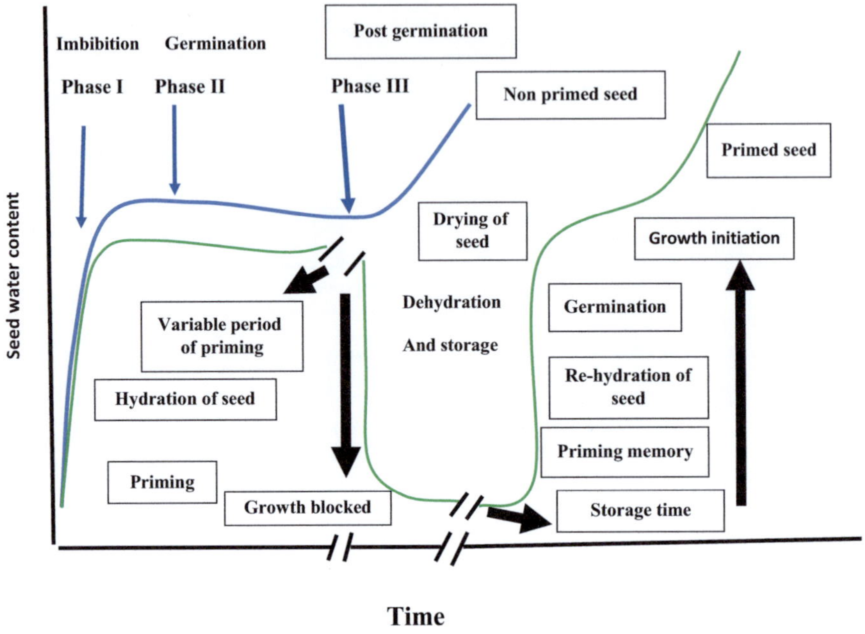

Fig. 2.2 Schematic representation of normal germination and seed priming process (Rajjou et al. 2012) (Phase I, imbibition phase; phase II, germination phase; phase III, post-germination phase)

germination (Anaytullah and Bose 2007), can break the dormancy, or may shorten the time of emergence with improved seedling vigor (Mondal et al. 2011). This leads to better crop stands and higher yields (Srivastava and Bose 2012).

Physiological, biochemical, and molecular changes during seed priming are represented in Fig. 2.3. During seed storage condition auto-oxidation of storage metabolites with time leads to lipid peroxidation, which can further cause membrane perturbance and loss of cellular compartmentalization. Long-term storage of seed causes dysfunction of cellular organelles and inactivation of enzyme and finally leads to genetic damage which can influence the seedling viability and vigor. During priming, phase I represented by the activation of priming memory, repairing DNA and mitochondria, respiration, and energy metabolism, ROS signaling and antioxidant, gene transcription and translation, cell cycle initiation, and induction of stress response gene such as LEA, DHY, AQP, and hormone signaling. Phase II is germination phase; in this priming memory is recruited upon second rehydration and protein synthesis by using new mRNA. Phase III is represented by the post-germination phase; in this phase mobilization of stored reserve, radical cell elongation events occur, and finally at the end of this stage, radicle emerges out by rupturing the seed coat (Chen and Arora 2013) (Fig. 2.3).

The pretreatment of seeds with priming agents facilitates the active absorption of ionic molecules with greater ATP availability and repair of deteriorated seed parts for reducing leakage of metabolites leading to faster embryo growth (Dahal et al. 1990). It also, reflected in greater cellular membrane integrity, counteraction of lipid

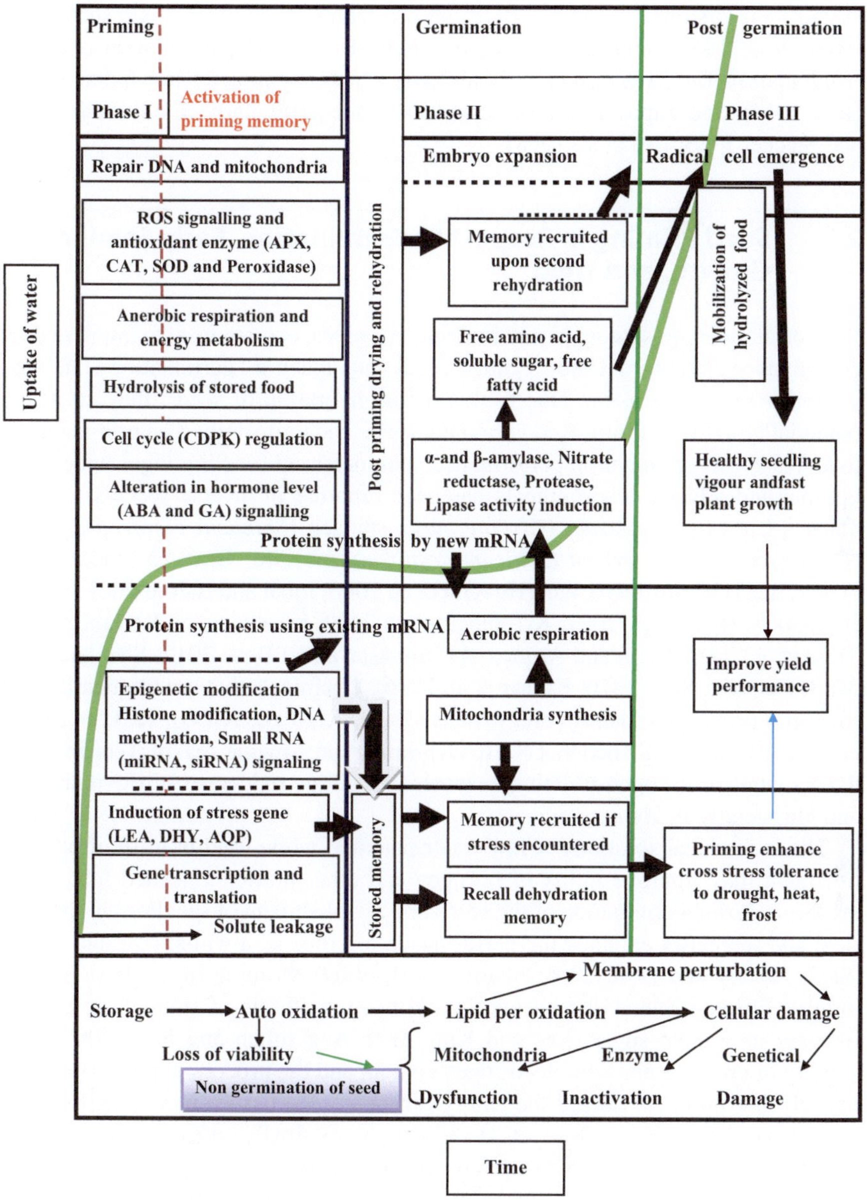

Fig. 2.3 Model for priming memory and its relation to abiotic stress signaling (Chen and Arora 2013)

peroxidation, and free radical chain reaction often are found to be directly correlated with the maintenance of viability and reduce moisture uptake by hydrated-dehydrated seed (Dollypan and Basu 1985), antipathogenic effects (Powell and Mathews 1986), repair of biochemical lesions by the cellular enzymatic repair system (Villiers and Edgcumbe 1975) and metabolic removal of toxic substances (Basu et al. 1973), counteraction of free radical and lipid peroxidation reactions (Rudrapal and Basu 1982), biochemical changes like enzyme activation (Sananda and Bose 2012), and improvement of germination rate particularly in old seeds (Gray and Steckel 1983; Lee et al. 1998).

2.5 Seed Priming in Respect to Germination, Early Seedling Growth, and Yield

Seed priming, a pre-sowing partial hydration of seeds, is often used to improve crop performance (Bradford 1986). Primed seeds thus reach a "germinating state" but without radicle protrusion. Due to this partial hydration of seeds, they adapt the potentiality to exhibit improved germination rate and uniformity. The priming leads to an enhanced germination performance, and increased seedling vigor under both optimal and adverse environments has been reported in diverse species, such as maize (*Zea mays*), soybean (*Glycine max*), spinach (*Spinacia oleracea*), pepper (*Capsicum annuum*), wheat (*Triticum aestivum*), mustard (Brassica juncea), sunflower, rice (*Oryza sativa*), etc. (Hussain et al. 2006; Iqbal and Ashraf 2007; Farooq et al. 2008; Bose et al. 2007; Krishnotar et al. 2009; Korkmaz and Korkmaz 2009; Zhou et al. 2009; Chen and Arora 2011; Srivastava and Bose 2012; Sheidaie et al. 2013; Pant and Bose 2016; Kumar et al. 2016). Chemicals that are frequently used for osmo-priming/hardening are, namely, NaCl, KH_2PO_4, K_3PO_4, KCl, etc. Bose et al. (1983a, b) established that $NaHCO_3$ priming (pre-sowing soaking) can improve the activities of nitrogen-utilizing enzymes like nitrate reductase, nitrite reductase, and nitrogenase in *Vigna mungo*.

These chemical solutions of low water potential while used in form of priming treatment may improve influx of nitrogen and other nutrients needed for protein synthesis during germination of the seeds and establishment of seedlings in the crop field and thereafter enhance the performance of crops as a whole (Du and Tuong 2002; Pandey and Bose 2006; Sharma et al. 2009). Priming of seeds in general induces different metabolic events like de novo synthesis of α-amylase with an increase in soluble sugar (Lee and Kim 2000; Anaytullah and Bose 2007), and activity of protease and soluble nitrogen content and the process of mobilization of hydrolyzing products toward the embryo from storage tissues in seeds (Bose and Srivastava 1980, 1982; Bose et al. 1982a, b, c) finally enhance the seedling establishment and seedling vigor, i.e., plant density, fertile tillers, test weight, number of grain per panicle, etc., compared with non-primed treated seeds (Bose and Mishra 1992; Du and Tuong 2002).

Among essential elements nitrogen plays the central role because it is one of the components of every enzyme/protein. However among nitrogen-containing salts, nitrate is known to act as dormancy breaking agent where ammonium salts are usually found less effective in this respect except in case of rice, and application of nitrate is observed to play an effective role in producing higher dry matter and grain yield (Bose et al. 1982a, b, c; Pandey and Bose 2006). Application of Mg salts as pre-sowing soaking treatment improves the protein of shoot and yield of mustard (Bose and Mishra 1997, 1999). Basra et al. (2004) and Sharma and Bose (2006) introduced a new technique for seed invigoration in which both seed hardening and osmo-conditioning were successfully integrated where seeds were hardened in various salt solutions instead of tap or distilled water; Basra et al. (2004) further concluded that osmohardening, using $CaCl_2$ solution (having an osmotic potential of −1.5 MPa), was best for vigor enhancement compared with other salts and simple hardening, whereas Sharma and Bose (2006) suggested $Mg(NO_3)_2$ is better than $K(NO_3)$ for this kind of seed treatment to wheat. Moeinzadeh et al. (2010) reported that bio-priming of sunflower seed with *Pseudomonas fluorescens* improves seed invigoration and seedling growth.

2.6 Seed Priming in Respect to Abiotic Stress

Different stressors, i.e., water deficit, heat, cold, salinity, etc., affect the crop yield. To counteract the effects of stress, plants undergo a process of stress acclimation via modulating various physiological and biochemical actions. Further, to feed the growing population of the world in future, there is a need to develop some techniques that can enhance yield in terms of grain even at stressed condition with not much application of nitrogen fertilizer, used commonly and save from nitrate pollution with minimum fertility loss and low cost, are the paramount challenges, tackled by plant scientists mainly seed physiologist. It has been noted that most of the types of stresses reduce the uptake of water in seed during germination. Further the reduced availability of water is observed to influence the process of the cell elongation; consequently it shows immense impact on the growth of the embryo followed by the seedling emergence at the time of seed germination.

Seed priming of rice varieties HUR-3022 and Sahabhagi Dhan with Mg $(NO_3)_2$ and K_2SO_4 was found to mitigate the inhibitory effect of PEG-6000 (creating the osmotic potential equivalent to –0.30 to –0.49 MPa) during seed germination. It is noted that in primed seeds the vigor index, germination index, and absolute water content percentage were increased as compared to non-primed PEG-treated seeds (Pant and Bose 2016). Therefore it seems to be an economically viable option for vertical intensification of yield potential in rice varieties. Salinity stress is known to trigger oxidative stress in plant tissues through the increase in reactive oxygen species (Apel and Hirt 2004). Chloroplasts are the major organelles producing the reactive oxygen species (ROS) such as the superoxide radical (O_2^-), hydrogen peroxide (H_2O_2), and singlet oxygen (${}_1O_2^*$) during photosynthesis (Asada 1992). The production of ROS can be particularly high when plants are exposed to salinity stress (Athar et al. 2008; Ashraf 2009). ROS cause chlorophyll degradation and membrane lipid peroxidation. So, malondialdehyde (MDA) accumulation as product of lipid peroxidation and chlorophyll retention are two oxidative stress indicators that are tested tools for determining salt tolerance in plants (Yildirim et al. 2008).

Srivastava et al. (2010) evaluated the effect of different seed priming methods to enhance the sodium chloride (NaCl) and polyethylene glycol-8000 (PEG-8000) stress tolerance in Indian mustard (*Brassica juncea* L.). Plant growth regulators salicylic acid (SA), ascorbic acid, and abscisic acid (ABA) while used as seed priming treatment to wheat seeds showed an improvement in germination and seedling growth in saline condition Afzal et al. (2006); SA(50 ppm) and ascorbic acid (50 ppm) primed seeds not only presented higher count for germination but also reduced the germination time and electrolyte leakage in saline and nonsaline conditions in primed seeds as compared to non-primed one, respectively; ABA was not found effective in this experiment. In presence of abiotic and biotic stresses, a number of plant proteins are formed, and their syntheses are activated by the plant hormones/PGR like SA and ABA (Jin et al. 2000). Therefore during the priming process, these chemicals (SA/ABA) may influence the synthesis of stress-regulating proteins in the seeds.

Heavy metals are known to inhibit the germination of field crops which resultantly decreases the elongation rates of roots and shoots, their dry mass, and the content of soluble proteins (Wang et al. 2003). Sethy and Ghosh (2013) studied about affectivity of various heavy metals like Pb, Ni, Cd, Co, Cr, and Hg on the events related to seed germination of various crops and established their toxic roles on the loss in productivity. Impact of $HgCl_2$ in respect to germination physiology of maize seed was studied by Bose et al. (2008) and observed that heavy metals start to show their inhibitory role from the first dynamic phase of plant's life, i.e., germination. In case of Hg it acts through the changes in the permeability of cell membrane via reacting with sulfhydryl (-SH-) groups with cations (Bose et al. 1983a, b); heavy metals also interfere through replacement of essential ions, oxidative stress, and having the affinity to react with phosphate groups of ATP or ADP. However, Kumar et al. (2016) reported that hydro(distilled) and hallow (Mg $(NO_3)_2$ and $Ca(NO_3)_2$ priming can mitigate the effects of heavy metal ($HgCl_2$) stress in wheat during the process of germination by improving the germination percentage, radical and plumule length, seedling emergence, soluble sugar content, and α-amylase activity in endosperm.

2.7 Seed Priming in Respect to Biotic Stress

Plants are sessile in nature and hence encountered with a number of biotic agents in their complete life cycle. These biotic agents may be bacteria, viruses, fungi, insects, nematodes, and protists. Presence of these biotic agents may influence the microenvironment of plant and also modulates the internal physiological and biochemical status of the system itself; consequently the yield potential of the cropping systems drops in the agricultural sector. But with the invent of modern technologies, the agriculture sector is continuously growing, breaking various yield barriers and enhancing crop productivity even in presence of various biotic agents, causing biotic stresses. Seed priming technology is used to overcome various biotic stresses, and for this purpose, some growth regulators are in use. Integration of microbial products, plant extracts, and some biotic agents when used either alone or in combination with some chemicals are referred as bio-priming.

Seed bio-priming is an important aspect of seed enhancement technology. Bio-priming is a type of seed priming that involves the coating of seed with various biocontrol agents such as *Pseudomonas chlororaphis*, *Trichoderma harzianum*, *P. fluorescens*, *B. subtilis*, *Streptomyces* sp., and *Gliocladium virens* (Callan et al. 1990; Nemec et al. 1996). Salicylic acid (SA) is a key molecule in the signal transduction pathway of biotic stress responses. Kuril (2010) observed that Kranti and Vardan varieties of mustard (*Brassica juncea* L.Cezern and Coss), while primed with salicylic acid, showed reduced percentage of disease index, caused by the fungus *Alternaria blight.*

El-Mohamedy et al. (2006) evaluated the efficacy of soil amendment with *Trichoderma harzianum* formulated on sugarcane bagasse and/or bio-priming seed treatment in controlling cowpea root rot pathogens under greenhouse and field conditions. The percentage of root rot diseases caused by *Fusarium solani, Rhizoctonia solani, and Macrophomina phaseolina* were reduced significantly. Nayaka et al. (2010) attempted the use of *T. harzianum* as seed treatment for the controlling maize ear rot and managing fumonisin (synthesized by *Fusarium verticillioides*) in maize seeds. Study showed that *T. harzianum* improves the seed germination and emergence, vigor index, plant height, test weight, yield and reduces the incidence of ear rot disease and the level of fumonisin.

Devi et al. 2013 showed that bio-priming with *Trichoderma harzianum* (NBAII-THIO) and *P. fluorescens* along with PGPRs (plant growth-promoting rhizobacteria) in cucumber (*Cucumis sativus* L.) enhances seedling vigor by increase germination percentage, shoot and root length, biocontrol efficiency, and lower level disease incidence. In other study, seed priming with *Trichoderma harzianum* (PGPFYCM-2; PGPFYCM-8; and PGPFYCM-14) promotes growth and induces resistance in sunflower against downy mildew caused by *Plasmopara halstedii*. Under field and greenhouse condition, susceptible sunflower cultivar improves vegetative and reproductive growth by improving NPK macronutrient uptake, plant height, early flowering, reduced crop duration, ear head size, and crop yield (Nagaraju et al. 2012). Also, Mastouri et al. (2010) showed that seed treatment with *Trichoderma harzianum* alleviates seed and seedling disease caused by *Pythium ultimum* and abiotic stress (osmotic, salinity, chilling, or heat stress) by overcoming physiological stress. They showed that bio-priming improves seed quality, overcomes oxidative damage, and elevates antioxidant system of plant.

Dual application of beneficial microorganisms used combination of one bacterial (*Pseudomonas chlororaphis* MA342 or *Pseudomonas fluorescens* CHA0) and one fungal isolate (*Clonostachys rosea* IK726d11 or *Trichoderma harzianum* T22) on onion and carrot as seed priming treatment; they observed that all microorganisms proliferated during the priming process on carrot seeds, and these treatments significantly affected the number of microorganisms recovered from the rhizosphere and ultimately improve the growth of root as well as plant (Bennett and Whipps 2008).

On-farm priming with hydro- and halo-priming reduces biotic stress tolerance, reduces disease severity, and increase yield. Many on-farm trials in countries like

India, Pakistan, and Bangladesh in many crops such as chick pea, mung bean, rice, pearl millet, and cowpea showed that on-farm priming improves yield by improving disease resistance traits and yield components (Jones et al.1995; Harris et al. 1999; Musa et al. 2001). An on-farm trial in Pakistan (Rashid et al. 2004a, b) showed that primed seeds of mung bean cv. NM 92 for 8 h in water resulted in a significant fivefold increase in grain yield relative to a non-primed crop. This was associated with a large difference in the severity of symptoms of mung bean yellow mosaic virus (MYMV) assessed using a visual scoring index.

Halo-priming is a pre-sowing soaking of seeds in salt solution, which enhance germination and seedling emergence uniformly under normal and adverse environmental conditions (Bose and Mishra 1999; Kausar and Ashraf 2003; Basra et al. 2005). The changes in the activities of phenylalanine ammonia lyase, chitinase, and beta-1,3 glucanase and phenolic content in groundnut were studied and observed an increase in phenolics content in leaves of SA pretreated and 5 days after inoculation of plants with *A. alternate* (Chitra et al. 2008).

Nitric oxide (NO) is a signaling molecule that takes part in pathophysiological and developmental processes and acts mainly against oxidative stress and also plays a role in plant-pathogen interactions. Potassium ferrocyanide, a structural analog of NO donor lacking NO moiety, failed to protect the pearl millet plants from downy mildew indicating a role for NO in induced host resistance reported by Manjunatha et al. (2008).

In presence of various stresses like salinity and drought, the plants generate defense mechanism via enhancing the contents of compatible solutes (proline, mannitol, malondialdehyde, soluble sugars, and quaternary ammonium), and improving the activities of protective enzymes, such as superoxide dismutase (SOD), peroxidase (POD), and catalase (CAT), are important indicators (Mittal and Dubey 1995; Bohnert and Shen 1999). Priming is noted to increase these factors which in turn enhance either crop's resistance or their tolerance capacity to drought, salinity, and late sowing stress, and the same is observed in case of biotic stresses (Wang and Shen 1991; Liao and Sun 1994; Kuril 2010; Kazemi and Eskandari 2012; Afzal et al. 2012). However SMP (solid matrix priming) in combination with *Trichoderma viride* can be successfully used to improve seedling emergence and productivity of okra under low temperatures (Pandita et al. 2010).

2.8 Impact of Seed Priming on Molecular Processes During Germination

It has been established that seed priming can increase seed germination, seedling vigor index, and germination potential (Hu et al. 2005); similar results were obtained by Pant and Bose (2016) with PEG-primed seeds of rice cultivars. PEG priming also shortens the time for seed emergence and enhanced percentage of germination under water-deficit condition. It also enhances or regulates the capacity of tolerance toward salt and chilling stress (Dursun and Ekinci (2010); Munir and Aftab (2009); Dong et al. (2013). Salah et al. (2015) studied about the influence of seed priming using polyethylene glycol on the physiological and molecular mechanism of rice cultivars of *Oryza sativa* under nano-ZnO stress. PEG priming significantly increased the level of photosynthetic pigments in presence of stress, but activities of superoxide dismutase (SOD), peroxidase (POD), catalase (CAT) enzymes, and melanoaldehyde contents were decreased in this situation. Expression of *APXa*, *APXb*, *CATa*, *CATb*, *CATc*, *SOD1*, *SOD2*, and *SOD3* genes were downregulated with PEG priming in respect to non-primed one under nano-ZnO stress. They concluded that PEG priming of rice can alleviate the toxic effect of nano-ZnO stress and improve the damaging level of leaf and roots.

During seed germination an extensive role is played by the water. Aquaporins has a putative role during the kinetic exchange of water which has been studied in *Arabidopsis*. Vander et al. (2006) observed their studies with microarrays carrying aquaporin gene-specific tags and antibodies raised against aquaporin subclasses that dry and young seedlings have two distinct aquaporin expressions. High and low expression of tonoplast intrinsic proteins (TIP) isoforms (TIP3;1, TIP3;2, and TIP5;1) and all the 13 plasma membrane aquaporins (PIPs) isoforms, respectively, were present in dry and germinating seeds, whereas aquaporins of TIP1, TIP2, and PIP subgroup expression are induced during seedling establishment. Proteomic analysis of the model plant Arabidopsis in unprimed and primed seeds during germination identified 1300 seed proteins by two-dimensional gels, in which an abundant change was observed in 74 proteins during germination, i.e., prior to radical emergence and radical protrusion step. This study further showed that during the dehydration process in priming of seeds, some new proteins have formed; one of them was cytosolic glyceraldehyde 3-phosphate dehydrogenase (Gallardo et al. 2001).

The physiological deterioration of seeds during storage and seed priming are closely associated with germination, hence related with plant growth and subsequent grain yields. Proteomics analysis based on the isobaric tandem mass-tag labeling using wheat seeds during different stages of artificial aging (45 °C; 50% relative humidity; 98%, 50%, 20%, and 1% germination rates) and priming (hydro-priming treatment) and observed that a total of 162 differentially expressed proteins (DEPs), identified during artificial aging, are mainly involved in metabolism, energy supply, and defense/stress responses; this indicated that seed deterioration leads to the incremental decomposition of the stored substance, which presented an inability to protect the seeds against aging (Lv et al. 2016). They also reported that upregulated proteins involved in seed aging are mainly enriched in ribosome, whereas the downregulated proteins are mainly accumulated in energy supply (starch and sucrose metabolism) and stress defense (ascorbate and aldarate metabolism), and proteins regarded as new markers of seed deterioration are hemoglobin 1, oleosin, agglutinin, and non-specific lipid-transfer proteins by using Kyoto Encyclopedia of Genes and Genomes (KEGG) analysis. During seed priming 531 DEPs are recognized in respect to non-primed seeds, and several upregulated DEPs are found to involve in energy supply (the processes are glycolysis, TCA cycle, and fatty acid oxidation) anabolic processes, like synthesis various amino acids and fats and finally the growth and division of cell. Through KEGG and protein-protein interaction analysis, it has been established that upregulated proteins in seed priming are mainly enriched in amino acid synthesis, stress defense (plant-pathogen interactions and ascorbate and aldarate metabolism), and energy supply (oxidative phosphorylation and carbon metabolism). Therefore, these studies open a channel to understand how seed priming helps in the maintenance of seed vigor and optimize germination enhancement treatments. This work adds new proteomic insights into protein changes, occurred during seed deterioration and priming.

2.9 Future Prospect

Seed priming technique is innovative, cheap, and easy to apply at farmer's field conditions. Also, many successful case studies were reported regarding seed priming in amelioration of biotic and abiotic stress till date. But, until now, complete molecular mechanism of seed priming during various stresses is unknown. Therefore, future goals of agricultural scientists are to identify novel genes, proteins, and transcription factors, which are expressed during stresses, and to improve our knowledge regarding crop behavior under drastic climatic change scenario. Then only, we can achieve our need of food security for everyone in near future.

References

Afzal I, Basra SMA, Farooq M, Nawaz A (2006) Alleviation of salinity stress in spring wheat by hormonal priming with ABA, salicylic acid and ascorbic acid. Int J Agric Biol 8:23. 1560-8530/2006/08-1-23-28http://.fspublishers.org

Afzal I, Butt A, Rehman HU, Basra SMA, Afzal A (2012) Alleviation of salt stress in fine aromatic rice by seed priming. Aust J Crop Sci 6:1401–1407

Anaytullah, Bose B (2007) Nitrate-hardened seeds increase germination, amylase activity and proline content in wheat seedlings at low temperature. Physiol Mol Biol Plants 13:199–207

Anaytullah, Srivastava AK, Bose B (2012) Impact of seed hardening treatment with nitrate salts on nitrogen and anti oxidant defense metabolisms in *Triticum aestivum* L under different sowing conditions. Vegetos 25:292–299

Apel K, Hirt H (2004) Reactive oxygen species: metabolism, oxidative stress, and signal transduction. Annu Rev Plant Biol 55:373–399

Araújo SDS, Paparella S, Dondi D, Bentivoglio A, Carbonera D, Balestrazzi A (2016) Physical methods for seed invigoration: advantages and challenges in seed technology. Front Plant Sci 7:646

Asada K (1992) Ascorbate peroxidase–a hydrogen peroxide-scavenging enzyme in plants. Physiol Plant 85:235–241

Ashraf M (2009) Biotechnological approach of improving plant salt tolerance using antioxidants as markers. Biotechnol Adv 27:84–93

Ashraf M, Foolad MR (2005) Pre-sowing seed treatment—a shotgun approach to improve germination, plant growth, and crop yield under saline and non-saline conditions. Adv Agron 88:223–271

Athar HUR, Khan A, Ashraf M (2008) Exogenously applied ascorbic acid alleviates salt-induced oxidative stress in wheat. Environ Exp Bot 63:224–231. https://doi.org/10.1016/j.envexpbot.2007.10.018

Basra SMA, Farooq M, Hafeez K, Ahmad N (2004) Osmohardening: a new technique for rice seed invigoration. Int Rice Res Notes 29:80–81

Basra SMA, Afzal I, Anwar S, Shafique M, Haq A, Majeed K (2005) Effect of different seed invigoration techniques on wheat (Triticum aestivum L.) seeds sown under saline and non-saline conditions. J Seed Technol 28:36–45

Basu RN, Chattopadhyay K, Pal P (1973) Maintenance of seed viability in rice (*Oryza sativa* L.) and Jute (*Corchorus capsularis* L. and *C. olitorius* L.). Indian Agric 18:76–79

Bennett AJ, Whipps JM (2008) Dual application of beneficial microorganisms to seed during drum priming. Appl Soil Ecol 38:83–89

Bishaw Z, Niane AA, Gan Y (2007) Quality seed production. Lentil:349–383

Bohnert HJ, Shen B (1999) Transformation and compatible solutes. Sci Hortic 78:237–260

Bose B, Mishra T (1992) Response of wheat seed to pre sowing seed treatment with Mg (NO_3) 2. Ann Agric Res 13:132–136

Bose B, Mishra T (1997) Effect of Mg salts on seed germination of Brassica sp. *Cruciferae Newsletter* (United Kingdom)

Bose B, Mishra T (1999) Influence of pre-sowing soaking treatment in *Brassica juncea* seed with Mg salt on growth, nitrate reductase activity, total protein content and yield responses. Physiol Mol Biol Plant 5:83–88

Bose B, Srivastava HS (1980) Proteolytic activity and nitrogen transfer in maize seed during imbibition. Biol Plant 22:414–419

Bose B, Srivastava HS (1982) Protein hydrolysis and nitrogen transfer during germination of maize seeds at various temperatures. Indian J Plant Physiol 25:141–148

Bose B, Tandon A (1991) Effect of magnesium nitrate on metabolism in germinating maize seeds. Indian J Plant Physiol 34:69–71

Bose B, Srivastava HS, Mathur SN (1982a) Effect of antibiotic on germination and protein activity of maize seeds. Indian J Plant Physiol 25:271–275

Bose B, Srivastava RC, Mukherjee D, Mathur SN (1982b) Multipurpose use of $NaHCO_3$ to *Vigna mungo*. In: Proceeding of the national symposium of biological nitrogen fixation, New Delhi, 298–309

Bose B, Srivastava HS, Mathur SN (1982c) Effect of some nitrogenous salt on nitrogen transfer and protease activity in germination of maize (*Zea Mays* L.) seeds. Biol Plant 24:89–95

Bose B, Srivastava HS, Mathur SN (1983a) Partial purification and properties of protease from maize seed endosperm. Beitage Biologie der Pflanzen 58:383–391

Bose B, Srivastava RC, Singh VK, Mathur SN (1983b) Sodium hydrogen carbonate: a potent electron donor for the nitrogen utilizing enzymes in Vigna mungo. Biochem Physiol Pflanz 178:443–448

Bose B, Kumar R, Kuril SK, Srivastava AK (2007) Hardening of mustard seeds with Mg(NO3)2 increases seed germination, vegetative growth, nitrogen assimilation and yield. Brassica 9:33–38

Bose B, Kumari S, Anaytullah SAK, Kuril SK, Singh PK (2008) Phyto-toxic role of mercuric chloride on germination physiology and seedling growth of maize (Zea mays L.) var. Jaunpuri. Indian J Plant Physiol 13(3):284–290

Bradford KJ (1986) Manipulation of seed water relations via osmotic priming to improve germination under stress conditions. Hort Sci (USA) 21:1105

Callan NW, Mathre D, Miller JB (1990) Bio-priming seed treatment for biological control of Pythium ultimum preemergence damping-off in sh-2 sweet corn. Plant Dis 74:368–372

Chen K, Arora R (2011) Dynamics of the antioxidant system during seed osmopriming, post-priming germination, and seedling establishment in spinach (Spinacia oleracea). Plant Sci 180:212–220

Chen K, Arora R (2013) Priming memory invokes seed stress-tolerance. Environ Exp Bot 94:33–45

Chitra K, Ragupathi N, Dhanalakshmi K, Mareeshwari P, Indra N, Kamalakannan A, Sankaralingam A, Rabindran R (2008) Salicylic acid induced systemic resistant on peanut against Alternaria alternata. Arch Phytopathol Plant Protect 41:50–56

Dahal P, Bradford KJ, Jones RA (1990) Effects of priming and endosperm integrity on seed germination rates of tomato genotypes II. J Exp Bot 41:1441–1453
Devi TS, Shivaprakash MK, Maina CC (2013) Efficacy of seed bio-priming in enhancing seedling vigour of cucumber (Cucumis sativus L.) under biotic stress conditions. Mysore J Agric Sci 47:107–111
Dieter F, Bouman F (1995) The seed: structure and function, Seed Development and Germination. Marcel Dekker, New York, pp 1–24
Dollypan, Basu RN (1985) Mid-storage and pre-sowing seed treatments for lettuce and carrot. Sci Hortic 33:1026–1027
Dong X, Bi H, Wu G, Ai X (2013) Drought-induced chilling tolerance in cucumber involves membrane stabilisation improved by antioxidant system. Int J Plant Prod 7:67–80
Du LV, Tuong TP (2002) Enhancing the performance of dry-seeded rice: effects of seed priming, seedling rate, and time of seedling. In: Direct seeding: research strategies and opportunities. International Research Institute, Manila. 241–256
Dursun A, Ekinci M (2010) Effects of different priming treatments and priming durations on germination percentage of parsley (*Petroselinum crispum* L.) seeds. Agric Sci 1:17–23
El-Mohamedy RSR, Abd Alla MA, Badiaa RI (2006) Soil amendment and seed bio-priming treatments as alternative fungicides for controlling root rot diseases on cowpea plants in Nobaria Province. Res J Agric Biol Sci 2:391–398
Farooq M, Basra SMA, Rehman H, Saleem BA (2008) Seed priming enhances the performance of late sown wheat (*Triticumaestivum* L.) by improving chilling tolerance. J Agron Crop Sci 194:55–60
Gallardo K, Job C, Groot SP, Puype M, Demol H, Vandekerckhove J, Job D (2001) Proteomic analysis of Arabidopsis seed germination and priming. Plant Physiol 126:835–848
Gray D, Steckel JR (1983) Seed quality in carrots: the effects of seed crop plant density, harvest date and seed grading on seed and seedling variability. J Hortic Sci 58:393–401
Guan YJ, Hu J, Wang XJ, Shao CX (2009) Seed priming with chitosan improves maize germination and seedling growth in relation to physiological changes under low temperature stress. J Zhejiang University-Sci B 10:427–433
Halmer P (2006) Seed technology and seed enhancement. In: XXVII international Horticultural Congress-IHC2006: international symposium on seed enhancement and seedling production 771, pp 17–26
Harris D, Joshi A, Khan PA, Gothkar P, Sodhi PS (1999) On-farm seed priming in semi-arid agriculture development and evaluation in maize, rice and chickpea in India using participatory methods. Exp Agric 35:15–29
Hu J, Zhu ZY, Song WJ, Wang JC, Hu WM (2005) Effect of sand priming on germination and field performance in direct-sown rice (*Oryza sativa* L.). Seed Sci Technol 33:243–248
Hussain M, Farooq M, Basra SM, Ahmad N (2006) Influence of seed priming techniques on the seedling establishment, yield and quality of hybrid sunflower. Int J Agric Biol 8:14–18
Iqbal M, Ashraf M (2007) Seed treatment with auxins modulates growth and ion partitioning in salt-stressed wheat plants. J Integr Plant Biol 49:1003–1015
Jin S, Chen CCS, Plant AL (2000) Regulation by ABA of osmotic stressinduced changes in protein synthesis in tomato roots. Plant Cell Environ 23: 51–60
Jones ES, Liu CJ, Gale MD, Hash CT, Witcombe JR (1995) Mapping quantitative trait loci for downy mildew resistance in pearl millet. TAG Theor Appl Genet 91:448–456
Kausar A, Ashraf M (2003) Alleviation of salt stress in pearl millet (Pennisetum glaucum (L.) R. Br.) through seed treatments. Agronomie 23:227–234
Kazemi K, Eskandari H (2012) Does priming improve seed performance under salt and drought stress? J Basic Appl Sci Res 2:3503–3507
Korkmaz A, Korkmaz Y (2009) Promotion by 5-aminolevulenic acid of pepper seed germination and seedling emergence under low-temperature stress. Sci Hortic 119:98–102
Krishnotar BB, Srivastava AK, Shahi JP (2009) Response of rabi maize crop to seed invigoration with magnesium nitrate and distilled water. Indian J Plant Physiol 14:71–77

Kumar M, Pant B, Mondal S, Bose B (2016) Hydro and halo priming: influenced germination responses in wheat Var-HUW-468 under heavy metal stress. Acta Physiol Plant 38:217

Kuril SK (2010) Influence of seed hardening with $Mg(NO_3)_2$ and salicylic acid on timely and late sown mustard (*Brassica juncea* L. Czern and Coss) varieties, Ph.D. thesis, BHU, India

Lee S-S, Kim J-H (2000) Total sugars, $\alpha $-amylase activity, and germination after priming of normal and aged rice seeds. Kor J Crop Sci 45:108–111

Lee SS, Kim JH, Hong SB, Yun SH, Park EH (1998) Priming effect of rice seeds on seedling establishment under adverse soil conditions. Korean J Crop Sci 43:194–198

Liao XR, Sun Q (1994) Cobaltous ion inhibits the inducement of mungbean draught resistance by seed priming. Seed 3:29–32

Lv Y, Zhang S, Wang J, Hu Y (2016) Quantitative proteomic analysis of wheat seeds during artificial ageing and priming using the isobaric tandem mass tag labeling. PLoS One 11:e0162851

Malnassy PG (1971) Physiological and biochemical studies on a treatment hastening the germination of seeds at low temperature, Ph.D. thesis, Rutgers University, New Brunswick

Manjunatha G, Raj SN, Shetty NP, Shetty HS (2008) Nitric oxide donor seed priming enhances defense responses and induces resistance against pearl millet downy mildew disease. Pestic Biochem Physiol 91:1–11

Mastouri F, Björkman T, Harman GE (2010) Seed treatment with Trichoderma harzianum alleviates biotic, abiotic, and physiological stresses in germinating seeds and seedlings. Phytopathology 100:1213–1221

Miller G, Suzuki N, Ciftci-Yilmaz S, Mittler R (2009) Reactive oxygen species homeostasis and signaling during drought and salinity stresses. Plant Cell Environ 33:453–467

Mittal R, Dubey RS (1995) Influence of sodium chloride salinity on polyphenol oxidase, indole 3-acetic acid oxidase and catalase activities in rice seedlings differing in salt tolerance. Trop Sci 35:141–149

Moeinzadeh A, Sharif-Zadeh F, Ahmadzadeh M, Tajabadi F (2010) Biopriming of sunflower ('Helianthus annuus' L.) seed with 'Pseudomonas fluorescens' for improvement of seed invigoration and seedling growth. Aust J Crop Sci 4:564

Mondal S, Vijai P, Bose B (2011) Role of seed hardening in rice variety Swarna (MTU 7029). Res J Seed Sci 4:157–165

Munir N, Aftab F (2009) The role of polyethylene glycol (PEG) pretreatment in improving sugarcane's salt (NaCl) tolerance. Turk J Bot 33:407–415

Musa AM, Harris D, Johansen C, Kumar JVDK (2001) Short duration chickpea to replace fallow after aman rice: the role of on-farm seed priming in the High Barind Tract of Bangladesh. Exp Agric 37:509–521

Nagaraju A, Sudisha J, Murthy SM, Ito SI (2012) Seed priming with Trichoderma harzianum isolates enhances plant growth and induces resistance against Plasmopara halstedii, an incitant of sunflower downy mildew disease. Australas Plant Pathol 41(6):609–620

Nayaka CS, Niranjana SR, Uday Shankar AC, Niranjan Raj S, Reddy MS, Prakash HS, Mortensen CN (2010) Seed biopriming with novel strain of Trichoderma harzianum for the control of toxigenic Fusarium verticillioides and fumonisins in maize. Arch Phytopathol Plant Protect 43:264–282

Nemec S, Datnoff LE, Strandberg J (1996) Efficacy of biocontrol agents in planting mixes to colonize plant roots and control root diseases of vegetables and citrus. Crop Prot 15:735–742

Pandey D, Bose B (2006) Influence of pre-sowing seed treatment with nitrate salts and different sowing dates on performance of mustard. Indian J Plant Physiol 11:261–265

Pandita VK, Anand A, Nagarajan S, Seth R, Sinha SN (2010) Solid matrix priming improves seed emergence and crop performance in okra. Seed Sci Technol 38:665–674

Pant B, Bose B (2016) Mitigation of the influence of PEG-6000 imposed water stress on germination of halo primed rice seeds. Int J Agric Environ Biotechnol 9:275

Powell AA, Mathews S (1986) Cell membranes and seed leachate conductivity in relation to the quality of seeds for sowing. J Seed Technol 10:81–100

Rajjou L, Duval M, Gallardo K, Catusse J, Bally J, Job C, Job D (2012) Seed germination and vigor. Annu Rev Plant Biol 63:507–533

Rashid A, Harris D, Hollington P, Ali S (2004a) On-farm seed priming reduces yield losses of mungbean (Vigna radiata) associated with mungbean yellow mosaic virus in the North West Frontier Province of Pakistan. Crop Prot 23:1119–1124

Rashid A, Harris D, Hollington P, Rafiq M (2004b) Improving the yield of mungbean Vigna radiata in the North West Frontiers province of Pakistan using on farm seed priming. Exp Agric 40:233–244

Rudrapal AB, Basu RN (1982) Use of chlorine and bromine in controlling mustard seed deterioration. Seed Res 9:188–191

Salah SM, Yajing G, Dongdong C, Jie L, Aamir N, Qijuan H, … Jin H (2015) Seed priming with polyethylene glycol regulating the physiological and molecular mechanism in rice (Oryza sativa L.) under nano-ZnO stress. Sci Rep:5

Sananda M, Bose B (2012) Kinetics studies on α-amylase extracted from germinating wheat endosperm of primed and non-primed seeds. Indian J Agric Biochem 25:137–141

Sethy SK, Ghosh S (2013) Effect of heavy metals on germination of seeds. J Natl Sci Biol Med 4:272–275

Sharma MK, Bose B (2006) Effect of seed hardening with nitrate salts on seedling emergence, plant growth and nitrate assimilation of wheat (*Triticumaestivum* L.). Physiol Mol Biol Plants 12:173–176

Sharma MK, Bose B, Shrivastava AK (2009) Effect of seed hardening with nitrate salts on physiological attributes at ear head emergence stage and yield of wheat (Triticum aestivum L.). Int J Agric Sci 5:439–442

Sheidaie S, Divsalar M, Oskouei B, Sadeghi H, Rezvani E (2013) Seed hydro-priming application to alleviation of water stress during germination in sunflower hybrids (Helianthus annuus L.). Int J Agric Crop Sci 5:155

Srivastava AK, Bose B (2012) Effect of nitrate seed priming on phenology, crop growth rate and yield attributes in rice (*Oryza sativa* L.). Vegetos 25:174–181

Srivastava AK, Lokhande VH, Patade VY, Suprasanna P, Sjahril R, D'Souza SF (2010) Comparative evaluation of hydro-, chemo-, and hormonal-priming methods for imparting salt and PEG stress tolerance in Indian mustard (Brassica juncea L.). Acta Physiol Plant 32:1135–1144

Taylor AG, Klein DE, Whitlow T (1988) SMP: solid matrix priming of seeds. Sci Hortic 37:1–11

Taylor AG, Allen PS, Bennett MA, Bradford KJ, Burris JS, Misra MK (1998) Seed enhancements. Seed Sci Res 8:245–256. https://doi.org/10.1017/S0960258500004141

Vander WC, Postaire O, Tournaire-Roux C, Boursiac Y, Maurel C (2006) Expression and inhibition of aquaporins in germinating Arabidopsis seeds. Plant Cell Physiol 47:1241–1250

Villiers TA, Edgcumbe DJ (1975) On the cause of seed determination in dry storage. Seed Sci Technol 3:761–774

Wang X, Shen B (1991) Effect of MET-soaked seed on drought resistance of rice seedlings. Acta Phytophysiol Sin 17:105–108. (in Chinese with English abstract)

Wang W, Vinocur B, Altman A (2003) Plant responses to drought, salinity and extreme temperatures towards genetic engineering for stress tolerance. Planta 218:1–14

Yildirim B, Yaser F, Ozpay T, Ozpay DT, Turkozu D, Terziodlu O, Tamkoc A (2008) Variations in response to salt stress among field pea genotypes (Pisum sativum sp. arvense L.). J Anim Vet Adv 7:907–910

Zhou ZS, Guo K, Elbaz AA, Yang ZM (2009) Salicylic acid alleviate mercury toxicity by preventing oxidative stress in roots of *Medicago sativa*. Environ Exp Bot 65:27–34

Seed Priming: An Emerging Technology to Impart Abiotic Stress Tolerance in Crop Plants

3

Shambhu Krishan Lal, Sudhir Kumar, Vijay Sheri, Sahil Mehta, Panditi Varakumar, Babu Ram, Bhabesh Borphukan, Donald James, Dhirendra Fartyal, and Malireddy K. Reddy

Abstract

Crop plants encounter a complex set of abiotic and biotic stresses very frequently. Abiotic stresses, being unavoidable, have major negative impact on crop production worldwide. These stresses such as inadequate and inconsistent rainfall, alkalinity, salinity, extreme temperature, and some other factors aren't only limiting crop yield but also seem to be inevitably worsening. Considering present situation, it is imperative to switch to some more sophisticated techniques that shall combat abiotic environmental challenges and improve crop yield efficiently. Among these, seed priming is a commonly utilized technology for enhancing seed vigor and stress tolerance. Seed priming involves the attainment of a specific physiological state by synthetic or natural compounds. Crop plants raised from primed seeds exhibit instant cellular response against abiotic stresses. Primed seed acquire resistance through various cellular and metabolic pathways which involves cascades of signaling networks. Studies, till date, have confirmed that primed seeds have several advantages over traditionally used methods which include uniform germination, reduction in germination and emergence time, and broad range of tolerance against disease and environmental stresses. Seed priming methods are widely used as an emerging technology to produce tolerant crop varieties against abiotic stresses.

S. K. Lal (✉)
International Centre for Genetic Engineering and Biotechnology, New Delhi, India

ICAR-Indian Institute of Agricultural Biotechnology, Ranchi, Jharkhand, India

S. Kumar
ICAR-Indian Institute of Agricultural Biotechnology, Ranchi, Jharkhand, India

V. Sheri · S. Mehta · P. Varakumar · B. Ram · B. Borphukan · D. James · D. Fartyal · M. K. Reddy
International Centre for Genetic Engineering and Biotechnology, New Delhi, India

© Springer Nature Singapore Pte Ltd. 2018
A. Rakshit, H. B. Singh (eds.), *Advances in Seed Priming*,
https://doi.org/10.1007/978-981-13-0032-5_3

Keywords
Abiotic stress · Seed priming · Seed germination · Biomarker · Sustainable agriculture

3.1 Introduction

In nature, plants are continuously exposed to number of environmental stresses during different developmental stages. The abiotic stress which adversely affects plant growth includes drought, extreme temperature, salinity, and other crucial factors (Zhao et al. 2007). In agriculture, abiotic stresses on standing crops lead to severe economic yield loss and affect plant health (Jakab et al. 2005). Drought, salinity, and extreme temperature cause osmotic stress on crop plants which create an imbalance at cellular, molecular, and physiological levels which ultimately leads to plant death (Xiong and Zhu 2002). Environmental stresses such as cold, radiation, light, and high temperature are also reported to affect the growth of the plants (Reyes and Cisneros-Zevallos 2007). In natural environment, crops are being exposed to various abiotic and biotic stresses, and to thrive well under such situations, crops have developed intricate mechanism to sense the signals and respond according to external environmental stimulus resulting in modulation of plant metabolites; activation of hormone signaling pathways that are regulated by plant hormones like abscisic acid, salicylic acid, jasmonate, and ethylene; and free radical generation (Fujita et al. 2006). In response to environmental and biotic stress, plant generates numerous volatile organic molecules that play key roles in important defense and signaling processes. Under stress conditions, the amount of emitted volatile organic compounds relies on stress tolerance, time duration, and severity of stresses (Niinemets 2009). Abiotic stresses like drought and salinity are complex traits, governed by a complete set of parameters. The improvement of crop plants through direct selection-based conventional breeding for drought and salinity stress appears to be quite complex (Cushman and Bohnert 2000; Flowers et al. 2000).

In agriculture, other innovative technologies such as plant tissue culture, seed priming, and genetic engineering could play a major role in increasing yield and productivity. Genetic engineering and genome editing technologies, developed for targeted insertion of foreign genes into desired crop cultivars, lead to the creation of designer crops (Gust et al. 2010). The transgenic crop has demerits of pleiotropic effect and gene silencing. Due to above limitation of available technology, we should think of some alternate technologies such as priming, mutagenesis, and tissue culture for combating abiotic stresses. Seed priming is a very simple, cost-effective, and sustainable approach which could be easily carried forward into the field.

3.2 Seed Priming

Priming involves prior exposure to elicitors which brings a cellular state that hinders the harmful effects of abiotic stress, and plants raised after priming are more tolerant of abiotic stress. Priming induces resistance upon pathogen attack and confers enhanced disease protection to the plant (Van Hulten et al. 2006). Seed priming involves soaking of seed in water or other solvents for a certain period that leads to changes in metabolic profile of the seed (Hussain et al. 2016).

Seed priming enhances the germination rate and uptake of nutrients and also controls seed-borne pathogens by influencing the pre-germination metabolic activities (Taylor and Harman 1990). It has the benefit of enhanced and uniform germination along with better yield and high vigor in floriculture and vegetable crops (Bruggink et al. 1999) and field crops (Kaur et al. 2005).

Primed seed exhibits significantly improved germination and greater germination uniformity and plant performance (Basra et al. 2005). Seed priming techniques such as halopriming, chemical priming, osmopriming, hormone priming, hydropriming, and nutrient priming are being used in rice for many environmental stresses (Jisha et al. 2013; Paparella et al. 2015). Seed priming technique could be adopted to enhance germination in soils that are contaminated with heavy metals (Sneideris et al. 2015). The imbibition of pepper seeds in water or other solution containing NaCl resulted in an increased germination percentage (Khan et al. 2009).

The major obstacle for practical application of primed seeds is storage and viability (McDonald 2000). This limitation could be addressed by knowing genes/markers for seed germination, and identified markers could be used for assessing the effect of priming on germination efficiency and seed vigor (Job et al. 2000). Genes/markers involved in rice seed priming were identified by comparing differential proteins between the dry and imbibed seed using two-dimensional electrophoresis (Cheng et al. 2017). Improved germination was reported in polyethylene glycol-6000 (PEG-6000) primed rapeseeds which, on germination, resulted in differential expression of 952 genes and 75 proteins (Kubala et al. 2015).

3.3 Physiological and Subcellular Basis of Seed Priming

Seed priming results in improved germination, vigor, and plant stand. Many previous studies have elucidated the seed priming mechanisms at various physiological, biochemical and cellular conditions using different kinds of spectrophotometric techniques in crop plants. Different proteins specifically appear during seed hydropriming and osmopriming, which are identified by MALDI-TOF in model plant *Arabidopsis thaliana* (Gallardo et al. 2004). There is accumulation of degradation product of B-subunit of 11S globulin and suggested enzyme involved in mobilization of storage protein during seed priming in sugar beet (Job et al. 2000). Activation of reserve mobilization enzymes, viz., carbohydrate and lipid mobilization enzymes, takes place during seed priming. Seed priming drives the synthesis of free radical scavenging enzymes, with expression of catalase and superoxide dismutase being increased. These enzymes protect the vital cell membrane and organs against oxidative stress induced during priming. Studies on gene expression profiling in osmoprimed seeds revealed genes are being expressed which are responsible for cellular metabolism (Soeda et al. 2005). Damage of DNA occurs inside the seed, which accumulates during seed aging, and is repaired to a great extent by accelerated hydration during early hours of germination (Thornton et al. 1993).

Comparing transcript profiling of rice seedlings under submergence stress reveals genes were differentially expressed in primed and non-primed seed and priming alleviated the harmful effect of submergence stress (Hussain et al. 2016). On constructing and sequencing small RNA libraries from water-soaked seed embryos under salt, drought, and control treatments, there is significant downregulation of six microRNA (miRNA) families and upregulation of one miRNA family, and further analysis revealed miRNA participate in regulation of plant hormone synthesis and play a crucial role in seed germination under environmental stress (Jian et al. 2016).

3.4 Role of Seed Priming Technique in Combating Abiotic Stress

3.4.1 Germination and Growth Promotion

Seed priming exhibited a promotive effect on seedling emergence, seedling growth, and plant performance in sesame (Shabbir et al. 2014). In sunflower priming with pre-optimized levels of priming reagents such as salicylic acid, H_2O_2, thiourea, ascorbic acid, gibberellic acid, sodium chloride, freezing, and thawing induces metabolic changes that lead to improvement in germination, reduced germination time, and promotive effect on seedling growth (Wahid et al. 2008). Hydration treatment and redrying increase germination efficiency and seedling emergence in many species of *Brassica* (Jett et al. 1996). Seed priming improved the germination and vigor of wheat crop that leads to 17% increase in grain yield than non-primed seed (Farooq et al. 2008). Priming with potassium nitrate and hydropriming resulted in better germination and enhanced plant performance in cotton (Rezaee et al. 2015). Seedling emergence involves interaction between soil matrix potential, soil aggregate size, and hydropriming in cotton and maize (Murungu et al. 2003). Seed primed with $CuSO_4$, $ZnSO_4$, and Na_2SO_4 significantly increased the germination in maize (Foti et al. 2008). Osmoriming and hydropriming increased the seedling emergence in sorghum (Moradi and Younesi 2009). Hydropriming increased the yield as a result of better germination and vigorous growth in mung bean (Rashid et al. 2004)

3.4.2 Salinity Stress

Under salinity stress, germination percentage and germination index decrease significantly, and also there is a decrease in chlorophyll a, chlorophyll b, and carotenoid content in maize. During germination, seed priming induces metabolic changes that help in better acclamation under salinity stress in maize (Sali et al. 2015). Seed primed with $CaCl_2$ followed by KCl induces salt tolerance in rice cultivar that is revealed by enhanced germination efficiency, seedling growth, and dry weight under saline medium (Afzal et al. 2012). High concentration of NaCl reduced the seed germination in wheat cultivars (Akbari et al. 2007). Seed treated with H_2O_2 showed improved salt tolerance in wheat cultivars (Wahid et al. 2007). Mustard seed primed with water, $CaCl_2$, and abscisic acid exhibited higher germination, and crop raised from primed seed contains high dry weight and chlorophyll content under common salt and PEG stress (Srivastava et al. 2010a). Thiourea supplementation in roots of *Brassica juncea* imparts tolerance against salt stress possibly by maintaining water homeostasis (Srivastava et al. 2010b). Hydropriming and KNO_3 treatment significantly improved germination and seedling growth under salinity and water deficit stress (Kaya et al. 2006). Halopriming enhances the growth under salt and drought stress in sugarcane cultivars (Patade et al. 2009). Halopriming counteracts salt stress by increasing the synthesis of antioxidant enzymes, H_2O_2 production, and accumulation of proline in mung bean (Saha et al. 2010). Soaked seed in water exhibited higher grain and straw yield than non-primed seed under saline-sodic condition in barley (Rashid et al. 2006).

3.4.3 Drought Stress

Seed priming with $CaCl_2$ improved seedling emergence, seedling establishment, plant height, tillers number, grain number, grain weight and yield under drought stress in wheat (Hussain et al. 2013). Seed priming with urea, KNO_3 increases germination efficiency, seedling growth, root length, proline and protein content under salinity and drought stress in maize hybrids (Anoshehet et al. 2011). Osmoprimed rice seed with KCl and $CaCl_2$ enhances seed vigour evidenced by improved germination, seedling emergence, kernel yield, number of fertile tiller, straw yield and harvest index. Rapid and uniform germination occurs due to elevated α-amylase activity and enhanced starch hydrolysis there by more sugar are available for embryo and seedling growth that leads to improvement in yield and quality (Farooq et al. 2006). The hydropriming was also used as a seed invigoration technique in maize inbred lines for effective seed germination in the presence of salinity and drought stress (Janmohammadi et al. 2008). The primed and non-primed rice seed performed differently under water deficit condition and the amount of proline and soluble protein are found higher in primed seed (Yuan-Yuan et al. 2010). Seed primed with water, PEG and KCl showed enhanced germination in winter wheat cultivar but priming doesn't benefitted field emergence (Ghana and Schillinger, 2003). Seed primed with water and mannitol showed more seedling growth as compared to non-primed seed under water deficit condition in chickpea (Kaur et al. 2002). Hydropriming increased the germination under water and temperature stress in cotton (Casenave and Toselli 2007).

3.4.4 Temperature Stress

During different reproductive stages of wheat, the heat stress poses a negative impact on final grain yield by affecting multiple traits like grain weight, seed size, tiller size and seed moisture content (Iqbal et al. 2017). Osmopriming in late sown wheat resulted into increased number of tillers, biological yield, and harvest index (Mustafa et al. 2017). Chilling stress or low-temperature stress also has a detrimental effect on normal functioning and biological system of crop plants. Cotton seeds primed with PEG-6000, KNO_3, KH_2PO_4, NaCl, and mannitol were evaluated for the best priming agent under cold stress. Priming with KNO_3 gives the best result for germination rate and 2 h priming with KNO_3 and then drying enables seed for cold tolerance and vigorous seedling (Cokkizgin and Bolek 2015). Maize seed primed with chitosan solutions exhibited enhanced germination, seedling growth, and plant biomass at low temperature (Guan et al. 2009). The water-soaked maize seeds were able to germinate below the optimum temperature and resulted in an improved crop growth and higher yield potential (Finch-Savage et al. 2004). In chickpea, hydropriming and osmopriming with different concentrations of PEG induce faster germination at temperature ranging from 5 to 32 °C (Elkoca et al. 2007) Figure 3.1 Diagramatic dipiction of seed priming induced plant acclimation to various environmental stresses (Tanou et al. 2012).

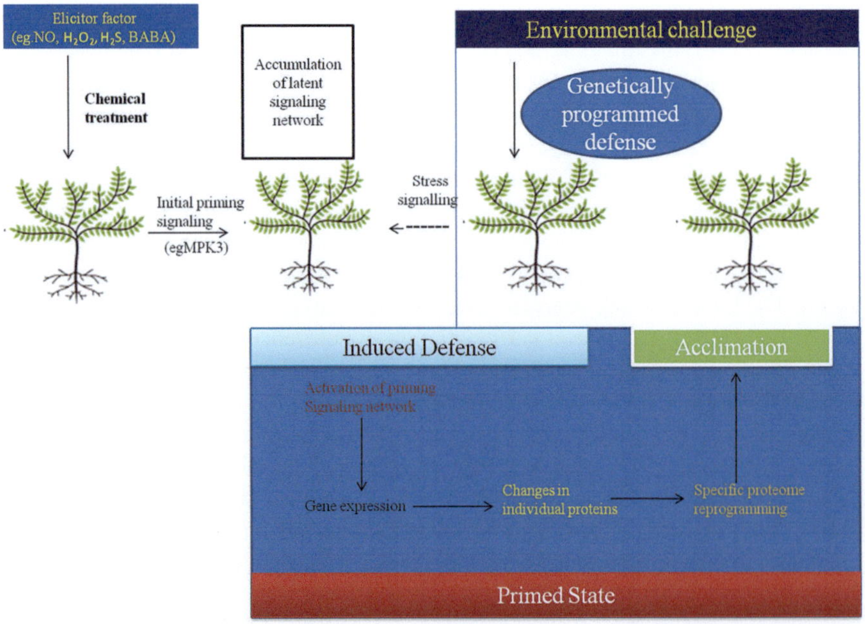

Fig. 3.1 Model depicting the acclimation of plant driven by seed priming in response to environmental stress

3.5 Conclusion

Seed priming emerges as a promising technology for combating abiotic stress in crops and alleviating the detrimental effects of abiotic stress without much affecting its fitness. Drought, salinity, submergence, and extreme temperature (heat, cold) are major challenges in agriculture. These abiotic stresses have similar effects at biochemical, cellular, and molecular levels, and somehow, they are interconnected and activate similar signaling cascades. It is established that seed priming affects signaling pathway at an early stage of growth and development in a crop that results in better tolerance against abiotic and biotic stress. Experimental results revealed that improved germination and vigorous seedling growth occur in primed seed by the mobilization of reserve food and activation of genes responsible for the synthesis of vital enzymes. Priming is also capable of repairing damage which occurs inside the seed. Seed priming has an effect on early stage of germination, and it modulates the DNA replication, transcription, and translation. There is a need to standardize suitable priming methods in different crops to combat abiotic stress in a sustainable manner.

References

Afzal I, Butt A, Rehman HU, Basra SMA, Afzal A (2012) Alleviation of salt stress in fine aromatic rice by seed priming. Aust J Crop Sci 6:1401

Akbari G, Sanavy SA, Yousefzadeh S (2007) Effect of auxin and salt stress (NaCl) on seed germination of wheat cultivars (*Triticum aestivum* L.). Pak J Biol Sci 10:2557–2561

Anosheh HP, Sadeghi H, Emam Y (2011) Chemical priming with urea and KNO3 enhances maize hybrids (*Zea mays* L.) seed viability under abiotic stress. J Crop Sci Biotechnol 14(4):289–295

Basra SMA, Farooq M, Tabassum R (2005) Physiological and biochemical aspects of seed vigour enhancement treatments in fine rice (*Oryza sativa* L.). Seed Sci Technol 33:623–628

Bruggink GT, Ooms JJJ, Vander Toorn P (1999) Induction of longevity in primed seeds. Seed Sci Res 9:49–53

Casenave EC, Toselli ME (2007) Hydropriming as a pre-treatment for cotton germination under thermal and water stress conditions. Seed Sci Technol 35:88–98

Cheng J, Wang L, Zeng P, He Y, Zhou R, Zhang H, Wang Z (2017) Identification of genes involved in rice seed priming in the early imbibition stage. Plant Biol 19:61–69

Cokkizgin H, Bolek Y (2015) Priming treatments for improvement of germination and emergence of cotton seeds at low temperature. Plant Breed Seed Sci 71:121–134

Cushman JC, Bohnert HJ (2000) Genomic approaches to plant stress tolerance. Curr Opin Plant Biol 3:117–124

Elkoca E, Haliloglu K, Esitken A, Ercisli S (2007) Hydro- and osmopriming improve chickpea germination. Acta Agric Scand Sect B Soil Plant Sci 57:193–200

Farooq M, Barsa SM, Wahid A (2006) Priming of field-sown rice seed enhances germination, seedling establishment, allometry and yield. Plant Growth Regul 49:285–294

Farooq M, Basra S, Rehman H, Saleem B (2008) Seed priming enhances the performance of late sown wheat (*Triticum aestivum* L.) by improving chilling tolerance. J Agron Crop Sci 194:55–60

Finch-Savage WE, Dent KC, Clark LJ (2004) Soak conditions and temperature following sowing influence the response of maize (*Zea mays* L.) seeds to on-farm priming (pre-sowing seed soak). Field Crop Res 90:361–374

Flowers TJ, Koyama ML, Flowers SA, Sudhakar C, Singh KP, Yeo AR (2000) QTL: their place in engineering tolerance of rice to salinity. J Exp Bot 51:99–106
Foti R, Aburenia K, Tigerea A, Gotosab J, Gerec J (2008) The efficacy of different seed priming osmotica on the establishment of maize (*Zea mays* L.) caryopses. J Arid Environ 72:1127–1130
Fujita M, Fujita Y, Noutoshi Y, Takahashi F, Narusaka Y, Yamaguchi-Shinozaki K, Shinozak K (2006) Crosstalk between abiotic and biotic stress responses: a current view from the points of convergence in the stress signaling networks. Curr Opin Plant Biol 9:436–442
Gallardo K, Job C, Groot SPC, Puype M, Demol H, Vandekerekhove J, Job D (2004) Proteomics of arabidospis seed germination and priming. In: Nichlolas G et al (eds) The biology of seeds: recent advances. CABI, Cambridge, MA, pp 199–209
Ghana SG, Schillinger WF (2003) Seed priming winter wheat for germination, emergence, and yield. Crop Sci 43:2135–2141
Guan YJ, Jin HU, Wang XJ, Shao CX (2009) Seed priming with chitosan improves maize germination and seedling growth in relation to physiological changes under low temperature stress. Zhejiang Univ Sci B 10(6):427–433
Gust AA, Brunner F, Nurnberger T (2010) Biotechnological concepts for improving plant innate immunity. Curr Opin Biotechnol 21:204–210
Hussain I, Ahmad R, Farooq M, Wahid A (2013) Seed priming improves the performance of poor quality wheat seed. Int J Agric Biol 15:1343–1348
Hussain S, Yin H, Peng S, Faheem A, Khan FA, Khan F, Sameeullah M, Hussain HA, Huang J, Kehui Cui K, Nie L (2016) Comparative transcriptional profiling of primed and non-primed rice seedlings under submergence stress. Front Plant Sci 7:1125
Iqbal M, Raja NI, Yasmeen F, Hussain M, Ejaj M, Shah MA (2017) Impacts of heat stress on wheat: a critical review. Adv Crop Sci Technol 5(1):01–09
Jakab G, Ton J, Flors V, Zimmerli L, Metraux JP, Mauch-Mani B (2005) Enhancing Arabidopsis salt and drought stress tolerance by chemical priming for its abscisic acid responses. Plant Physiol 139:267–274
Janmohammadi M, Moradi Dezfuli P, Sharifzadeh F (2008) Seed invigoration techniques to improve germination and early growth of inbred line of maize under salinity and drought stress. Gen Appl Plant Physiol 34:215–226
Jett LW, Welbaum GE, Morse RD (1996) Effects of matric and osmotic priming treatments on broccoli seed germination. J Am Soc Hortic Sci 121:423–429
Jian H, Jia Wang J, Wang T, Wei L, Li J, Liu L (2016) Identification of rapeseed micro RNAs involved in early stage seed germination under salt and drought stresses. Front Plant Sci 7:658
Job D, Capron I, Job C, Dacher F, Corbineau F, Come D (2000) Identification of germination-specific protein markers and their use in seed priming technology. Black M, Bradford KJ, Vazguez-Ramos J Seed biology: advances and applications, CAB International, Wallingford, 449–459
Jisha KC, Vijayakumari K, Puthur JT (2013) Seed priming for abiotic stress tolerance: an overview. Acta Physiol Plant 35:1381–1396
Kaur S, Gupta AK, Kaur N (2002) Effect of osmo- and hydropriming of chickpea on seedling growth and carbohydrate metabolism under water deficit stress. Plant Growth Regul 37:17–22
Kaur S, Gupta AK, Kaur N (2005) Seed priming increases crop yield possibly by modulating enzymes of sucrose metabolism in chickpea. J Agron Crop Sci 19:81–87
Kaya MD, Okcu G, Atak M, Cikili Y, Kolsarici O (2006) Seed treatments to overcome salt and drought stress during germination in sunflower (*Helianthus annuus* L.). Eur J Agron 24:291–295
Khan HA, Ayub CM, Pervez MA, Bilal RM, Shahid MA, Ziaf K (2009) Effect of seed priming with NaCl on salinity tolerance of hot pepper (*Capsicum annuum* L.) at seedling stage. Soil Environ 28:81–87
Kubala S, Garnczarska M, Wojtyla L, Clippe A, Kosmala A, Zmienko A, Lutts S, Quinet M (2015) Deciphering priming-induced improvement of rapeseed (*Brassica napus* L.) germination through an integrated transcriptomic and proteomic approach. Plant Sci 231:94–113
McDonald MB (2000) Seed priming. In: Black M, Bewley JD (eds) Seed technology and its biological basis. Sheffield Academic Press, Sheffield, pp 287–325

Moradi A, Younesi O (2009) Effects of osmo- and hydropriming on seed parameters of grain sorghum (*Sorghum bicolor* L.). Aust J Basic Appl Sci 3:1696–1700

Murungu FS, Nyamugafata P, Chiduza C, Clark LJ, Whalley WR (2003) Effects of seed priming, aggregate size and soil matric potential on emergence of cotton (*Gossypium hirsutum* L.) and maize (*Zea mays* L.). Soil Tillage Res 74:161–168

Mustafa HSB, Mahmood T, Ullah A, Sharif A, Bhatti AN, Muhammad Nadeem M, Ali R (2017) Role of seed priming to enhance growth and development of crop plants against biotic and abiotic stresses. Bull Biol Allied Sci Res 2:1–11

Niinemets U (2009) Mild versus severe stress and BVOCs: thresholds, priming and consequences. Trends Plant Sci 15:145–153

Patade VY, Sujata B, Suprasanna P (2009) Halopriming imparts tolerance to salt and PEG induced drought stress in sugarcane. Agric Ecosyst Environ 134:24–28

Paparella S, Araújo SS, Rossi G, Wijayasinghe M, Carbonera D, Balestrazzi A (2015) Seed priming: state of the art and new perspectives. Plant Cell Rep 34:1281–1293

Rashid A, Harris D, Hollington P, Ali S (2004) On-farm seed priming reduces yield losses of mung bean (*Vigna radiata*) associated with mung bean yellow mosaic virus in the North West Frontier Province of Pakistan. Crop Prot 23:1119–1124

Rashid A, Hollington PA, Harris D, Khan P (2006) On-farm seed priming for barley on normal, saline and saline–sodic soils in North West Frontier Province, Pakistan. Eur J Agron 24:276–281

Reyes LF, Cisneros-Zevallos L (2007) Electron-beam ionizing radiation stress affects on mango fruit (*Mangifera indica* L.) antioxidant constituents before and during post harvest storage. J Agric Food Chem 55:6132–6139

Rezaee S, Moghaddam MRR, Bazrgar AB (2015) Cotton seed germination as affected by salinity by and priming. Indian J Fundam Appl Life Sci 5:312–318

Saha P, Chatterjee P, Biswas AK (2010) NaCl pretreatment alleviates salt stress by enhancement of antioxidant defense system and osmolyte accumulation in mung bean (*Vigna radiata* L. Wilczek). Indian J Exp Biol 48:593–600

Sali A, Rusinovci I, Fetahu S, Gashi B, Simeonovska E, Rozman L (2015) The effect of salt stress on the germination of maize (*Zea mays* L.) seeds and photosynthetic pigments. Acta Agric Slov 105:85–94

Shabbir I, Ayub M, Tahir M, Bilal M, Tanveer A, Hussain M, Afzal M (2014) Impact of priming techniques on emergence and seedling growth of sesame (*sesamum indicum* l.) genotypes. Scientia 1:92–96

Sneideris LC, Gavassi MA, Campos ML, D'Amico-Damiao V, Carvalcho RF (2015) Effects of hormonal priming on seed germination of pigeon pea under cadmium stress. Ann Braz Acad Sci 87(3):1847–1852

Soeda Y, Konings MCJM, Vorst O, Van Houwelingen AMML, Stoopen GM, Maliepaard CA, Kodde J, Bino RJ, Groot SPC, Van der Geest AHM (2005) Gene expression programs during *Brassica oleracea* seed maturation, osmopriming, and germination are indicators of progression of the germination process and the stress tolerance level. Plant Physiol 137:354–368

Srivastava AK, Lokhande VH, Patade VY, Suprasanna P, Sjahril R, D'Souza SF (2010a) Comparative evaluation of hydro-, chemo-, and hormonal priming methods for imparting salt and PEG stress tolerance in Indian mustard (*Brassica juncea* L.). Acta Physiol Plant 32(1135):1144

Srivastava AK, Suprasanna P, Srivastava S, D'Souza SF (2010b) Thiourea mediated regulation in the expression profile of aquaporins and its impact on water homeostasis under salinity stress in *Brassica juncea* roots. Plant Sci 178:517–522

Tanou G, Fotopoulos V, Molassiotis A (2012) Priming against environmental challenges and proteomics in plants: update and agricultural perspectives. Front Plant Sci 3:216.

Taylor AG, Harman GE (1990) Concepts and technologies of selected seed treatments. Annu Rev Phytopathol 28:321–329

Thornton JM, Collins ARS, Powell AA (1993) The effect of aerated hydration on DNA synthesis in embryos of *Brassica oleracea* L. Seed Sci Res 3:195–1999

Van Hulten M, Pelser M, van Loon LC, Pieterse CMJ, Ton J (2006) Costs and benefits of priming for defense in Arabidopsis. Proc Natl Acad Sci U S A 103:5602–5607

Wahid A, Perveen M, Gelani S, Shahzad MA, Basra SMA (2007) Pretreatment of seed with H_2O_2 improves salt tolerance of wheat seedlings by alleviation of oxidative damage and expression of stress proteins. J Plant Physiol 164:283–294

Wahid A, Noreen A, Basra SM, Gelani S, Farooq M (2008) Priming-induced metabolic changes in sunflower (*Helianthus annuus*) achenes improve germination and seedling growth. Bot Stud 49:343–350

Xiong L, Zhu JK (2002) Molecular and genetic aspects of plant responses to osmotic stress. Plant Cell Environ 25:131–139

Yuan-Yuan S, Yong-Jian S, Ming-Tian W, Xu-Yi LI, Guo X, Rong HU, Jun MA (2010) Effects of seed priming on germination and seedling growth under water stress in rice. Acta Agron Sin 36:1931–1940

Zhao TJ, Liu Y, Yan YB, Feng F, Liu WQ, Zhou HM (2007) Identification of the amino acids crucial for the activities of drought responsive element binding factors (DREBs) of *Brassica napus*. FEBS Lett 581:3044–3050

Recent Advances in Abiotic Stress Tolerance of Plants Through Chemical Priming: An Overview

4

Muhammad Arslan Ashraf, Ali Akbar, Sajjad Hassan Askari, Muhammad Iqbal, Rizwan Rasheed, and Iqbal Hussain

Abstract

Plants under natural conditions often face multiple environmental constraints in terms of submergence, temperature extremes, salinity, and drought stress. Plant growth and productivity is negatively influenced by these abiotic stresses. Presently, a variety of approaches are being used to overcome abiotic stresses in plants. Recently, seed soaking with various priming agents has emerged as a promising strategy to induce tolerance in plants against abiotic stresses. In seed priming, seeds are treated with synthetic or natural compounds prior to germination so as to initiate specific physiological state in plants. Seed priming could also be defined as physiological state which enables plants to more quickly respond to abiotic stresses. Plants raised from seeds treated with various priming agents tend to show greater abiotic stress tolerance over unprimed seeds. Induction of abiotic stress tolerance through priming is an intricate process that involves various metabolic events. Primed seeds show early and uniform germination and seedling emergence. Seed priming enables plants to survive adverse environmental conditions and gives appropriate yield. In this book chapter, we have discussed a wide range of chemical which are extensively being used for seed priming.

Keyword

Seed preconditioning · Chemical priming · Hormonal priming · Redox priming · Hydropriming

M. A. Ashraf (✉) · A. Akbar · S. H. Askari · M. Iqbal · R. Rasheed · I. Hussain
Department of Botany, Government College University Faisalabad, Faisalabad, Pakistan

© Springer Nature Singapore Pte Ltd. 2018

A. Rakshit, H. B. Singh (eds.), *Advances in Seed Priming*,
https://doi.org/10.1007/978-981-13-0032-5_4

4.1 Introduction

Stress is called disturbance in equilibrium which produces changes in physiological parameters, and due to stress plant's chemical and physiological changes occur (Gaspar et al. 2002). Plants are growing in continuously under varying environmental conditions which were stressful and unfavorable for plant growth. These stressful circumstances are biotic stress (herbivores, pathogens attack) and abiotic stresses such as cold, heat, heavy metal, nutrient deficiency, salinity, and drought (Fedoroff et al. 2010). The main abiotic strains include heat, cold, salinity, and drought which produced adverse effects on biomass production and yield (Kaur et al. 2008; Thakur et al. 2010), and this creates food insecurity throughout the world. Drought, temperature, and salinity stress produced dehydration and reduced growth of plant and development (Jaleel et al. 2009; Thakur et al. 2010). Abiotic stresses induce excessive (reactive oxygen species) ROS generation. Among ROS, H_2O_2 functions as signal transduction but also reduces photosynthesis and cause cellular damage. The effect of ROS is reduced by antioxidant mechanism, which in turn reduced the effect of stress (Baxter et al. 2013).

4.2 Salinity Stress

Salinity is one of the problems of semiarid and arid regions which reduced the crop yield. Near about one third of the overall word land is affected by salinity (Flowers and Colmer 2008). It is reported that about 65% yield is reduced in wheat because of salinity stress (Ahmad 2012). Excessive buildup of salinity induced osmotic potential lowers the water potential which makes water inaccessible to plants (Munns et al. 2006). Chinnusamy et al. (2005) reported that accumulation of Na^+ and Cl^- cause ion toxicity and ion imbalance which reduce the uptake of other nutrients and reduced growth.

Salinity stress reduces the water potential, and this condition limits plants to uptake water and nutrients from land. Under this situation plants show two types of responses. Firstly, due to excess accumulation of Na^+ in roots, plants face osmotic stress, and second condition involves the increase of ionic stress due to imbalance in essential nutrients (Munns et al. 2006). It is reported that high salt concentration (Na^+ and Cl^-) in soil generate competition between essential nutrients in plants (Munns et al. 2006). High level of Na^+ reduced the influx of K^+ ion which disturbs enzymatic activity, osmotic balance, and stomatal functions. Furthermore, extreme level of Na^+ affects physiological attributes such as fruiting, flowering seed growth, and seed germination (Parihar et al. 2015; Singh et al. 2013). It is reported that ion toxicity in cell increases ROS levels (Parida and Das 2005; Parihar et al. 2015). It is reported by Aldesuquy et al. (2014) that irrigation of wheat with seawater reduced the photosynthetic pigments, light reaction, photosynthesis, and leaf area. In an investigation, it was reported that buildup of salts in barley decreased the photosystem II and photosynthetic activity (Kalaji et al. 2011). Salinity significantly affects plant growth and yield.

4.3 Drought Stress

Drought is termed as duration without substantial rainfall. The existing water reduces, and loss of water occurs through evaporation or transpiration under drought conditions (Jaleel et al. 2007a, b). Shortage of water causes drought that is the most dangerous hazard to food security all over the world. It was the main cause of starvations of past time. Because water supply is limited all over the world, food demand becomes a major problem (Somerville and Briscoe 2001). The harshness of drought stress is unpredictable because it is due to various circumstances like evaporation demand, moisture quality of soil, rainfall distribution, and its occurrence (Farooq et al. 2009). The primary impact of drought on plants is poor establishment and reduced germination (Harris et al. 2002). It was reported by Okçu et al. (2005) that drought stress severely affected the early seedling and germination of five cultivars of pea plants. In another investigation by Kaya et al. (2006), it was studied that drought stress reduced seed stand and its germination. Furthermore, it was examined that under drought stress caused by polyethylene glycol, root length was improved, but it reduced dry and fresh weight of root and shoot, hypocotyl length, and germination potential in alfalfa species (Zeid and Shedeed 2006). It was studied that plant development and growth in vegetative phase was greatly reduced under drought stress in rice (Manickavelu et al. 2006). It was examined by Samarah (2005) that grains and grain weight, spikes per plants, and number of tillers were reduced under drought stress in barley (*Hordeum vulgare*). Nam et al. (2001) reported that drought stress reduced about 40–55% seed yield in pigeon pea. Water relation of plant was disturbed due to disturbance in transpiration rate, canopy temperature, leaf temperature, stomatal resistance, water potential of leaf, and leaf water content. Due to accumulation of dry matter, relative water content of wheat plant leaves was reduced (Siddique et al. 2000). It was reported by Siddique et al. (2000) that relative water content was reduced in stressed plants as compared to nonstressed one.

It was studied by Nerd and Nobel (1991) that total water content was reduced to 57% in cladode (*Opuntia ficus-indica*) under drought stress. Egilla et al. (2005) reported that drought stress reduced the stomatal conductance, turgor potential, transpiration, and relative water content in *Hibiscus rosa-sinensis*. It was concluded that under water stress condition its water use efficiency was greater than nonstressed plants; they correlate it with closure of stomata and reduction of transpiration as reported by Abbate et al. (2004). It was reported that under drought stress the water use efficiency increased under well-watered conditions in lucerne (*Medicago sativa*) as investigated by Lazaridou and Noitsakis (2003). Water availability was reduced due to drought stress which restricts the uptake of nutrient and reduced tissue concentration in plants (McWilliams 2003). Different plant species showed different responses under drought stress; generally moisture stress showed decrease in P and increase in N, but K was unaffected (Garg 2003). McWilliams (2003) reported that K and N uptake was restricted in cotton under drought stress. Similarly, PO_3^{-3} and P uptake was also reduced under drought in plant tissues because moisture accessibility is low as reported by Peuke et al. (2002). It was investigated by Wahid et al. (2005) that drought stress reduced the efficiency of photosynthesis which in turn decreased food production and senescence of premature leaves and disturbed photosynthetic machinery and leaf expansion. It was also investigated that drought stress affects the photosynthetic machinery (Anjum et al. 2003) which disturbed the production of photosynthetic components and pigments (Fu and Huang 2001) and reduced crop yield by damaging Calvin cycle function (Monakhova and Chernyad'ev 2002). It was investigated by Bota et al. (2004) that drought stress decreased photosynthesis by decreasing the enzymatic activity of Rubisco. Plants undergo changes in different physiological or biochemical (antioxidant defense, cell membrane stability, plant growth regulators, osmotic adjustment, morphological (escape, avoidance, and phenotypic flexibility) and molecular mechanisms (aquaporins, stress protein), and other approaches to tolerate drought and induction of drought resistance through seed priming with plant growth regulators, osmoprotectants, and other important priming agents is also encourgaed (Farooq et al. 2009).

4.4 Heavy Metal Stress

Heavy metal was defined as a metal which have a gravity higher than 4 (Nagajyoti et al. 2010) based on dictionaries and encyclopedia. Nagajyoti et al. (2010) reported that heavy metal phytotoxicity in plants varies with taxonomical groups, and they are mainly derived naturally from geochemical material. Pollution of heavy metals may arise naturally (forest fires, spontaneous combustion, volcanic outgassings, surface mineralization, etc.) or may occur by human action (agriculture and mining). Man produced heavy metal contamination through agricultural activities like use of pesticides, automobiles, municipal incinerators, generator stations, mills, mines, metal smelters, fertilizers, and refineries (El-Ramady et al. 2015). Heavy metals were categorized in to two groups named as nonessential and essential nutrients for plant normal growth and development. The essential micronutrients were Cu, Zn, Ni, Mo, Mn, Fe, and Co which work as electron transfer, redox reaction, and other metabolic procedures, whereas nonessential nutrients were Hg, Cd, Pb, Cr, and As, and these were much toxic for plants (Sebastiani et al. 2004; Rai et al. 2004). Metals related with bioactivity were divided into two groups based on physiochemical property: (I) redox metals, i.e., Mn, Cr, Fe, and Cu, and (II) non-redox metals like Zn, Al, Hg, Ni, and cadmium (Valko et al. 2005; Jozefczak et al. 2012). The first grouped metal members generate oxidative injury in plants through two methods: one was Haber-Weiss method and second was Fenton reaction, and this leads to production of reactive oxygen species (ROS) in plants which produced homeostasis disruption in cell, breakage of DNA strand, protein defragmentation, damage to the photosynthetic apparatus, and ultimately cell death (Schutzendubel and Polle 2002; Flora 2009). Furthermore, second group metal members indirectly influence plant through various mechanisms like depletion of glutathione and sulfhydryl group binding (Valko et al. 2005), inhibiting the activity of antioxidants and produced ROS which damage plant (Bielen et al. 2013).

It is well documented that Cr is a toxic agent for the growth and development of plants (Panda and Choudhury 2005; Mohanty and Patra 2013). It is reported by Singh et al. (2013) that chromium also caused pollution in environment. Chromium is found in two forms, i.e., hexavalent Cr^{6+} and trivalent Cr^{3+}. Chromium metal is not uptaken directly or absorbed by the plants; it enters into cell through specific ion carriers, e.g., iron or sulfate (Singh et al. 2013; Gajalakshmi et al. 2012). Chromium is mostly accumulated in root part instead of other parts of plants (Kumar and Maiti 2013), because it is immobilized in plant cell vacuole so mostly it builds up in roots of the plant (Oliveira 2012; Nematshahi et al. 2012). Chromium mostly showed effect in early stages of plants and reduced the seed dry matter and stem development (Nematshahi et al. 2012). It is reported by Shanker et al. (2005) that chromium stress constrains the elongation and cell division in roots and reduced the length and also shortens the root system. In another report it was investigated that chromium toxicity disturbed cell cycle in roots and decreased shoot length by restricting the nutrient and water absorption process (Srivastava and Jain 2011). It was investigated that seed of cowpea (*Vigna sinensis* L.) showed significant reduction in germination characteristics by reducing total sugar and amylase activity when treated with different amount of chromium (Cr^{6+}) (Nath et al. 2008). It was reported that chromium interfered with other metal ions and buildup of extensive range of various nutrients, i.e., Ca, Mn, P, K, and Fe in root and shoot of plant and reduced plant growth (Samantaray et al. 1998).

Plants have various interrelated and sophisticated mechanisms to combat heavy metal stress. First line of defense against heavy metal toxicity was physical barriers; i.e., trichome tissues, thick cuticle, mycorrhizal association, and cell wall were activated when plant confronted with heavy metal stress (Hall 2002; Wong et al. 2004; Harada et al. 2010). It was reported that trichomes worked as heavy metal storage organs to detoxify the effect of metals by secreting secondary metabolites (Lee et al. 2002; Hauser 2014).

Under heavy metal stress, plant produced low-molecular-weight protein known as chelators or metallochaperones, i.e., mugineic acid, spermine, putrescine, nicotianamine, metallothionein, phytochelatin, organic acid, and glutathione or some cellular exudates, i.e., heat shock protein, protons, phenolics, flavonoids, and various amino acids like histidine, proline and hormones, i.e., ethylene, jasmonic acid, and salicylic acid (Viehweger 2014; Dalvi and Bhalerao 2013; Sharma and Dietz 2006). When plant was unable to combat with heavy metal stress with the above-mentioned strategies, then plant produced reactive oxygen species (Mourato et al. 2012). To alleviate the adverse effects of ROS radicals, plants have defensive system in terms of antioxidants such as glutathione (GSH), guaiacol peroxidase (GPX), ascorbate peroxidase (APX), catalase (CAT), superoxide dismutase (SOD), proline, tocopherol, alkaloids, carotenoids, and other phenolic compounds such as lignin, tannin, and flavonoids. These all scavenged ROS radicals (Sharma et al. 2012; Michalak 2006; Rastgoo et al. 2011)

4.5 Temperature Stress

When temperature was raised from the threshold level which damaged plant growth and its development, then it is called as heat stress. When this temperature raised up to 10–15 °C above ambient temperature, then heat shock occurs (Wahid et al. 2007a, b). High-temperature stress is the cause of decline in crop yield all over the world (Hall 2001). High-temperature stress directly damages the crop by denaturing and aggregating the protein, and indirect injury includes protein degradation and inhibition and degradation of enzymes of mitochondria and chloroplast, and its membrane integrity was also lost (Howarth 2005). These damages produced reactive oxygen species and toxic compounds and reduced ionic influx which leads to growth reduction (Howarth 2005). High-temperature stress causes diverse and adverse effects on plant development, growth, yield, and physiological processes (Hasanuzzaman et al. 2012, 2013). Under high temperature physiological attributes were badly disturbed such as fruit damage, root and shoot inhibition, leaf scrolling and senescence, and abscission which in turn become a cause of decrease in crop productivity (Vollenweider and Günthardt-Goerg 2005). High stress exposure induced molecular level disturbance in plants such as changing gene expression and transcripts accumulation, which lead to synthesis of stress-related genes (Iba 2002). High temperature reduces the dry weight of crop by decreasing plant growth net assimilation rate of shoot as investigated by Wahid et al. (2007a, b). Barnabás et al. (2008) reported that due to high temperature respiration and photosynthesis of plant were disturbed which reduced plant productivity. The initial impact of heat stress on plant was the degradation of enzymes and altered chloroplast protein structure as examined by Ahmad et al. (2010). It was reported that heat stress decreased respiration and photosynthetic activity as well as deterioration of chlorophyll (Todorov et al. 2003; Zhang et al. 2010). It was reported by Adams et al. (2001) that plant induced different mechanisms to tolerate heat stress, such as induction of long- and short-term changes in morphological and molecular changes like cooling, transpiration, leaf orientation, and changing of lipid membrane composition.

4.6 Priming Agents

There are a number of priming agents and/or methods including solid matric priming, redox priming, biological priming, hormonal priming, chemical priming, osmopriming, hydropriming, etc. Seed priming increases germination and growth especially under environmental constraints (Jisha et al. 2013). However, the degree of efficacy of different priming agents varies with plant species and diverse environmental conditions (Iqbal and Ashraf 2005).

4.7 Seed Priming with Nutrients

Seed priming with nutrients is an innovative practice that holds positive influences in terms of enhanced nutrient supply (Al-Mudaris and Jutzi 1999). Seeds are soaked in solutions with limited nutrients, whereas for comparison seeds are also soaked in water (Arif et al. 2008).There exists evidence in the literature that plants with better nutrient supply possess greater potential to tolerate abiotic stress (Marschner 2011). Among different mineral nutrients, potassium is more prominent because it enables plants to survive under different environmental stresses (Cakmak 2005). In this context, seed soaking in Zn^{2+} improved yield in wheat and chickpea (Arif et al. 2007). Ascorbic acid is vitamin C and acts as an important antioxidant. Ascorbic acid is also a promising priming agent (Jisha et al. 2013). A compelling evidence is available in the literature that advocated the requirement of greater endogenous levels of ascorbate for maintaining antioxidant defense system so as to protect plants from oxidative damage (Zhou et al. 2009). For instance, pretreatment with ascorbic acid increased germination in *Agropyron elongatum* when subjected to salinity stress (Tavili et al. 2009). Likewise, salt tolerance in maize has been induced through seed soaking in distilled water or $CaCl_2$, KCl, and NaCl (200 $meq{\cdot}L^{-1}$). Plants raised from primed and unprimed seeds were subjected to varying concentrations of NaCl salinity (0, 100, and 200 mM) for 14 days. The results indicated the positive effects of all priming agents on salt tolerance of maize. However, seed priming with $CaCl_2$ was more effective in the induction of salinity tolerance in maize in terms of improved germination percentage and plant biomass. Plants raised from primed seeds also had elevated cellular levels of Ca^{2+}, K^+, and Na^+. Chloride contents were greater in maize plants raised from seeds soaked in NaCl and KCl (Ashraf and Rauf 2001). In a study, seed soaking with KCl, $CaCl_2$, and NaCl (100 mM) alleviated the negative effects of salt stress in wheat under field conditions. Of different priming agents, priming with $CaCl_2$ was found to be more effective because of greater plant biomass and grain yield in two wheat cultivars (Inqlab-91 and MH-97). Seed soaking in $CaCl_2$ also mitigated the inhibitory effects of salinity on hormonal balance in wheat cultivars. In wheat cultivar, namely, MH-97, seed pretreatment with $CaCl_2$ enhanced endogenous levels of salicylic acid (SA) and decreased leaf ABA (abscisic acid) contents under saline or nonsaline conditions. Conversely, $CaCl_2$ priming in cv. Inqlab-91 increased IBA (indolebutyric acid) and (indoleacetic acid) contents. Furthermore, seed-soaking treatment did not cause any change in cellular levels of polyamines. However, plants from seeds treated with $CaCl_2$ had lower levels of spermidine in both wheat cultivars under salinity stress. In addition, seed priming with KCl improved plant growth in Inqlab-91, whereas plants of MH-97 wheat cultivar remained unaffected under salinity stress. Interestingly, NaCl priming mitigated salt-induced decline in auxins (IBA and IAA) in wheat plants under salinity stress. NaCl priming decreased putrescine and SA and increased leaf ABA contents in wheat cultivar Inqlab-91 under salinity stress. The results suggested that all priming agents were effective in improving plant growth, but the influence of seed priming on hormonal balance was not the same in two wheat cultivars (Iqbal et al. 2006). Iqbal and Ashraf (2007) studied the influence of seed soaking in $CaCl_2$, KCl, and NaCl (100 mM) in salt tolerance of wheat cultivars, namely, MH-97 and Inqlab-91. Both unprimed and primed seeds were grown in field with 15 dS $m^{-1}$of salinity. Seed soaking in $CaCl_2$ was better followed by KCl and NaCl in mitigating the inhibitory effects of salinity on photosynthesis, plant growth, and grain yield. Endogenous levels of Na^+ were significantly lower in plants raised from seeds treated with different priming agents. Despite the positive effect of priming agents on concentrations of ions and photosynthesis, the influence of priming agents was variable in two wheat cultivars.

A study by Anosheh et al. (2011) studied the effect of seed priming with KNO_3 and urea on seedling growth, germination, and proline and protein content in four maize hybrids under salt and drought stress. Plants were subjected to severe and moderate salt and drought stress and control (without stress). Seed priming with urea improved germination percentage, germination rate, and seedling length, whereas seed soaking with KNO_3 increased root length. Seed priming with either of priming agents had no influence on proline contents. Authors concluded that priming with urea and KNO_3 improved abiotic stress tolerance of maize hybrids and recommended that priming could increase plant growth of maize under environmental constraints. Likewise, Jafar et al. (2012) studied the performance of wheat cultivars (MH-97 and SARC-1) under salinity stress. These authors soaked seeds of two wheat cultivars in solutions containing $CaCl_2$, ascorbate, kinetin, and salicylic acid (50 mg L^{-1}) for 12 h. Authors also soaked seeds in distilled water for comparison. In this experiment, plants from unsoaked seeds acted as control plants. Among different priming agents, plants raised from seeds treated with $CaCl_2$ were more effective in terms of stand establishment, harvest index, 1000-grain weight, grains per spike, and number of fertile tillers. These results with $CaCl_2$ seed priming were closely followed by ascorbate in both wheat cultivars. There was decrease in Na^+ and increase in K^+ contents in wheat plants raised from seeds treated with priming agents, and this effect was more pronounced with $CaCl_2$ priming. Plants raised from seeds soaked in $CaCl_2$ had greater values for activities of protease and α-amylase, total soluble protein, and phenolic contents, and this trend was closely followed by seed priming with ascorbate. Authors concluded that priming with $CaCl_2$ is more economical from benefit-to-cost ratio viewpoint. Furthermore, authors found greater yield in wheat plants raised from seeds treated with $CaCl_2$, whereas kinetin was inferior in this context. Likewise, seed priming with potassium nitrate significantly influenced germination percentage and emergence in two soybean cultivars (Gorgan-3 and Sahar) under salinity stress. Cultivar Gorgan-3 performed better over cv. Sahar. Seed priming with KNO_3 increased emergence, germination percentage, plumule and radical length, plant dry weight, leaf area, plant height, and seedling dry weight under salinity stress as compared to plants raised from unprimed seeds (Ahmadvand et al. 2012).

4.8 Osmoconditioning

Osmopriming or osmoconditioning is the seed soaking in solutions with low water potential (Jisha et al. 2013). In osmopriming, degree and rate of imbibition is restricted through the exposure of seeds to low external water potential. Osmopriming is analogous to lengthy initial imbibition of seeds that induces metabolic activities even before germination. Osmopriming enables researchers to study alteration in seeds from physiologically inactive dry state to physiologically active hydrated state (Chen and Arora 2011). Researchers employ a wide range of chemicals in order to generate solutions with low water potential. In this context, PEG (polyethylene glycol) is nontoxic with large molecular size compound that decreases water potential of the solution without influencing seeds during soaking (Thomas et al. 2000). Furthermore, $MgSO_4$, KH_2PO_4, K_3PO_4, KCl, KNO_3, mannitol, NaCl, $CaCl_2$, etc. are other chemicals used to create solutions with low water potential (Jisha et al. 2013). Osmopriming financially, technically, and methodologically is more challenging as compared to hydropriming (Moradi and Younesi 2009) as osmopriming yields easier and faster results and economical than water conservation systems. In addition, osmopriming provides farmers with very smart substitute to improve crop yield (Foti et al. 2008).Osmopriming can maintain the integrity of plasma membrane and gives better germination percentage over hydropriming. This effect of osmopriming is attributed to prolonged precise seed hydration (Jett et al. 1996). Osmopriming of rice seeds increased starch hydrolysis to improve sugar availability of embryo and produced strong seedling growth. Furthermore, osmopriming also improved yield and yield quality parameters (Jisha et al. 2013).

In Bromus seeds, osmoconditioning with PEG increased germination percentage under drought and salt stress (Tavili et al. 2011). PEG osmopriming of *Trifolium alexandrinum* seeds increased seedling growth and germination percentage (Rouhi et al. 2010). PEG osmopriming also produced positive results in soybean (Khalil et al. 2001). Osmopriming of spinach seeds with PEG increased germination percentage, stimulated antioxidant defense system, and thereby induced tolerance to spinach plants (Chen and Arora 2011). Similarly, greater germination percentage was noted in plants raised from seed soaked in water for 12 h or osmoprimed with PEG for 24 h under cold soil environment (Elkoca et al. 2007). Likewise, osmopriming in adequate concentration of PEG improved seedling growth and germination as compared to hydropriming in rice (Sun et al. 2010). In another study, germination varied differentially in *Festuca ovina*, *Agropyron elongatum*, *osmopriming*, *Festuca arundinacea*, and *Bromus inermis* due to seed priming with PEG or hydropriming. Authors concluded that osmopriming with PEG was superior to hydropriming (Rouhi et al. 2011).

Plants produced from seeds osmoprimed with NaCl increase osmotic adjustment ability of plants due to greater endogenous Na^+ and Cl^- contents in roots of primed seeds. In addition, plants raised from primed seeds had greater organic acids and sugar contents of leaves as compared to non-primed seeds (Cayuela et al. 1996). Moreover, seeds soaked in NaCl increase the chances of direct germination from these seeds when sown under saline conditions (Cuartero and Fernández-Muñoz 1998). The better performance of plants from osmoprimed seeds with NaCl could have been due to accumulation of germination metabolites, metabolic repair courses, and osmotic adjustment during treatments (Haghpanah et al. 2009). Likewise, sugarcane osmoconditioning with NaCl is an attractive pregermination exercise to enable plants to tolerate drought and salinity stress (Patade et al. 2009). Seed osmopriming with NaCl enhanced seedling emergence, seed germination, and plant growth in canola and chickpea (Farhoudi and Sharifzadeh 2006; Sarwar et al. 2006). In a study, NaCl seed priming stimulated antioxidant system and decreased lipid peroxidation in *Momordica charantia* and *Zea mays* (Yeh et al. 2005; Randhir and Shetty 2005). Similarly, salt priming increased salinity tolerance of wheat in terms of improved metabolite reserves, seedling vigor, and K^+ and Ca^{2+} contents along with decrease in Na^+ contents (Afzal et al. 2005). Seed osmopriming with lower dose of NaCl alleviated the inhibitory effects of salinity on mung bean plants in terms of greater values for osmolytes accumulation, chlorophyll contents, and better antioxidant defense system and osmotic adjustment (Saha et al. 2010). Similarly, melon plants produced from seeds treated with salt primed seeds had greater tolerance to salinity stress. Moreover, osmopriming with NaCl also aided in plant establishment in saline medium. Furthermore, salinity tolerance of melon plants is attributed to increased levels of Ca^{2+} and K^+. Plants raised from primed seeds also accumulated more organic solutes that improved osmotic adjustment (Sivritepe et al. 2005). In another study, osmopriming with NaCl also improved salinity tolerance of sunflower through increasing cellular levels of Ca^{2+} and K^+. Plants also showed better osmotic adjustment in terms of greater proline accumulation (Bajehbaj 2010). In a medicinal plant, *Silybum marianum* priming with GA3 and NaCl increased germination percentage and plant biomass under salinity stress. Authors recommended seed priming with NaCl to get better yield under environmental constraints (Sedghi et al. 2010).

Osmopriming with mannitol and hydropriming alleviated the inhibitory effects of salinity and drought on plant growth in chickpea. Plants produced from seeds soaked in different concentrations of mannitol (2% and 4%) improved biomass and length of shoot and roots as compared to plants from non-primed seeds under saline conditions (Kaur et al. 2005). Likewise, hydropriming and mannitol osmopriming increased seedling growth and germination in alfalfa under salinity stress. Plants from seeds primed with mannitol had higher activities of antioxidant enzymes (SOD, POD, and CAT) and minimal electrolyte leakage and malondialdehyde (MDA) contents. Authors concluded that mannitol priming could prevent alfalfa plants from salinity stress damages (Amooaghaie 2011).

A number of other chemical are also used for the purpose of osmopriming. For instance, osmopriming with KNO_3 increased germination in water melon (Demir and Mavi 2004). Similarly, seed soaked in KNO_3 enhanced germination index, germination percentage, shoot and root fresh and dry masses, and lengths in tomato (Nawaz et al. 2011). Likewise, priming of rice seeds with $CaCl_2$, KCl, and ascorbate improved germination percentage and seedling growth. Plants obtained from primed seeds showed higher length and plant fresh and dry masses. Plants from primed seeds had significant alterations in calcium and nitrogen homeostasis of seedlings that was correlated with improved α-amylase activity and decreased levels of reducing sugars (Jisha et al. 2013). Similarly, osmopriming with KNO_3 increased seedling emergence in tomato (Ozbingol et al. 1998), sorghum (Moradi and Younesi 2009), and leek (Brocklehurst et al. 1984). In another study, seed osmopriming with KNO_3 in sunflower increased seedling vigor and seedling growth and increased germination. Seed soaking with KNO_3 also improved seedling growth in cucumber (Singh and Rao 1993; Ghassemi-Golezani and Esmaeilpour 2008).

4.9 Hydropriming

Hydropriming is a nontoxic, economical, and simple technique that improved plant growth and production under environmental constraints in terms of improved seedling establishment and osmotic adjustment (Kaur et al. 2002). In hydropriming, seeds are soaked in distilled water at room temperature, and seed imbibition determines length of hydropriming (Kaya et al. 2006). It is evident from the literature that soaked seeds must be properly dried because inadequate drying is harmful (Thomas et al. 2000). Soaked seeds are brought to their original weight through proper drying under shade (Jisha et al. 2013). Hydropiming initiates physiochemical alterations in seeds prior to germination (Basra et al. 2003). In addition, protoplast in hydroprimed seeds has greater permeability to nutrients and water, lesser viscosity, and greater resistance to dehydration (Jisha et al. 2013). Plants produced from hydroprimed seeds had greater water uptake which is positively associated with seedling growth (Yagmur and Kaydan 2008). Evidence is available in the literature which indicates that hydropriming is advantageous over non-primed seeds. Plants from seeds soaked in water showed three- to fourfold increase in plant growth as compared to plants from untreated seeds under drought conditions (Kaur et al. 2002). Hydropriming is inexpensive and simple method to improve abiotic stress tolerance of plants (Jisha et al. 2013). Some researchers have also found hydropriming as an excellent method to produce better germination and seedling establishment in different grain crops (Abebe and Modi 2009).

Seed soaking in water improved germination, seedling establishment, and yield in different crops under contrasting environmental conditions (Rashid et al. 2006). Likewise, hydropriming improved seed germination in onion (Caseiro et al. 2004). Hydropriming also improved germination percentage in cauliflower (Jisha et al. 2013). Hydropriming was most effective among various priming agents for mustard (Srivastava et al. 2010a). Similarly, hydropriming in barley is recommended to the farmers in NWFP of Pakistan and in other regions of the world with same environmental conditions. Hydropriming improved seedling emergence and growth in melon (Sung and Chiu 1995). Furthermore, hydropriming also increased germination percentage in coriander (Rithichai et al. 2009), pyrethrum (Li et al. 2011), *Allium porrum* (Ashraf and Bray 1993), and desert cacti (Dubrovsky 1996).

Hydropriming resulted in increased seedling establishment and germination percentage in mung bean plants as compared to that in non-primed seeds. Plants produced from hydroprimed seeds yielded 80% greater above ground plant biomass, 415% additional grain yield, and 264% additional pod yield, whereas non-primed seeds were inferior in this regard (Rashid et al. 2004). Likewise, hydropriming coupled with proline increased germination percentage and plant growth in *Vigna radiata* under temperature stress (Posmyk and Janas 2007). Authors concluded that plants pretreated with proline had minimal lipid peroxidation that in turn maintained the integrity of membranes under chilling stress. Moreover, proline also acts as carbon and nitrogen source and thereby seedling regeneration and growth (Posmyk and Janas 2007).

4.10 Hormonal Priming

The use of plant hormones and other plant growth regulators as seed presowing treatment can improve plant growth under stressful conditions (Jisha et al. 2013). Of various plant hormones, ABA (abscisic acid) mediates plant responses to a number of abiotic stresses such as osmotic, low-temperature, and drought stress (Fujita et al. 2006). Apart from the involvement of ABA in the upregulation in the expression of stress-related genes, exogenous application of ABA improves salinity tolerance of plants (Xiong and Zhu 2002). Evidence is present in the literature that indicates the wide range of genes induced by ABA (Jisha et al. 2013). Among ABA-induced genes, some genes encode different signal transduction pathways including transcription factors, protein kinases/phosphatase, and putative receptors that could improve salinity tolerance of plants (Xiong and Zhu 2002). Mustard plants produced from ABA-primed seeds showed higher rate of seed germination as compared to non-primed seeds (Srivastava et al. 2010a). ABA-primed seeds of canola showed greater germination percentage as compared to non-primed seeds under low-temperature, drought, or salt stress (Gao et al. 2002).

There are a number of reports in the literature where seed soaking in GA3 (gibberellic acid) has been shown to improve germination (Khan et al. 2002; Jisha et al. 2013). In a study, on cotton and faba beans, GA3 enhanced Ca and K when applied as presowing seed treatment (Harb 1992). Akbari et al. (2007) studied seed priming with auxin in wheat crop where auxin priming increased hypocotyl dry weight, hypocotyl length, and seedling fresh and dry weight. Similarly, priming with plant growth regulator GA3 and IAA improved germination in pyrethrum under nonsaline regime (Bisht et al. 2009). Ascorbic acid priming gave better results in wheat over other priming agents such as GA3, kinetin, ascorbic acid, and salicylic acid (Khan et al. 2011). Likewise, seed priming with salicylic acid enhanced germination in fennel under low water potential (Farahbakhsh 2012). Seed soaking with different growth regulators including ethephon, kinetin, thiourea, and fusicoccin mitigated the negative effects of salt stress on germination in a halophyte, namely, *Salicornia utahensis*, while nitrate, betaine, and proline priming did not significantly influence germination at varying salinity levels (Gul and Khan 2004). Likewise, priming with 3 μM methyl jasmonate and 3% KNO_3 enhanced germination in dormant seeds of *Amaranthus cruentus* L. (Tiryaki et al. 2005). In another study, seed priming with ABA and GA increased activities of antioxidant enzymes over plants from unprimed seeds under drought conditions (Eisv and et al. 2010). Seed priming with vitamin and hormone improved root architecture (Shakirova et al. 2003). Priming with growth regulators might increase cell division in apical meristem of root and thereby increase plant growth. Furthermore, hormonal priming cause increase in tissue levels of cytokinin and IAA which in turn improve plant growth (Shakirova et al. 2003).

4.11 Chemical Priming

Literature indicates the use of diverse chemicals in different crop species. Tolerance in crop plants has been reported in response to seed priming with synthetic or natural compounds such as chitosan, choline, paclobutrazol, putrescine, ethanol, KH_2PO_4, $ZnSO_4$, $CuSO_4$, selenium, and butenolide (Jisha et al. 2013; Demir et al. 2012).

4.12 Seed Priming with Chitosan

Chitosan, a large polysaccharide cation, is extracted from seafood waste materials. Seed preconditioning with chitosan increased tolerance in plants against a number of diseases, and moreover, increased yield quality, IAA and GA3 contents, lipase activity, and germination percentage (Shao et al. 2005; Jisha et al. 2013). For instance, chitosan treatment in wheat improved tolerance against various diseases and improved seed quality (Bhaskara Reddy et al. 1999). Likewise, Ruan and Xue (2002) reported increase in seed germination and stress tolerance in rice plants in response to chitosan. Similarly, chitosan priming improved seedling vigor in maize (Shao et al. 2005). In another study, Guan et al. (2009) reported that seed priming with chitosan increased germination and plant growth in maize under low-temperature stress.

4.13 Seed Priming with Choline

Supplementation of choline induces tolerance in plants against salinity stress by improving cellular levels of glycine betaine (Su et al. 2006). Choline is an integral component in the biosynthesis of glycine betaine in plants (Sakamoto and Murata 2002). In addition, choline is an important component of phosphatidylcholine which is a conspicuous part of membrane phospholipids in plants (Rathinasabapathi 2000). In another study, seed priming with choline chloride improved salinity tolerance ability in wheat plants (Salama et al. 2011).

4.14 Paclobutrazol, Putrescine, and Ethanol

A study by Farooq et al. (2006) reported better germination and leaf score as a result of seed priming with 1% or 5% ethanol. Similarly, seed priming with putrescine improved seed germination and growth as a result of better antioxidant system in tobacco under chilling stress (Xu et al. 2011). Likewise, seed priming with paclobutrazol in *Catharanthus roseus* resulted in better salt tolerance in terms of stimulated antioxidant system (Jaleel et al. 2007a, b). In another study, seed preconditioning with paclobutrazol improved modulated growth and key physiochemical processes in turf grass under drought stress. However, this effect of paclobutrazol is associated with plant species and levels of paclobutrazol (Shahrokhi et al. 2011).

According to Shahrokhi et al. (2011), paclobutrazol seed priming in *F. arundinacea* L. (turf grass) affected growth and physiological traits of plants during drought stress, but it related to concentration of paclobutrazol and the nature of the cultivar.

4.15 Seed Preconditioning with $CuSO_4$, KH_2PO_4, $ZnSO_4$, and Selenium

Selenium improves plant growth and stress tolerance potential of plants when applied in lower concentrations (Mirza et al. 2010). Likewise, priming with selenium protected plants from temperature-induced oxidative injury by stimulating the antioxidant defense system of plants (Chen and Sung 2001). Seed preconditioning with $ZnSO_4$ and $CuSO_4$ enhanced seed germination as well as seed emergence in maize plants (Foti et al. 2008). Likewise, seed priming with KH_2PO_4 improved germination and seedling growth in wheat (Korkmaz and Pill 2003). It has been reported that seed priming with lesser levels of KNO_3 and K_2HPO_4 gave better results as compared to greater levels of these priming agents (Sarwar et al. 2006). Likewise, seed priming with KH_2PO_4 enhanced seedling growth and germination percentage in triticale (Yagmur and Kaydan 2008).

4.16 Butenolide

Butenolide is another important chemical priming agent that improves seedling vigor and growth in diverse plant species (Demir et al. 2012). Butenolide is extracted from burnt cellulose and plant-based smoke. Plants raised from seeds treated with butenolide had lower chances of pathogen attack mainly due to rapid emergence of seedling. Moreover, seed treatment with smoke also protects seeds from pathogen attack in seedbed (Kulkarni et al. 2006). Similarly, treatment of this compound produced better results in tomato grown under low-temperature conditions (Jain et al. 2006).

4.17 Redox Priming

Cellular redox state regulates important processes that mediate growth and development under stressful environment. Plants undergo change in redox status due to external stimuli, and the degree of change in redox status of plants depends upon the concentration, duration, and nature of stimulus (Miller et al. 2009). Plants decrease the intensity of stress damage through the maintenance of reduced redox state (Mittler 2002). It has been reported in the literature that foliar application and seed soaking in a number of thiol compounds especially thiourea improved stress tolerance and plant productivity in different plant species, e.g., wheat and mustard (Sahu et al. 2005; Jisha et al. 2013). In another study by Srivastava et al. (2008), seed treatment with thiourea maintained integrity of mitochondria in seedlings under salinity stress that negatively influenced mitochondrial integrity in *Brassica juncea* (Srivastava et al. 2008). Thiourea pretreatment also improved signaling pathways in *Brassica juncea* under salinity stress (Srivastava et al. 2010b).

Reactive nitrogen species like NO and reactive oxygen species (ROS) mediate a large number of cellular processes under stressful conditions including drought, pathogen attack, and heat and cold stress (Hancock et al. 2011). In another study by Yadav et al. (2011), seed soaking in H_2O_2 improved chilling tolerance in capsicum. Seed priming with H_2O_2 enhanced chilling tolerance in *B. juncea* L., and this effect of H_2O_2 priming was further aggravated when combined with 24-EBL (24-epibrassinolide) (Jisha et al. 2013). Similarly, seed treatment with 24-EBL mitigated the inhibitory effects of H_2O_2 by stimulating the antioxidant enzymes such as SOD, CAT, POD, and APX (Kumar et al. 2010). In another study, seed soaking in H_2O_2 mitigated the adverse effects of salinity stress in wheat (Wahid et al. 2007a, b). Similarly, H_2O_2 seed priming enhanced drought tolerance in maize plants (Ashraf et al. 2015b). Likewise, seed priming with NO donors increased germination and seedling growth in pearl millet (Manjunatha et al. 2008). Moreover, H_2O_2 functioned as signaling molecule and thereby improved seed germination through the induction of alterations at hormonal, transcriptomic, and proteomic levels (Barba-Espín et al. 2012). Glutathione, an important compound of the antioxidant pathways in plants, mediates redox signaling networks which directly affect plant growth and defense (Maughan and Foyer 2006). Similarly, seed priming with antioxidant compounds such as tocopherol, glutathione, and ascorbic acid improved seedling vigor in sunflower under low-temperature stress (Draganic and Lekic 2012). Cysteine is not directly involved in redox priming but significantly affects synthesis of GSH (Barba-Espín et al. 2012). Furthermore, seed priming with cysteine alleviated injury due to gamma radiation, and cysteine caused more conspicuous effects in the elongation of primary root (Reddy and Smith 1978). Contrarily, seed priming with L-cysteine failed to prevent methyl methanesulfonate-induced damages to barley seeds except for little protection at higher concentrations (Rubluo 1982).

4.18 Proline As Promising Priming Agent for Plants Under Abiotic Stress

Plants often face multitude of abiotic stresses due to their sessile nature. UV radiation, nutrient deficiency, trace metal toxicity, temperature extremes, water-limited conditions, and salinity are among foremost environmental limitations to crop production worldwide. Plants possess a variety of cellular mechanisms to mitigate negative effects of abiotic stress. For example, a wide range of compatible solutes is accumulated in plants. Of compatible solutes, amino acids such as proline are accumulated in substantial amount. There evidence is still not obvious to explain the relationship of proline accumulation with abiotic stress tolerance of plants. However, as a general consensus among plant researchers, proline accumulation is taken as beneficial for plants particularly after stress recovery (Verbruggen and Hermans 2008; Kaur and Asthir 2015). According to an estimate, plants accumulate 80% proline under abiotic stress in comparison to 5% under normal conditions. This greater accumulation of proline under stress condition is attributed to enhanced biosynthesis and limited degradation of this biomolecule (Szabados and Savouré 2010; Kaur et al. 2017). Proline has been documented as a multifunctional amino acid with diverse functions under stress conditions. Protection of cellular functions through detoxification of ROS, maintenance of membrane integrity, proteins, and subcellular structure are the major functions displayed by this important biomolecule. Functional variation in the metabolism of proline is largely due to compartmentalization of proline degradation and biosynthesis in mitochondria, chloroplast, and cytosol. Proline is involved in the maintenance of redox balance and homeostasis via dissolving additional reduction potential due to saturated electron transport chain under environmental constraint (Kaur and Asthir 2015). Proline plays important role in maintaining plant growth particularly after stress recovery because proline degradation is associated with oxidative respiration where it directs energy to recommence growth. Proline protects plants from damage of trace metals through scavenging ROS, but it may also act as protein-compatible hydrotrope (Busch et al. 2014) thereby mitigating cytoplasmic acidosis and preserving suitable $NADP^+$/NADPH adjusted with the cellular metabolism (Filippou et al. 2014). Proline impacts plant growth and development through its action as metabolic signal mediating metabolite reservoir (Verbruggen and Hermans 2008). Proline also provides protection to plants through the induction of stress-related proteins (Kaur and Asthir 2015; Kaur et al. 2017).

Plants subjected to abiotic stress conditions experience growth retardation. Conversely, supplemented proline improved plant growth through osmoprotection under salinity stress (Roy et al. 1993; Yancey 1994; Savvides et al. 2016). Exogenous application of proline in lower concentration significantly alleviated the inhibitory effects of salt stress in rice (Roy et al. 1993). Similarly, addition of proline to the culture medium increased fresh weight and decreased membrane damage in terms of lipid peroxidation in *Arachis hypogea*. Conversely, when greater levels of proline were added to the medium, it imposed no beneficial effects (Jain et al. 2001). In another study by Ehsanpour and Fatahian (2003), addition of proline to the saline medium improved dry mass and proline levels in *Medicago sativa*. Likewise, treatment of premature maize embryos with proline induced somatic embryogenesis (Duncan and Widholm 1987). Presowing seed treatment of wheat with proline increased growth and grain yield under drought conditions (Kamran et al. 2009). A study of Ali et al. (2008) reported improved uptake of N, P, Ca^{2+}, and K^{+} in maize plants treated with proline under water-deficit conditions. Likewise, added proline also mitigated negative effects of Cd on cultured tobacco (Islam et al. 2009). In a study by Hua-long et al. (2014), presowing seed treatment with proline increased rice germination under salinity stress. Furthermore, proline priming also increased relative germination energy and relative germination rate under salinity stress. Likewise, proline priming also stimulated amylase activities in rice under salinity stress. These authors concluded that presowing seed treatment with proline effectively alleviated negative effects of salinity on rice germination. In another study, seed priming with proline significantly alleviated the inhibitory effects of drought in wheat under water-deficit conditions (Kamran et al. 2009). Similarly, proline scavenged superoxide radicals by improving the SOD activity in *Solanum nigrum* plants under cadmium toxicity (Xu et al. 2009). In another study, Agami (2014) recorded the influence of presowing seed treatment on growth, leaf anatomy, antioxidant enzyme, electrolyte leakage, relative water content, proline, and chlorophyll in *Hordeum vulgare* L. under salinity stress. Plants grown under saline conditions (100 and 200 mM NaCl) had decrease in growth, soluble sugars, chlorophyll, and relative water content along with significant changes in leaf anatomy. However, proline presowing seed treatment mitigated negative effects of proline and enhanced above stated attributes. Furthermore, salinity stress increased activities of antioxidant enzymes (POX, CAT, and SOD) and endogenous levels of proline and electrolyte leakage. There was a marked variation in leaf anatomy and antioxidant enzymes in plants raised from proline-treated seeds under nonsaline or saline conditions. Likewise, Sadak and Mostafa (2015) studied the impact of seed soaking with proline on growth, key biochemical attributes, and yield in sunflower plants under seawater salinity. Salt stress caused reduction in plant growth, chlorophyll contents, yield, and yield quality. Conversely, seawater salinity increased proline contents, free amino acids, total soluble carbohydrates, and phenolics. Plants raised from proline-treated seeds had greater tolerance to salinity in terms of better growth, chlorophyll content, proline, total soluble carbohydrates, and free amino acids. Authors concluded that priming with proline alleviated the inhibitory effects of salinity on sunflower through improving endogenous levels of chlorophyll and osmoprotectants. Likewise, presowing seed treatment with proline mitigated the negative effects of temperature extremes in wheat plants in terms of improved germination and α-amylase activity (Sultana et al. 2000).

4.19 Glycine Betaine As Potential Priming Agent

A plethora of organisms accumulate GB under abiotic stresses (Chen and Murata 2002; Giri 2011). Plants with natural potential of GB production can grow well under saline and water-limited conditions (Chen and Murata 2008; Giri 2011). Compatible solutes are small organic molecules highly soluble in water and are not toxic even at higher concentrations (Giri 2011). GB is known to function as compatible solute (Chen and Murata 2008). A number of halotolerant plants accumulate substantial amount of GB in plastids and chloroplasts. GB accumulation is also very high in plants subjected to abiotic stress (Ashraf and Foolad 2007). There existed a positive association between the endogenous levels of GB and degree of abiotic stress tolerance in plants (Rhodes and Hanson 1993). GB protects and stabilizes the quaternary structure of complex proteins and thereby maintains the integrity of cell membranes at high-temperature and salinity stress (Papageorgiou and Murata 1995). The survey of literature showed that plants treated with added GB or plants engineered with GB biosynthesis genes had significant tolerance to environmental constrains (Ashraf and Foolad 2007; Chen and Murata 2008; Giri 2011). Of various compatible solutes, GB provides effective protection to plants against multitude of environmental constraints (Sakamoto and Murata 2000, 2002). Literature showed the effectiveness of this biomolecule in the induction of tolerance against a wide range of stresses during different ontogenic phases (Sulpice et al. 2003; Park et al. 2004).

Literature survey showed that exogenous application of GB markedly mitigated the negative influences of abiotic stresses in plethora of plant species where GB enhanced growth and yield (Chen and Murata 2008). Plants readily take up GB when applied exogenously as foliar spray (Park et al. 2006). Plants can also efficiently take up this biomolecule from roots (Park et al. 2003). Plants normally produce reactive oxygen species (ROS) as products of different metabolic products under nonstress conditions (Chen and Murata 2008). Plants hold an efficient defensive antioxidant system that scavenges ROS and thereby decrease cellular levels of ROS (Chen and Murata 2011). Abiotic stresses such as salinity, drought, and heavy metal stress induce oxidative damage to plants (Ashraf et al. 2015a). Addition of hydroxyl radicals to *Arabidopsis* roots induced concentration-dependent efflux of K^+. However, growth medium with GB (5 mM) significantly reduced efflux of K^+ (Cuin and Shabala 2007). In addition, tomato plants treated with exogenously applied GB markedly decreased the cellular levels of H_2O_2 under chilling stress (Park et al. 2006). GB is not directly involved in the scavenging of ROS, but GB stimulates the activities of antioxidant enzymes or reduces ROS generation through an unknown mechanism (Giri 2011; Chen and Murata 2008). In a study, turfgrass seeds were primed in 5–50 mM GB in order to evaluate the effect of GB on temperature, salinity, and drought stress during germination. It was observed that abiotic stresses more drastically affected daily germination percentage than final germination percentage. *Lolium perenne* L. exhibited better tolerance to temperature, salinity, and drought stress followed by *Festuca arundinacea* Schreb and *Agrostis palustris*. Conversely, Bermuda grass, zoysia grass, and Kentucky bluegrass were found sensitive to abiotic stress. Seed priming with 10 mM GB in Bermuda grass and Kentucky bluegrass enhanced germination percentage under different abiotic stresses. Furthermore, authors concluded that other grasses did not respond significantly to seed-soaking treatment with GB. The results suggested that impact of GB priming is plant, type of abiotic stress, and GB concentration dependent (Zhang et al. 2014).

4.20 Conclusion

Seed priming is an attractive approach to counteract the negative effects of stress in plants against a variety of abiotic stresses such as salinity, drought, heavy metal, and temperature stress. Seed priming is realistic, effective, and smart choice for better plant production. Oxidative stress, temperature extremes, salinity, and drought are associated and often induce similar type of damage. As a result, these multiple abiotic stresses introduce similar kinds of cellular responses and signaling pathways. Seed priming stimulates these signaling pathways earlier and improves plant defense responses. The precise process that governs priming is not well known. However, it is assumed that inactive signaling proteins accumulate in primed cells. When plants from primed seeds face abiotic stresses, such inactive signaling proteins are activated which amplify the signal transduction and result in more intricate response of plants under stress. There is increasing evidence in the literature where seed priming has been proposed as one of the possible solutions to counteract the negative effects of abiotic stresses. However, there are still many gaps where more input is required from researchers around the globe. For instance, whether seed priming also influences changes in physiochemical processes of plants at different growth stages or not. Seed priming agents work the same for different plant species or not. How far important physiochemical processes are linked with seed priming. In nutshell, seed priming acts as important yardstick for the induction of tolerance in plants against wide range of abiotic stresses.

References

Abbate PE, Dardanelli JL, Cantarero MG, Maturano M, Melchiori RJM, Suero EE (2004) Climatic and water availability effects on water-use efficiency in wheat. Crop Sci 44(2):474–483

Abebe AT, Modi A (2009) Hydro-priming in dry bean (*Phaseolus vulgaris* L.). Res J Seed Sci 2(2):23–31

Adams S, Cockshull K, Cave C (2001) Effect of temperature on the growth and development of tomato fruits. Ann Bot 88(5):869–877

Afzal I, Basra SMA, Ahmad N, Farooq M (2005) Optimization of hormonal priming techniques for alleviation of salinity stress in wheat (*Triticum aestivum* L.)

Agami RA (2014) Applications of ascorbic acid or proline increase resistance to salt stress in barley seedlings. Biol Plant 58(2):341–347. https://doi.org/10.1007/s10535-014-0392-y

Ahmad S (2012) Oxidative stress and antioxidant defenses in biology. Springer

Ahmad A, Diwan H, Abrol YP (2010) Global climate change, stress and plant productivity. In: Pareek A, Sopory SK, Bohnert HJ (eds) Abiotic stress adaptation in plants: physiological, molecular and genomic foundation. Springer, Dordrecht, pp 503–521. https://doi.org/10.1007/978-90-481-3112-9_23

Ahmadvand G, Soleimani F, Saadatian B, Pouya M (2012) Effects of seed priming on germination and emergence traits of two soybean cultivars under salinity stress. J Basic Appl Sci Res 3:234–241

Akbari G, Sanavy S, Yousefzadeh S (2007) Effect of auxin and salt stress (NaCl) on seed germination of wheat cultivars (*Triticum aestivum* L.). Pak J Biol Sci PJBS 10(15):2557–2561

Aldesuquy H, Baka Z, Mickky B (2014) Kinetin and spermine mediated induction of salt tolerance in wheat plants: leaf area, photosynthesis and chloroplast ultrastructure of flag leaf at ear emergence. Egypt J Basic Appl Sci 1(2):77–87

Ali Q, Ashraf M, Shahbaz M, Humera H (2008) Ameliorating effect of foliar applied proline on nutrient uptake in water stressed maize (*Zea mays* L.) plants. Pak J Bot 40(1):211–219

Al-Mudaris MA, Jutzi SC (1999) The influence of fertilizer-based seed priming treatments on emergence and seedling growth of *Sorghum bicolor* and *Pennisetum glaucum* in pot trials under greenhouse conditions. J Agron Crop Sci 182(2):135–142. https://doi.org/10.1046/j.1439-037x.1999.00293.x

Amooaghaie R (2011) The effect of hydro and osmopriming on alfalfa seed germination and antioxidant defenses under salt stress. Afr J Biotechnol 10(33):6269–6275

Anjum F, Yaseen M, Rasul E, Wahid A, Anjum S (2003) Water stress in barley (*Hordeum vulgare* L.). II. Effect on chemical composition and chlorophyll contents. Pak J Agric Sci 40:45–49

Anosheh HP, Sadeghi H, Emam Y (2011) Chemical priming with urea and KNO3 enhances maize hybrids (*Zea mays* L.) seed viability under abiotic stress. J Crop Sci Biotechnol 14(4):289–295. https://doi.org/10.1007/s12892-011-0039-x

Arif M, Waqas M, Nawab K, Shahid M (2007) Effect of seed priming in Zn solutions on chickpea and wheat. Afr Crop Sci Pro 8:237–240

Arif M, Jan MT, Marwat KB, Khan MA (2008) Seed priming improves emergence and yield of soybean. Pak J Bot 40(3):1169–1177

Ashraf M, Bray C (1993) DNA synthesis in osmoprimed leek (*Allium porrum* L.) seeds and evidence for repair and replication

Ashraf M, Foolad MR (2007) Roles of glycine betaine and proline in improving plant abiotic stress resistance. Environ Exp Bot 59(2):206–216. https://doi.org/10.1016/j.envexpbot.2005.12.006

Ashraf M, Rauf H (2001) Inducing salt tolerance in maize (*Zea mays* L.) through seed priming with chloride salts: growth and ion transport at early growth stages. Acta Physiol Plant 23(4):407–414. https://doi.org/10.1007/s11738-001-0050-9

Ashraf M, Iqbal M, Hussain I, Rasheed R (2015a) Physiological and biochemical approaches for salinity tolerance. Managing salt tolerance in plants: molecular and genomic perspectives, 79

Ashraf MA, Rasheed R, Hussain I, Iqbal M, Haider MZ, Parveen S, Sajid MA (2015b) Hydrogen peroxide modulates antioxidant system and nutrient relation in maize (*Zea mays* L.) under water-deficit conditions. Arch Agron Soil Sci 61(4):507–523. https://doi.org/10.1080/03650340.2014.938644

Bajehbaj AA (2010) The effects of NaCl priming on salt tolerance in sunflower germination and seedling grown under salinity conditions. Afr J Biotechnol 9(12):1764–1770

Barba-Espín G, Hernández JA, Diaz-Vivancos P (2012) Role of H2O2 in pea seed germination. Plant Signal Behav 7(2):193–195. https://doi.org/10.4161/psb.18881

Barnabás B, Jäger K, Fehér A (2008) The effect of drought and heat stress on reproductive processes in cereals. Plant Cell Environ 31(1):11–38

Basra SMA, Pannu IA, Afzal I (2003) Evaluation of seedling vigor of hydro and matriprimed wheat (*Triticum aestivum* L.) seeds. Int J Agric Biol 5(2):121–123

Baxter A, Mittler R, Suzuki N (2013) ROS as key players in plant stress signalling. J Exp Bot 65(5):1229–1240

Bhaskara Reddy M, Arul J, Angers P, Couture L (1999) Chitosan treatment of wheat seeds induces resistance to Fusarium graminearum and improves seed quality. J Agric Food Chem 47(3):1208–1216

Bielen A, Remans T, Vangronsveld J, Cuypers A (2013) The influence of metal stress on the availability and redox state of ascorbate, and possible interference with its cellular functions. Int J Mol Sci 14(3):6382–6413

Bisht C, Badoni A, Vashishtha R, Nautiyal M (2009) Photoperiodic effect on seed germination in pyrethrum (Chrysanthemum cinerariaefolium vis.) under the influence of some growth regulators. J Am Sci 5:147–150

Bota J, Medrano H, Flexas J (2004) Is photosynthesis limited by decreased Rubisco activity and RuBP content under progressive water stress? New Phytol 162(3):671–681

Brocklehurst PA, Dearman J, Drew RLK (1984) Effects of osmotic priming on seed germination and seedling growth in leek. Sci Hortic 24(3):201–210. https://doi.org/10.1016/0304-4238(84)90103-1

Busch S, Lorenz CD, Taylor J, Pardo LC, McLain SE (2014) Short-range interactions of concentrated proline in aqueous solution. J Phys Chem B 118(49):14267–14277

Cakmak I (2005) The role of potassium in alleviating detrimental effects of abiotic stresses in plants. J Plant Nutr Soil Sci 168(4):521–530. https://doi.org/10.1002/jpln.200420485

Caseiro R, Bennett M, Marcos-Filho J (2004) Comparison of three priming techniques for onion seed lots differing in initial seed quality. Seed Sci Technol 32(2):365–375

Cayuela E, Pérez-Alfocea F, Caro M, Bolarín MC (1996) Priming of seeds with NaCl induces physiological changes in tomato plants grown under salt stress. Physiol Plant 96(2):231–236. https://doi.org/10.1111/j.1399-3054.1996.tb00207.x

Chen K, Arora R (2011) Dynamics of the antioxidant system during seed osmopriming, post-priming germination, and seedling establishment in Spinach (*Spinacia oleracea*). Plant Sci 180(2):212–220. https://doi.org/10.1016/j.plantsci.2010.08.007

Chen THH, Murata N (2002) Enhancement of tolerance of abiotic stress by metabolic engineering of betaines and other compatible solutes. Curr Opin Plant Biol 5(3):250–257. https://doi.org/10.1016/S1369-5266(02)00255-8

Chen THH, Murata N (2008) Glycinebetaine: an effective protectant against abiotic stress in plants. Trends Plant Sci 13(9):499–505. https://doi.org/10.1016/j.tplants.2008.06.007

Chen TH, Murata N (2011) Glycinebetaine protects plants against abiotic stress: mechanisms and biotechnological applications. Plant Cell Environ 34(1):1–20

Chen CC, Sung JM (2001) Priming bitter gourd seeds with selenium solution enhances germinability and antioxidative responses under sub-optimal temperature. Physiol Plant 111(1):9–16. https://doi.org/10.1034/j.1399-3054.2001.1110102.x

Chinnusamy V, Jagendorf A, Zhu J-K (2005) Understanding and improving salt tolerance in plants. Crop Sci 45(2):437–448

Cuartero J, Fernández-Muñoz R (1998) Tomato and salinity. Sci Hortic 78(1):83–125. https://doi.org/10.1016/S0304-4238(98)00191-5

Cuin TA, Shabala S (2007) Compatible solutes reduce ROS-induced potassium efflux in Arabidopsis roots. Plant Cell Environ 30(7):875–885. https://doi.org/10.1111/j.1365-3040.2007.01674.x

Dalvi AA, Bhalerao SA (2013) Response of plants towards heavy metal toxicity: an overview of avoidance, tolerance and uptake mechanism. Ann Plant Sci 2(09):362–368

Demir I, Mavi K (2004) The effect of priming on seedling emergence of differentially matured watermelon (*Citrullus lanatus* (Thunb.) Matsum and Nakai) seeds. Sci Hortic 102(4):467–473. https://doi.org/10.1016/j.scienta.2004.04.012

Demir I, Ozuaydın I, Yasar F, Van Staden J (2012) Effect of smoke-derived butenolide priming treatment on pepper and salvia seeds in relation to transplant quality and catalase activity. S Afr J Bot 78:83–87. https://doi.org/10.1016/j.sajb.2011.05.009

Draganic I, Lekic S (2012) Seed priming with antioxidants improves sunflower seed germination and seedling growth under unfavorable germination conditions. Turk J Agric For 36(4):421–428

Dubrovsky JG (1996) Seed hydration memory in Sonoran Desert cacti and its ecological implication. Am J Bot 83:624–632

Duncan DR, Widholm JM (1987) Proline accumulation and its implication in cold tolerance of regenerable maize callus. Plant Physiol 83(3):703–708

Egilla J, Davies F, Boutton T (2005) Drought stress influences leaf water content, photosynthesis, and water-use efficiency of *Hibiscus rosa-sinensis* at three potassium concentrations. Photosynthetica 43(1):135–140

Ehsanpour AA, Fatahian N (2003) Effects of salt and proline on *Medicago sativa* callus. Plant Cell Tissue Organ Cult 73(1):53–56. https://doi.org/10.1023/a:1022619523726

Eisvand H, Tavakkol-Afshari R, Sharifzadeh F, Maddah Arefi H, Hesamzadeh Hejazi S (2010) Effects of hormonal priming and drought stress on activity and isozyme profiles of antioxidant enzymes in deteriorated seed of tall wheatgrass (*Agropyron elongatum* Host). Seed Sci Technol 38(2):280–297

Elkoca E, Haliloglu K, Esitken A, Ercisli S (2007) Hydro- and osmopriming improve chickpea germination. Acta Agric Scand Sect B Soil Plant Sci 57(3):193–200. https://doi.org/10.1080/09064710600914087

El-Ramady H, Abdalla N, Alshaal T, Domokos-Szabolcsy E, Elhawat N, Prokisch J, Sztrik A, Fári M, El-Marsafawy S, Shams MS (2015) Selenium in soils under climate change, implication for human health. Environ Chem Lett 13(1):1–19

Farahbakhsh H (2012) Germination and seedling growth in un-primed and primed seeds of Fenel as affected by reduced water potential induced by NaCl. Int Res J App Basic Sci 3(4):737–744

Farhoudi R, Sharifzadeh F (2006) The effects of NaCl priming on salt tolerance in canola (*Brassica napus* L.) seedlings grown under saline conditions. Indian J Crop Sci 1(1–2):74–78

Farooq M, Basra SM, Rehman H, Mehmood T (2006) Germination and early seedling growth as affected by pre-sowing ethanol seed treatments in fine rice. Int J Agric Biol 8:19–22

Farooq M, Wahid A, Kobayashi N, Fujita D, Basra S (2009) Plant drought stress: effects, mechanisms and management. Agron Sustain Dev 29(1):185–212

Fedoroff N, Battisti D, Beachy R, Cooper P, Fischhoff D, Hodges C, Knauf V, Lobell D, Mazur B, Molden D (2010) Radically rethinking agriculture for the 21st century. Science 327(5967):833–834

Filippou P, Bouchagier P, Skotti E, Fotopoulos V (2014) Proline and reactive oxygen/nitrogen species metabolism is involved in the tolerant response of the invasive plant species *Ailanthus altissima* to drought and salinity. Environ Exp Bot 97:1–10

Flora SJ (2009) Structural, chemical and biological aspects of antioxidants for strategies against metal and metalloid exposure. Oxidative Med Cell Longev 2(4):191–206

Flowers TJ, Colmer TD (2008) Salinity tolerance in halophytes. New Phytol 179(4):945–963

Foti R, Abureni K, Tigere A, Gotosa J, Gere J (2008) The efficacy of different seed priming osmotica on the establishment of maize (*Zea mays* L.) caryopses. J Arid Environ 72(6):1127–1130. https://doi.org/10.1016/j.jaridenv.2007.11.008

Fu J, Huang B (2001) Involvement of antioxidants and lipid peroxidation in the adaptation of two cool-season grasses to localized drought stress. Environ Exp Bot 45(2):105–114

Fujita M, Fujita Y, Noutoshi Y, Takahashi F, Narusaka Y, Yamaguchi-Shinozaki K, Shinozaki K (2006) Crosstalk between abiotic and biotic stress responses: a current view from the points of convergence in the stress signaling networks. Curr Opin Plant Biol 9(4):436–442. https://doi.org/10.1016/j.pbi.2006.05.014

Gajalakshmi S, Iswarya V, Ashwini R, Divya G, Mythili S, Sathiavelu A (2012) Evaluation of heavy metals in medicinal plants growing in Vellore District. Eur J Exp Biol 2(5):1457–1461

Gao Y-P, Bonham-Smith PC, Gusta LV (2002) The role of peroxiredoxin antioxidant and calmodulin in ABA-primed seeds of *Brassica napus* exposed to abiotic stresses during germination. J Plant Physiol 159(9):951–958. https://doi.org/10.1078/0176-1617-00782

Garg B (2003) Nutrient uptake and management under drought: nutrient-moisture interaction. Curr Agric 27(1/2):1–8

Gaspar T, Franck T, Bisbis B, Kevers C, Jouve L, Hausman J-F, Dommes J (2002) Concepts in plant stress physiology. Application to plant tissue cultures. Plant Growth Regul 37(3):263–285

Ghassemi-Golezani K, Esmaeilpour B (2008) The effect of salt priming on the performance of differentially matured cucumber (*Cucumis sativus*) seeds. Notulae Botanicae Hortic Agrobotanici Cluj-Napoca 36(2):67

Giri J (2011) Glycinebetaine and abiotic stress tolerance in plants. Plant Signal Behav 6(11):1746–1751. https://doi.org/10.4161/psb.6.11.17801

Guan Y-J, Hu J, Wang X-J, Shao C-X (2009) Seed priming with chitosan improves maize germination and seedling growth in relation to physiological changes under low temperature stress. J Zhejiang Univ Sci B 10(6):427–433. https://doi.org/10.1631/jzus.B0820373

Gul B, Khan MA (2004) Effect of growth regulators and osmotica in alleviating salinity effects on the germination of *Salicornia utahensis*. Pak J Bot 35(5; SPI):885–894

Haghpanah A, Younesi O, Moradi A (2009) The effect of priming on seedling emergence of differentially matured sorghum (*Sorghum bicolor* L.) seeds. J Appl Sci Res 5:729–732

Hall A (2001) Crop developmental responses to temperature, photoperiod, and light quality. In: Crop response to environment, pp 83–87

Hall J (2002) Cellular mechanisms for heavy metal detoxification and tolerance. J Exp Bot 53(366):1–11

Hancock JT, Neill SJ, Wilson ID (2011) Nitric oxide and ABA in the control of plant function. Plant Sci 181(5):555–559. https://doi.org/10.1016/j.plantsci.2011.03.017

Harada E, Kim J-A, Meyer AJ, Hell R, Clemens S, Choi Y-E (2010) Expression profiling of tobacco leaf trichomes identifies genes for biotic and abiotic stresses. Plant Cell Physiol 51(10):1627–1637

Harb E (1992) Effect of soaking seeds in some growth regulators and micronutrients on growth, some chemical constituents and yield of faba bean and cotton plants. Bull Fac Agric Cairo Univ (Egypt) 43:429–452

Harris D, Tripathi R, Joshi A (2002) On-farm seed priming to improve crop establishment and yield in dry direct-seeded rice. In: Direct seeding: research strategies and opportunities. International Research Institute, Manila, pp 231–240

Hasanuzzaman M, Hossain MA, da Silva JAT, Fujita M (2012) Plant response and tolerance to abiotic oxidative stress: antioxidant defense is a key factor. In: Crop stress and its management: perspectives and strategies. Springer, pp 261–315

Hasanuzzaman M, Nahar K, Fujita M, Ahmad P, Chandna R, Prasad M, Ozturk M (2013) Enhancing plant productivity under salt stress: relevance of poly-omics. In: Salt stress in plants. Springer, pp 113–156

Hauser M-T (2014) Molecular basis of natural variation and environmental control of trichome patterning. Front Plant Sci 5:320

Howarth C (2005) Genetic improvements of tolerance to high temperature. In: Ashraf M, Harris PJC (eds) Abiotic stresses–plant resistance through breeding and molecular approaches. The Haworth Press, New York, pp 277–300

Hua-Long L, Han-Jing S, Jing-Guo W, Yang L, De-Tang Z, Hong-Wei Z (2014) Effect of seed soaking with exogenous proline on seed germination of rice under salt stress. J Northeast Agric Univ (Engl Ed) 21(3):1–6. https://doi.org/10.1016/S1006-8104(14)60062-3

Iba K (2002) Acclimative response to temperature stress in higher plants: approaches of gene engineering for temperature tolerance. Annu Rev Plant Biol 53(1):225–245

Iqbal M, Ashraf M (2005) Changes in growth, photosynthetic capacity and ionic relations in spring wheat (*Triticum aestivum* L.) due to pre-sowing seed treatment with polyamines. Plant Growth Regul 46(1):19–30. https://doi.org/10.1007/s10725-005-5901-8

Iqbal M, Ashraf M (2007) Seed preconditioning modulates growth, ionic relations, and photosynthetic capacity in adult plants of hexaploid wheat under salt stress. J Plant Nutr 30(3):381–396. https://doi.org/10.1080/01904160601171330

Iqbal M, Ashraf M, Jamil A, Ur-Rehman S (2006) Does seed priming induce changes in the levels of some endogenous plant hormones in hexaploid wheat plants under salt stress? J Integr Plant Biol 48(2):181–189. https://doi.org/10.1111/j.1744-7909.2006.00181.x

Islam MM, Hoque MA, Okuma E, Banu MNA, Shimoishi Y, Nakamura Y, Murata Y (2009) Exogenous proline and glycinebetaine increase antioxidant enzyme activities and confer tolerance to cadmium stress in cultured tobacco cells. J Plant Physiol 166(15):1587–1597. https://doi.org/10.1016/j.jplph.2009.04.002

Jafar MZ, Farooq M, Cheema MA, Afzal I, Basra SMA, Wahid MA, Aziz T, Shahid M (2012) Improving the performance of wheat by seed priming under saline conditions. J Agron Crop Sci 198(1):38–45. https://doi.org/10.1111/j.1439-037X.2011.00485.x

Jain M, Mathur G, Koul S, Sarin N (2001) Ameliorative effects of proline on salt stress-induced lipid peroxidation in cell lines of groundnut (*Arachis hypogaea* L.). Plant Cell Rep 20(5):463–468. https://doi.org/10.1007/s002990100353

Jain N, Kulkarni MG, van Staden J (2006) A butenolide, isolated from smoke, can overcome the detrimental effects of extreme temperatures during tomato seed germination. Plant Growth Regul 49(2):263–267. https://doi.org/10.1007/s10725-006-9136-0

Jaleel CA, Gopi R, Manivannan P, Panneerselvam R (2007a) Responses of antioxidant defense system of *Catharanthus roseus* (L.) G. Don. to paclobutrazol treatment under salinity. Acta Physiol Plant 29(3):205–209. https://doi.org/10.1007/s11738-007-0025-6

Jaleel CA, Manivannan P, Sankar B, Kishorekumar A, Sankari S, Panneerselvam R (2007b) Paclobutrazol enhances photosynthesis and ajmalicine production in *Catharanthus roseus*. Process Biochem 42(11):1566–1570

Jaleel CA, Riadh K, Gopi R, Manivannan P, Ines J, Al-Juburi HJ, Chang-Xing Z, Hong-Bo S, Panneerselvam R (2009) Antioxidant defense responses: physiological plasticity in higher plants under abiotic constraints. Acta Physiol Plant 31(3):427–436

Jett LW, Welbaum GE, Morse RD (1996) Effects of matric and osmotic priming treatments on broccoli seed germination. J Am Soc Hortic Sci 121(3):423–429

Jisha KC, Vijayakumari K, Puthur JT (2013) Seed priming for abiotic stress tolerance: an overview. Acta Physiol Plant 35(5):1381–1396. https://doi.org/10.1007/s11738-012-1186-5

Jozefczak M, Remans T, Vangronsveld J, Cuypers A (2012) Glutathione is a key player in metal-induced oxidative stress defenses. Int J Mol Sci 13(3):3145–3175

Kalaji HM, Bosa K, Kościelniak J, Żuk-Gołaszewska K (2011) Effects of salt stress on photosystem II efficiency and CO 2 assimilation of two Syrian barley landraces. Environ Exp Bot 73:64–72

Kamran M, Shahbaz M, Ashraf M, Akram NA (2009) Alleviation of drought-induced adverse effects in spring wheat (*Triticum aestivum* L.) using proline as a pre-sowing seed treatment. Pak J Bot 41(2):621–632

Kaur G, Asthir B (2015) Proline: a key player in plant abiotic stress tolerance. Biol Plant 59(4):609–619. https://doi.org/10.1007/s10535-015-0549-3

Kaur S, Gupta AK, Kaur N (2002) Effect of osmo- and hydropriming of chickpea seeds on seedling growth and carbohydrate metabolism under water deficit stress. Plant Growth Regul 37(1):17–22. https://doi.org/10.1023/a:1020310008830

Kaur S, Gupta AK, Kaur N (2005) Seed priming increases crop yield possibly by modulating enzymes of sucrose metabolism in chickpea. J Agron Crop Sci 191(2):81–87. https://doi.org/10.1111/j.1439-037X.2004.00140.x

Kaur G, Athar M, Alam MS (2008) *Quercus infectoria* galls possess antioxidant activity and abrogates oxidative stress-induced functional alterations in murine macrophages. Chem Biol Interact 171(3):272–282

Kaur D, Grewal S, Kaur J, Singh S (2017) Differential proline metabolism in vegetative and reproductive tissues determine drought tolerance in chickpea. Biol Plant 61(2):359–366

Kaya MD, Okçu G, Atak M, Çıkılı Y, Kolsarıcı Ö (2006) Seed treatments to overcome salt and drought stress during germination in sunflower (*Helianthus annuus* L.). Eur J Agron 24(4):291–295. https://doi.org/10.1016/j.eja.2005.08.001

Khalil SK, Mexal JG, Murray LW (2001) Germination of soybean seed primed in aerated solution of polyethylene glycol 8000. Online J Biol Sci 1(3):105–107

Khan MA, Gul B, Weber DJ (2002) Improving seed germination of *Salicornia rubra* (Chenopodiaceae) under saline conditions using germination-regulating chemicals. West N Am Nat 62:101–105

Khan MB, Gurchani MA, Hussain M, Freed S, Mahmood K (2011) Wheat seed enhancement by vitamin and hormonal priming. Pak J Bot 43(3):1495–1499

Korkmaz A, Pill W (2003) The effect of different priming treatments and storage conditions on germination performance of lettuce seeds. Eur J Hortic Sci 68:260–265

Kulkarni MG, Sparg SG, Light ME, Van Staden J (2006) Stimulation of rice (*Oryza sativa* L.) seedling vigour by smoke-water and butenolide. J Agron Crop Sci 192(5):395–398. https://doi.org/10.1111/j.1439-037X.2006.00213.x

Kumar A, Maiti SK (2013) Availability of chromium, nickel and other associated heavy metals of ultramafic and serpentine soil/rock and in plants. Int J Emerg Technol Adv Eng 3(2):256–268

Kumar M, Sirhindi G, Bhardwaj R, Kumar S, Jain G (2010) Effect of exogenous H 2 O 2 on antioxidant enzymes of *Brassica juncea* L. seedlings in relation to 24-epibrassinolide under chilling stress

Lazaridou M, Noitsakis B (2003) The effect of water deficit on yield and water use efficiency of lucerne. In: Optimal forage systems for animal production and the environment. Proceedings of the 12th symposium of the European Grassland Federation, Pleven, Bulgaria, 26–28 May 2003. Bulgarian Association for Grassland and Forage Production (BAGFP), pp 344–347

Lee S, Moon JS, Domier LL, Korban SS (2002) Molecular characterization of phytochelatin synthase expression in transgenic Arabidopsis. Plant Physiol Biochem 40(9):727–733

Li J, Yin LY, Jongsma MA, Wang CY (2011) Effects of light, hydropriming and abiotic stress on seed germination, and shoot and root growth of pyrethrum (*Tanacetum cinerariifolium*). Ind Crop Prod 34(3):1543–1549. https://doi.org/10.1016/j.indcrop.2011.05.012

Manickavelu A, Nadarajan N, Ganesh S, Gnanamalar R, Babu RC (2006) Drought tolerance in rice: morphological and molecular genetic consideration. Plant Growth Regul 50(2–3):121–138

Manjunatha G, Raj SN, Shetty NP, Shetty HS (2008) Nitric oxide donor seed priming enhances defense responses and induces resistance against pearl millet downy mildew disease. Pestic Biochem Physiol 91(1):1–11. https://doi.org/10.1016/j.pestbp.2007.11.012

Marschner H (2011) Marschner's mineral nutrition of higher plants. Academic, London

Maughan S, Foyer CH (2006) Engineering and genetic approaches to modulating the glutathione network in plants. Physiol Plant 126(3):382–397. https://doi.org/10.1111/j.1399-3054.2006.00684.x

McWilliams D (2003) Drought strategies for cotton, cooperative extension service circular 582, College of Agriculture and Home Economics. New Mexico State University, Las Cruces

Michalak A (2006) Phenolic compounds and their antioxidant activity in plants growing under heavy metal stress. Pol J Environ Stud 15(4):523–530

Miller G, Honig A, Stein H, Suzuki N, Mittler R, Zilberstein A (2009) Unraveling Δ1-pyrroline-5-carboxylate-proline cycle in plants by uncoupled expression of proline oxidation enzymes. J Biol Chem 284(39):26482–26492

Mirza H, Hossain MA, Fujita M (2010) Selenium in higher plants: physiological role, antioxidant metabolism and abiotic stress tolerance. J Plant Sci 5(4):354–375

Mittler R (2002) Oxidative stress, antioxidants and stress tolerance. Trends Plant Sci 7(9):405–410. https://doi.org/10.1016/S1360-1385(02)02312-9

Mohanty M, Patra HK (2013) Effect of ionic and chelate assisted hexavalent chromium on mung bean seedlings (*Vigna radiata* L. wilczek. var k-851) during seedling growth. J Stress Physiol Biochem 9(2):232–241

Monakhova O, Chernyad'ev I (2002) Protective role of kartolin-4 in wheat plants exposed to soil draught. Appl Biochem Microbiol 38(4):373–380

Moradi A, Younesi O (2009) Effects of osmo-and hydro-priming on seed parameters of grain sorghum (*Sorghum bicolor* L.). Aust J Basic Appl Sci 3(3):1696–1700

Mourato M, Reis R, Martins LL (2012) Characterization of plant antioxidative system in response to abiotic stresses: a focus on heavy metal toxicity. In: Advances in selected plant physiology aspects. InTech

Munns R, James RA, Läuchli A (2006) Approaches to increasing the salt tolerance of wheat and other cereals. J Exp Bot 57(5):1025–1043

Nagajyoti P, Lee K, Sreekanth T (2010) Heavy metals, occurrence and toxicity for plants: a review. Environ Chem Lett 8(3):199–216

Nam N, Chauhan Y, Johansen C (2001) Effect of timing of drought stress on growth and grain yield of extra-short-duration pigeonpea lines. J Agric Sci 136(2):179–189

Nath K, Singh D, Shyam S, Sharma YK (2008) Effect of chromium and tannery effluent toxicity on metabolism and growth in cowpea (*Vigna sinensis* L. Saviex Hassk) seedling. Res Environ Life Sci 1(3):91–94

Nawaz A, Amjad M, Pervez MA, Afzal I (2011) Effect of halopriming on germination and seedling vigor of tomato. Afr J Agric Res 6(15):3551–3559

Nematshahi N, Lahouti M, Ganjeali A (2012) Accumulation of chromium and its effect on growth of (*Allium cepa* cv. Hybrid). Eur J Exp Biol 2:969–974

Nerd A, Nobel PS (1991) Effects of drought on water relations and nonstructural carbohydrates in cladodes of *Opuntia ficus-indica*. Physiol Plant 81(4):495–500

Okçu G, Kaya MD, Atak M (2005) Effects of salt and drought stresses on germination and seedling growth of pea (*Pisum sativum* L.). Turk J Agric For 29(4):237–242

Oliveira H (2012) Chromium as an environmental pollutant: insights on induced plant toxicity. J Bot 2012:1–8

Ozbingol N, Corbineau F, Come D (1998) Responses of tomato seeds to osmoconditioning as related to temperature and oxygen. Seed Sci Res 8(3):377–384

Panda S, Choudhury S (2005) Chromium stress in plants. Braz J Plant Physiol 17(1):95–102

Papageorgiou GC, Murata N (1995) The unusually strong stabilizing effects of glycine betaine on the structure and function of the oxygen-evolving photosystem II complex. Photosynth Res 44(3):243–252. https://doi.org/10.1007/bf00048597

Parida AK, Das AB (2005) Salt tolerance and salinity effects on plants: a review. Ecotoxicol Environ Saf 60(3):324–349

Parihar P, Singh S, Singh R, Singh VP, Prasad SM (2015) Effect of salinity stress on plants and its tolerance strategies: a review. Environ Sci Pollut Res 22(6):4056–4075

Park E-J, Jeknic Z, Sakamoto A, Denoma J, Murata N, Chen TH (2003) Genetic engineering of cold-tolerant tomato via glycinebetaine biosynthesis. Cryobiol Cryotechnol 49:77–85

Park E-J, Jeknić Z, Sakamoto A, DeNoma J, Yuwansiri R, Murata N, Chen THH (2004) Genetic engineering of glycinebetaine synthesis in tomato protects seeds, plants, and flowers from chilling damage. Plant J 40(4):474–487. https://doi.org/10.1111/j.1365-313X.2004.02237.x

Park E-J, Jeknic Z, Chen TH (2006) Exogenous application of glycinebetaine increases chilling tolerance in tomato plants. Plant Cell Physiol 47(6):706–714

Patade VY, Bhargava S, Suprasanna P (2009) Halopriming imparts tolerance to salt and PEG induced drought stress in sugarcane. Agric Ecosyst Environ 134(1):24–28. https://doi.org/10.1016/j.agee.2009.07.003

Peuke A, Schraml C, Hartung W, Rennenberg H (2002) Identification of drought-sensitive beech ecotypes by physiological parameters. New Phytol 154(2):373–387

Posmyk MM, Janas KM (2007) Effects of seed hydropriming in presence of exogenous proline on chilling injury limitation in *Vigna radiata* L. seedlings. Acta Physiol Plant 29(6):509–517. https://doi.org/10.1007/s11738-007-0061-2

Rai V, Vajpayee P, Singh SN, Mehrotra S (2004) Effect of chromium accumulation on photosynthetic pigments, oxidative stress defense system, nitrate reduction, proline level and eugenol content of *Ocimum tenuiflorum* L. Plant Sci 167(5):1159–1169

Randhir R, Shetty K (2005) Developmental stimulation of total phenolics and related antioxidant activity in light- and dark-germinated corn by natural elicitors. Process Biochem 40(5):1721–1732. https://doi.org/10.1016/j.procbio.2004.06.064

Rashid A, Harris D, Hollington P, Ali S (2004) On-farm seed priming reduces yield losses of mungbean (*Vigna radiata*) associated with mungbean yellow mosaic virus in the North West Frontier Province of Pakistan. Crop Prot 23(11):1119–1124. https://doi.org/10.1016/j.cropro.2004.04.002

Rashid A, Hollington PA, Harris D, Khan P (2006) On-farm seed priming for barley on normal, saline and saline–sodic soils in North West Frontier Province, Pakistan. Eur J Agron 24(3):276–281. https://doi.org/10.1016/j.eja.2005.10.006

Rastgoo L, Alemzadeh A, Afsharifar A (2011) Isolation of two novel isoforms encoding zinc- and copper-transporting P1B-ATPase from Gouan (*Aeluropus littoralis*). Plant Omics J 4(7):377–383

Rathinasabapathi B (2000) Metabolic engineering for stress tolerance: installing osmoprotectant synthesis pathways. Ann Bot 86(4):709–716

Reddy CS, Smith JD (1978) Effects of delayed post treatment of gamma-irradiated seed with cysteine on the growth of *Sorghum bicolor* seedlings. Environ Exp Bot 18(4):241–243. https://doi.org/10.1016/0098-8472(78)90050-3

Rhodes D, Hanson A (1993) Quaternary ammonium and tertiary sulfonium compounds in higher plants. Annu Rev Plant Biol 44(1):357–384

Rithichai P, Sampantharat P, Jirakiattikul Y (2009) Coriander (*Coriandrum sativum* l.) seed quality as affected by accelerated aging and subsequent hydropriming. Asian J Food Agro Ind 1:217–221

Rouhi HR, Afshari RT, Moosavi SA, Gharineh MH (2010) Effects of osmopriming on germination and vigour traits of bersim clover (*Trifolium alexandrinum* L.). Notulae Sci Biol 2(4):59

Rouhi HR, Aboutalebian MA, Sharif-Zadeh F (2011) Effects of hydro and osmopriming on drought stress tolerance during germination in four grass species. Int J Agrisience 1(2):107–114

Roy D, Basu N, Bhunia A, Banerjee SK (1993) Counteraction of exogenous L-proline with NaCl in salt-sensitive cultivar of rice. Biol Plant 35(1):69. https://doi.org/10.1007/bf02921122

Ruan S, Xue Q (2002) Effects of chitosan coating on seed germination and salt-tolerance of seedling in hybrid rice (*Oryza sativa* L.). Zuo Wu Xue Bao 28(6):803–808

Rubluo A (1982) The effect of L-cysteine on presoaked barley seeds treated with methyl methanesulfonate. Experientia 38(3):326–327. https://doi.org/10.1007/bf01949371

Sadak MS, Mostafa HA (2015) Physiological role of pre-sowing seed with proline on some growth, biochemical aspects, yield quantity and quality of two sunflower cultivars grown under seawater salinity stress. Sci Agric 9(1):60–69

Saha P, Chatterjee P, Biswas AK (2010) NaCl pretreatment alleviates salt stress by enhancement of antioxidant defense system and osmolyte accumulation in mungbean (*Vigna radiata* L. Wilczek)

Sahu M, Kumawat S, D'souza S, Ramaswamy N, Singh G (2005) Sulphydryl bioregulator technology for increasing mustard production. Research Bulletin RAU-BARC, pp 1–52

Sakamoto A, Murata N (2000) Genetic engineering of glycinebetaine synthesis in plants: current status and implications for enhancement of stress tolerance. J Exp Bot 51(342):81–88

Sakamoto A, Murata N (2002) The role of glycine betaine in the protection of plants from stress: clues from transgenic plants. Plant Cell Environ 25(2):163–171

Salama KH, Mansour MM, Hassan NS (2011) Choline priming improves salt tolerance in wheat (*Triticum aestivum* L.). Aust J Basic Appl Sci 5:126–132

Samantaray S, Rout GR, Das P (1998) Role of chromium on plant growth and metabolism. Acta Physiol Plant 20(2):201–212

Samarah NH (2005) Effects of drought stress on growth and yield of barley. Agron Sustain Dev 25(1):145–149

Sarwar N, Yousaf S, Jamil FF (2006) Induction of salt tolerance in chickpea by using simple and safe chemicals. Pak J Bot 38(2):325

Savvides A, Ali S, Tester M, Fotopoulos V (2016) Chemical priming of plants against multiple abiotic stresses: mission possible? Trends Plant Sci 21(4):329–340

Schutzendubel A, Polle A (2002) Plant responses to abiotic stresses: heavy metal-induced oxidative stress and protection by mycorrhization. J Exp Bot 53(372):1351–1365

Sebastiani L, Scebba F, Tognetti R (2004) Heavy metal accumulation and growth responses in poplar clones Eridano (Populus deltoides× maximowiczii) and I-214 (P.× euramericana) exposed to industrial waste. Environ Exp Bot 52(1):79–88

Sedghi M, Nemati A, Amanpour-Balaneji B, Gholipouri A (2010) Influence of different priming materials on germination and seedling establishment of milk thistle (*Silybum marianum*) under salinity stress. World Appl Sci J 11(5):604–609

Shahrokhi M, Tehranifar A, Hadizadeh H, Selahvarzi Y (2011) Effect of drought stress and paclobutrazol-treated seeds on physiological response of *Festuca arundinacea* L. master and *Lolium perenne* L. barrage. J Biol Environ Sci 5(14):77–85

Shakirova FM, Sakhabutdinova AR, Bezrukova MV, Fatkhutdinova RA, Fatkhutdinova DR (2003) Changes in the hormonal status of wheat seedlings induced by salicylic acid and salinity. Plant Sci 164(3):317–322

Shanker AK, Cervantes C, Loza-Tavera H, Avudainayagam S (2005) Chromium toxicity in plants. Environ Int 31(5):739–753

Shao C, Hu J, Song W, Hu W (2005) Effects of seed priming with chitosan solutions of different acidity on seed germination and physiological characteristics of maize seedling. J Zhejiang Univ (Agric Life Sci) 31(6):705–708

Sharma SS, Dietz K-J (2006) The significance of amino acids and amino acid-derived molecules in plant responses and adaptation to heavy metal stress. J Exp Bot 57(4):711–726

Sharma P, Jha AB, Dubey RS, Pessarakli M (2012) Reactive oxygen species, oxidative damage, and antioxidative defense mechanism in plants under stressful conditions. J Bot 2012:1–26

Siddique M, Hamid A, Islam M (2000) Drought stress effects on water relations of wheat. Bot Bull Acad Sin 41:35–39

Singh B, Rao G (1993) Effects of chemical soaking of sunflower (*Helianthus annuus* L.) seed on vigour index. Indian J Agric Sci 2:232–233

Singh HP, Mahajan P, Kaur S, Batish DR, Kohli RK (2013) Chromium toxicity and tolerance in plants. Environ Chem Lett 11(3):229–254

Sivritepe HÖ, Sivritepe N, Eriş A, Turhan E (2005) The effects of NaCl pre-treatments on salt tolerance of melons grown under long-term salinity. Sci Hortic 106(4):568–581. https://doi.org/10.1016/j.scienta.2005.05.011

Somerville C, Briscoe J (2001) Genetic engineering and water. American Association for the Advancement of Science

Srivastava S, Jain R (2011) In-situ monitoring of chromium cytotoxicity in sugarcane. J Environ Biol 32(6):759

Srivastava A, Ramaswamy N, Mukopadhyaya R, Jincy MC, D'Souza S (2008) Thiourea modulates the expression and activity profile of mtATPase under salinity stress in seeds of *Brassica juncea*. Ann Bot 103(3):403–410

Srivastava AK, Lokhande VH, Patade VY, Suprasanna P, Sjahril R, D'Souza SF (2010a) Comparative evaluation of hydro-, chemo-, and hormonal-priming methods for imparting salt and PEG stress tolerance in Indian mustard (*Brassica juncea* L.). Acta Physiol Plant 32(6):1135–1144. https://doi.org/10.1007/s11738-010-0505-y

Srivastava AK, Suprasanna P, Srivastava S, D'Souza SF (2010b) Thiourea mediated regulation in the expression profile of aquaporins and its impact on water homeostasis under salinity stress in *Brassica juncea* roots. Plant Sci 178(6):517–522. https://doi.org/10.1016/j.plantsci.2010.02.015

Su J, Hirji R, Zhang L, He C, Selvaraj G, Wu R (2006) Evaluation of the stress-inducible production of choline oxidase in transgenic rice as a strategy for producing the stress-protectant glycine betaine. J Exp Bot 57(5):1129–1135

Sulpice R, Tsukaya H, Nonaka H, Mustardy L, Chen THH, Murata N (2003) Enhanced formation of flowers in salt-stressed Arabidopsis after genetic engineering of the synthesis of glycine betaine. Plant J 36(2):165–176. https://doi.org/10.1046/j.1365-313X.2003.01873.x

Sultana N, Ikeda T, Mitsui T (2000) GA3 and proline promote germination of wheat seeds by stimulating α-Amylase at unfavorable temperatures. Plant Prod Sci 3(3):232–237

Sun Y-Y, Sun Y-J, Wang M-T, Li X-Y, Guo X, Hu R, Ma J (2010) Effects of seed priming on germination and seedling growth under water stress in rice. Acta Agron Sin 36(11):1931–1940. https://doi.org/10.1016/S1875-2780(09)60085-7

Sung JM, Chiu KY (1995) Hydration effect on seedling emergence strength of watermelon seeds differing in ploidy. Plant Sci 110(1):21–26. https://doi.org/10.1016/0168-9452(95)04183-U

Szabados L, Savouré A (2010) Proline: a multifunctional amino acid. Trends Plant Sci 15(2):89–97. https://doi.org/10.1016/j.tplants.2009.11.009

Tavili A, Zare S, Enayati Z (2009) Hydropriming, ascorbic and salicylic acid influence on germination of agropyron elongation host. Seeds under salt stress

Tavili A, Zare S, Moosavi SA, Enayati A (2011) Effects of seed priming on germination characteristics of Bromus species under salt and drought conditions. Am Eurasian J Agric Environ Sci 10(2):163–168

Thakur P, Kumar S, Malik JA, Berger JD, Nayyar H (2010) Cold stress effects on reproductive development in grain crops: an overview. Environ Exp Bot 67(3):429–443

Thomas U, Varughese K, Thomas A, Sadanandan S (2000) Seed priming for increased vigour, viability and productivity of upland rice. Leisa India 4:14

Tiryaki I, Korkmaz A, Nas M, Ozbay N (2005) Priming combined with plant growth regulators promotes germination and emergence of dormant *Amaranthus cruentus* L. seeds. Seed Sci Technol 33(3):571–579

Todorov D, Karanov E, Smith A, Hall M (2003) Chlorophyllase activity and chlorophyll content in wild and mutant plants of *Arabidopsis thaliana*. Biol Plant 46(1):125–127

Valko M, Morris H, Cronin M (2005) Metals, toxicity and oxidative stress. Curr Med Chem 12(10):1161–1208

Verbruggen N, Hermans C (2008) Proline accumulation in plants: a review. Amino Acids 35(4):753–759. https://doi.org/10.1007/s00726-008-0061-6

Viehweger K (2014) How plants cope with heavy metals. Bot Stud 55(1):35

Vollenweider P, Günthardt-Goerg MS (2005) Diagnosis of abiotic and biotic stress factors using the visible symptoms in foliage. Environ Pollut 137(3):455–465

Wahid A, Rasul E, Rao R, Iqbal R (2005) Photosynthesis in leaf, stem, flower and fruit. Handb Photosynth 2:479–497

Wahid A, Gelani S, Ashraf M, Foolad MR (2007a) Heat tolerance in plants: an overview. Environ Exp Bot 61(3):199–223

Wahid A, Perveen M, Gelani S, Basra SMA (2007b) Pretreatment of seed with H2O2 improves salt tolerance of wheat seedlings by alleviation of oxidative damage and expression of stress proteins. J Plant Physiol 164(3):283–294. https://doi.org/10.1016/j.jplph.2006.01.005

Wong HL, Sakamoto T, Kawasaki T, Umemura K, Shimamoto K (2004) Down-regulation of metallothionein, a reactive oxygen scavenger, by the small GTPase OsRac1 in rice. Plant Physiol 135(3):1447–1456

Xiong L, Zhu JK (2002) Molecular and genetic aspects of plant responses to osmotic stress. Plant Cell Environ 25(2):131–139. https://doi.org/10.1046/j.1365-3040.2002.00782.x

Xu J, Yin H, Li X (2009) Protective effects of proline against cadmium toxicity in micropropagated hyperaccumulator, *Solanum nigrum* L. Plant Cell Rep 28(2):325–333. https://doi.org/10.1007/s00299-008-0643-5

Xu S, Hu J, Li Y, Ma W, Zheng Y, Zhu S (2011) Chilling tolerance in *Nicotiana tabacum* induced by seed priming with putrescine. Plant Growth Regul 63(3):279–290. https://doi.org/10.1007/s10725-010-9528-z

Yadav PV, Kumari M, Ahmed Z (2011) Seed priming mediated germination improvement and tolerance to subsequent exposure to cold and salt stress in capsicum. Res J Seed Sci 4(3):125–136

Yagmur M, Kaydan D (2008) Alleviation of osmotic stress of water and salt in germination and seedling growth of triticale with seed priming treatments. Afr J Biotechnol 7(13):2156–2162

Yancey PH (1994) Compatible and counteracting solutes. In: Cellular and molecular physiology of cell volume regulation, pp 81–109

Yeh Y, Chiu K, Chen C, Sung J (2005) Partial vacuum extends the longevity of primed bitter gourd seeds by enhancing their anti-oxidative activities during storage. Sci Hortic 104(1):101–112

Zeid I, Shedeed Z (2006) Response of alfalfa to putrescine treatment under drought stress. Biol Plant 50(4):635–640

Zhang JX, Wang C, Yang CY, Wang JY, Chen L, Bao XM, Zhao YX, Zhang H, Liu J (2010) The role of Arabidopsis AtFes1A in cytosolic Hsp70 stability and abiotic stress tolerance. Plant J 62(4):539–548

Zhang Q, Rue K, Mueller J (2014) The effect of glycinebetaine priming on seed germination of six turfgrass species under drought, salinity, or temperature stress. Hortscience 49(11):1454–1460

Zhou ZS, Guo K, Elbaz AA, Yang ZM (2009) Salicylic acid alleviates mercury toxicity by preventing oxidative stress in roots of *Medicago sativa*. Environ Exp Bot 65(1):27–34. https://doi.org/10.1016/j.envexpbot.2008.06.001

Seed Priming Technology in the Amelioration of Salinity Stress in Plants

5

Aditya Banerjee and Aryadeep Roychoudhury

Abstract

A large proportion of the global cultivable land is inflicted by saline conditions. Several popular plants and staple crops cannot be cultivated on these vast stretches of land due to their susceptibility to salt stress. Crops growing under such suboptimal conditions exhibit deteriorated physiological development and compromised yields. Several agro-biotechnology-supported programmes are available to enhance plant salt tolerance. Among them, seed priming or 'pretreatment' is the most acceptable one from the point of biosafety and socio-economic views. Seed priming provides an abiotic stress-like condition to the dormant seed. It partially reprogrammes the seed metabolome so that it experiences such suboptimal condition and can better adapt to salt stress. Partial hydration of the seed during priming weakens the endosperm, channelizes the energy reserves, makes the seed ready for radicle protrusion (germination) and recharges the entire antioxidant machinery. This chapter provides an insight into the multiple mechanisms via which seed priming with various inorganic as well as endogenous agents can ameliorate salinity stress-related damages across multiple plant species.

Keywords

Antioxidative machinery · Protective agents · Seed priming · Salinity stress · Salt tolerance

A. Banerjee · A. Roychoudhury (✉)
Post Graduate Department of Biotechnology, St. Xavier's College (Autonomous), Kolkata, West Bengal, India

© Springer Nature Singapore Pte Ltd. 2018

A. Rakshit, H. B. Singh (eds.), *Advances in Seed Priming*,
https://doi.org/10.1007/978-981-13-0032-5_5

5.1 Introduction

Abiotic stresses like salinity, drought, heavy metal toxicity, irradiation, etc. lead to large-scale crop losses throughout the world. Among these environmental challenges, salt stress is most prevalent in arid, semiarid and coastal regions and spreads easily in the irrigated lands (Munns and Tester 2008). The FAO report (2011) highlights that 60–80 million hectares of land are inflicted by salt. This ultimately will lead to the loss of about 50% of the cultivable lands by the twenty-first century. High salt content in the soil or the irrigated water directly interferes with seed germination and seedling growth, thus making most plants susceptible to this kind of abiotic stress (Hubbard et al. 2012). Salt stress delays the advent of germination in susceptible plant seeds (Thiam et al. 2013). An interesting contradiction has been noted in the development of plants in response to low and high salt concentrations (Khan and Weber 2008). It was seen that whereas low salt levels promote seed dormancy, high salt concentrations directly inhibit seed germination. However, both these stress inductions ultimately decrease the germination rate and thus lead to phenotypically retarded development (Khan and Weber 2008). Several crops like *Oryza sativa*, *Zea mays*, *Brassica oleracea*, *Abelmoschus esculentus*, *Vigna unguiculata*, *Apium graveolens*, *Foeniculum vulgare*, *Petroselinum crispum*, *Raphanus sativus*, *Ipomoea aquatica*, *Silybum marianum*, *Lactuca sativa*, *Glycine max*, etc. are reportedly sensitive to a gradient of salt concentrations (Banerjee and Roychoudhury 2016a; Basu and Roychoudhury 2014; Ibrahim 2016).

Esechie (1995) showed that the top 10 cm layer of the soil accumulates higher salt levels than the lower layers. Seeds of cultivated crops are usually sown in this top layer. High evapotranspiration in plants growing in the arid environments results in water loss and accumulation of salt around the roots. This retards translocation and crucial physiological processes (Bernstein and Hayward 1958). Hence, novel strategies are required to ameliorate salt stress in developing crop plants. Transgenic technology has often been adopted to generate genetically modified (GM) plants overexpressing a target gene which confers stress tolerance. However, this technology faces several biosafety issues across multiple countries, and hence such GM plants cannot be popularly marketed. Thus researchers have designed a novel technology called 'seed priming' where an inorganic chemical solution or an endogenous osmoprotectant or 'eliciting factor' is purified and used as the pretreating agent to make the seeds tolerant to future stress exposures (Tanou et al. 2012). In this technology, the seeds are hydrated in a prescribed solution containing the optimum concentration of the 'eliciting factor' and then dried. This improves germination, triggers multiple epigenetic alterations and up-regulates genes encoding stress-responsive transcription factors (TFs) (Farooq et al. 2009; Bruce et al. 2007). The treated seeds reportedly exhibit higher germination and seedling emergence rates under stress conditions in comparison to the non-treated seeds (Sharma et al. 2014). Studies show that seed priming can even improve crop productivity under optimum conditions (Jisha et al. 2013). The popularity of seed priming lies in its easy usage, low cost and lesser environmental risk (Ibrahim 2016).

5.2 Salinity and Seed Germination

Salt stress primarily increases the soil osmotic potential which results in constrained water and solvent uptake via roots (Daszkowska-Golec 2011). The osmotic balance in the plant gets disrupted due to generation of reactive oxygen species (ROS) like hydroxyl radicals, superoxides and hydrogen peroxides (Das and Roychoudhury 2014). Massive oxidative stress caused by Na^+ and Cl^- toxicity jeopardizes macromolecular structures and membrane integrity and even affects embryo development. Physiological processes like photosynthesis, growth, respiration and flowering are severely inhibited by salt stress (Roychoudhury and Chakraborty 2013). The overall systemic deterioration leads to cellular apoptosis coupled with the degeneration of membrane lipids, enzymes and nucleic acids (Banerjee and Roychoudhury 2017a). Peroxidation of membrane lipids produces malondialdehyde (MDA), an important stress marker in plants. Such MDA levels sharply increase in salt-sensitive plants exposed to stress (Das and Roychoudhury 2014).

Salinity-induced ROS accumulation triggers the up-regulation of *osmotic stress responsive* (*OR*) genes and their upstream transcription factors (TFs) in a cultivar-dependent fashion (Roychoudhury et al. 2013; Banerjee and Roychoudhury 2017b). The *OR* gene products confer tolerance in specific cultivars of the crops exposed to salt stress. The salt-tolerant cultivars exhibit higher expression of antioxidant enzymes like superoxide dismutase (SOD), peroxidase (POX), catalase (CAT), glutathione reductase (GR), etc. Most of these enzymes restore the cellular oxidative equilibrium by operating through the ascorbate-glutathione cycle (Anjum et al. 2015). Recent studies have also highlighted the massive histone modifications, DNA methylation and chromatin remodelling occurring in signature genomic regions of the plants exposed to salinity (Banerjee and Roychoudhury 2017c). The transposon-associated differentially methylated regions (DMRs) in IR-64 (stress susceptible), Pokkali (salt tolerant) and Nagina 22 (drought tolerant) rice cultivars were closely related to the transcript abundance of the protein-coding genes (Garg et al. 2015). However, close association of the hypermethylated silenced heterochromatin with the small RNAs (smRNAs) was noted (Banerjee et al. 2016). This clarified the existence of a crosstalk among the chromatin methylation status, gene expression and smRNA abundance during salt stress response in rice.

5.3 Seed Physiology and Priming

Seed hydration triggers germination via three stages: imbibition, lag phase and radicle protrusion through the testa (Ibrahim 2016). Priming promotes partial hydration of seeds. This effectively accelerates pregermination metabolism but is not enough to facilitate the transition of a dormant seed towards complete germination (Paparella et al. 2015). Hence priming converts a metabolically naive seed into a quasi-metabolically active unit. However, such quasi-metabolic state does not support the complete emergence of the radicle.

Table 5.1 The different applied seed priming techniques

Seed priming technique	Nature of the treatment
Hydropriming	Water treated
Osmopriming	Aqueous solution of osmolytes like polyamines, etc.
Halopriming	Inorganic salt solution
Hormone priming	Phytohormones like abscisic acid (ABA), salicylic acid (SA), etc.
Hardening	Hydrated seeds are redried
Solid matrix	Solid materials like ground Leonardite Shale (Agro-Lig), etc. mixed with water in known proportions
Humidification and stratification	Seed processing like cold and moist treatments to promote faster germination after sowing
Physical	Irradiation, heat, etc.

The next crucial phase is the post-priming redrying (or drying back) of the seeds to restore their relative moisture content back to the initial control levels. Redrying of the primed seeds at the correct stage is extremely important for seed storage, preserving seed longevity and tolerance towards abiotic stresses (Ratikanta 2011). It has been reported that the partially hydrated seeds in the imbibition or lag phase tolerate redrying without significant physiological deteriorations (Rajjou et al. 2012). However, seeds with already emerged radicles if redried usually exhibit compromised seed vigour (Rajjou et al. 2012). The rate of redrying also regulates seed viability in due course (Gurusinghe and Bradford 2001). Bruggink et al. (1999) stated that the drying back of the primed seeds should be performed slowly as this improves seed longevity and tolerance to desiccation.

5.4 Seed Priming Techniques

The classification of the priming techniques varies with the chemical nature of the priming agent. Eight different priming techniques are usually reported (Ibrahim 2016). They have been highlighted in Table 5.1. Out of the different priming strategies, hydro-, osmo-, halo- and hormone priming are the most popular (Paparella et al. 2015; Maiti and Pramanik 2013). Depending on the technique to be used, other variable parameters also require standardization to gain optimum ameliorative results. These variables include water potential, priming duration, temperature, seed vigour, cultivar and post-priming storage conditions (Maiti and Pramanik 2013).

5.5 Priming-Induced Alterations Which Ameliorate Salt Stress in Susceptible Plants

Priming promotes embryo swelling and accelerates the development of immature embryos. The partial hydration state reduces the physical resistance of the endosperm, improves physiological parameters and leaches out the chemical inhibitors of germination (Bewley et al. 2013). Sadeghi et al. (2011) reported that priming modifies the seed metabolic balance as a result of which germination and seedling development is more rapid even under suboptimal saline conditions. Such stress tolerance is facilitated by metabolome reprogramming and generation of 'priming memory' in seeds (Pastor et al. 2013). 'Priming memory' is supposedly epigenetic signatures etched within the seed genome during the stress-like conditions created as a result of seed priming (Banerjee and Roychoudhury 2017c). Such epigenetic alterations in the chromatin architecture lead to the overexpression of several stress-responsive genes like *late embryogenesis abundant* (*LEA*), whose protein products confer tolerance towards salt stress (Banerjee and Roychoudhury 2016a; Roychoudhury et al. 2007).

Sharma et al. (2015) showed accelerated germination in the primed seeds. Such improvements could be attributed to specific germination-associated genes which get up-regulated in the primed seeds (Sharma et al. 2015). Several antioxidant genes also exhibit increased expression as the entire metabolic equilibrium of the seed is altered after optimum priming (Sadeghi et al. 2011). Such antioxidants promote seed germination and seedling development by scavenging the toxic ROS and lowering oxidative stress under saline conditions (Kubala et al. 2015). Salt stress imposes large-scale oxidative stress in the plant. If uncontrolled, this can lead to chromosomal damages, protein degradation and metabolite leakage (Netondo et al. 2004). Oxidative stress-induced membrane peroxidation triggers the accumulation of MDA which inhibits the activities of crucial enzymes (Younesi and Moradi 2015). Priming reportedly reverses these degenerative effects of salt stress and facilitates early replication, transcription and chromosomal repair (Roychoudhury and Chakraborty 2013).

The abiotic stress tolerance generated by seed priming is conferred via the synchronization of several physiological, biochemical, systemic, cellular and molecular modulations (Siri et al. 2013). The metabolome reprogramming enables mobilization of energy reserves via endosperm weakening and promotes the expansion and initial development of the dormant embryo (Chen and Arora 2011). This boosts the germination potential of the seed. The activities of several enzymes which facilitate reserve mobilization are enhanced. These are essentially proteases, lyases and amylases (Varier et al. 2010). Proper seedling development is allowed by inducing cell division, elongation, plasma membrane fluidity and stress-responsive proteins like the heat shock proteins (HSPs) and LEAs. Reports have shown alterations in H^+/ATPase activities and even in the transcriptome and proteome of the primed seeds (Ibrahim 2016). Stress tolerance in the primed seeds is also mediated by an increased potential in protein synthesis and post-translational modifications and by maintaining the optimum quotient for the translational turnover (Kubala et al. 2015).

Bakht et al. (2011) reported that seed priming efficiently eliminated the harmful Na^+ and Cl^- ions via activating membrane efflux pumps. On the contrary, the active uptake of inorganic ions facilitates the accumulation of K^+ and Ca^{2+} ions which in turn lowers the cellular osmopotential and promotes water uptake under saline conditions. Apart from these beneficial effects, K^+ ions balance membrane potential and turgor, whereas Ca^{2+} ions maintain the cellular morphology and integrity and mask the growth inhibitory effects of Na^+ ions (Summart et al. 2010; Gobinathan et al. 2009).

A large number of inorganic and organic solutes have been isolated from plants which mediate osmotic adjustments and confer salt tolerance. Solutes like proline (Pro), glycine betaine, free amino acids, soluble sugars, etc. undergo accumulation in the seeds and seedlings after osmopriming. These solutes might also be used as the priming agents to ameliorate salt susceptibility in plants (Roychoudhury and Chakraborty 2013). A chronological representation of the significant priming reagents used across several plant species to generate salt tolerance is presented in Table 5.2. Antioxidant enzymes like SOD, CAT and peroxidase (POX) also exhibit increased ROS scavenging upon seed priming (Nawaz et al. 2012). Compatible solutes like polyamines [putrescine (Put^{2+}), spermidine (Spd^{3+}) and spermine (Spm^{4+})] maintain cellular osmolarity and membrane integrity by chelating out the toxic Na^+ ions (Paul and Roychoudhury 2016; Roychoudhury et al. 2008). Similar antioxidative effects are conferred by seed priming using ascorbic acid and glutathione (Roychoudhury et al. 2012). Imbibition with the universal stress hormone, abscisic acid (ABA), generates a 'stress memory' in the seeds and makes them salt tolerant (Roychoudhury et al. 2009). Priming also induces the accumulation of photoprotective pigments like anthocyanin which exhibit ROS scavenging and plant protection (Banerjee and Roychoudhury 2016b). Overall, the priming strategies utilized to generate salt tolerance reduce MDA content and optimize ROS levels via accumulation of multivariant antioxidants and protective proteins (Nawaz et al. 2012).

5.6 Conclusion and Future Perspectives

Priming is a biologically safe and cheap crop expansion technology which modifies the seed metabolome and makes the tissue ready to tolerate suboptimal conditions like salinity. From the mechanism of stress amelioration by several priming agents (Table 5.2), it can be summarized that they recharge the antioxidant machinery and up-regulate multiple stress-responsive genes (Paul and Roychoudhury 2017). This promotes seed development and germination even under adversely saline conditions. Seed priming is also economically cheap since a small volume of priming solution is sufficient for seed imbibition, and this solution can even be reused. In spite of the huge potential of this technology, little information regarding its molecular mechanisms actually exists. One such perspective is the epigenomic basis of 'stress memory', which is required to be unravelled. Precise concentrations of the priming agents are extremely important for agronomic purposes as unusually high concentrations can cause irreversible damages to the developing seeds. Thus, future investigations revolving around the molecular and metabolomic platforms in this field shall bear credible impacts.

Table 5.2 Some seed priming approaches adopted to ameliorate salinity stress across susceptible plant species

Type	Agent	Concentration/dose	Treatment time	Plant Species	Salt tolerance	Mechanism	References
Halopriming	KNO_3	0.25 mM	–	*Silybum marianum*	^	Improved germination indices and peroxidase activity	Zavariyan et al. (2015)
	Silicon priming (Na_2SiO_3)	30 mM	–	*Triticum aestivum*	^	Accumulation of Ca^{2+} and K^+ ions	Azeem et al. (2015)
	KCl	10 mM	36 h	*Capsicum annuum*	^	Increased proline accumulation	Aloui et al. (2014)
	$CaCl_2$	2 mM	24 h	*Cucumis sativus*	^	Increased proline accumulation	Joshi et al. (2013)
	NaCl	300 mM	24 h	*Solanum lycopersicum*	^	Increased gibberellin content promoting endosperm weakening and germination	Nakaune et al. (2012)
	NaCl	100 mM	36 h	*Cucumis melo*	^	Increased proline, soluble carbohydrates and antioxidants	Farhoudi et al. (2011)
	KNO_3	−1.0 MPa	24 h	*Helianthus annuus*	^	Better osmoregulation	Bajehbaj (2010)
	NaCl	1 mM	–	*Capsicum annuum*	^	Generated 'stress memory'	Khan et al. (2009)
	KCl, KNO_3, $CaCl_2.2H_2O$, $Ca(NO_3)_2.4H_2O$	Varying concentration	12 h	*Triticum aestivum*	No change	–	Ashraf and Iram (2002)
	Mixed salt	–	–	*Oryza sativa*	^	Increased activities of amylases, root dehydrogenase and shoot catalase	Chang-Zheng et al. (2002)
	$CaCl_2$	10–60 mmol L^{-1}	12 h	*Gossypium hirsutum*	˅	Species-specific effect; low germination	Xiao-Fang et al. (2000)
	KNO_3 $CaCl_2$	–	–	*Cajanus cajan*	^	Accumulation of proteins, free amino acids and soluble sugars	Jyotsna and Srivastava (1998)

(continued)

Table 5.2 (continued)

Type	Agent	Concentration/dose	Treatment time	Plant Species	Salt tolerance	Mechanism	References
Osmopriming	Spermidine	5 mM	8 h	*Oryza sativa*	^	Up-regulation of several stress-responsive genes and transcription factors; enhanced expression of membrane Na^+ efflux pumps	Paul et al. (2017)
	Glycine betaine	10 mM	24 h	*Capsicum annuum*	^	Decreased MDA levels due to higher accumulation of proline	Roychoudhury and Banerjee (2016) and Korkmaz and Şirikçi (2011)
	Nitric oxide	75 μM (Na-nitroprusside)	24 h	*Jatropha curcas*	^	Increased accumulation of glutathione and ascorbate in the endosperm-embryo axis; higher activities of catalase 1, catalase 2 and glutathione reductases (GR1 and GR2)	Gadelha et al. (2017)
	H_2O_2	–	–	*Cakile maritima, Eutrema salsugineum*	^	Redox balance was restored	Ellouzi et al. (2017)
	Spermine	2.5 mM	8 h	*Oryza sativa*	^	Osmo-protection	Paul and Roy choudhury (2016)
	β-amino butyric acid	1 mM	6 h	*Vigna radiata*	^	Increased accumulation of proline, total carbohydrates, total protein; enhanced activities of nitrate reductase, superoxide dismutase and guaiacol peroxidase	Jisha and Puthur (2016)
	Trehalose	10 mM	48 h	*Oryza sativa*	^	Enhanced the activities of multiple antioxidant enzymes	Mostofa et al. (2015)
	Ascorbic acid	Variable	–	*Triticum durum*	^	Proteins associated with metabolism, energy, disease, defence and storage showed increased abundance	Fercha et al. (2014)
	Ascorbic acid	0.5 mM	24 h	*Cucurbita pepo*	^	Increased activities of catalase and peroxidase	Fazlali et al. (2013)
	Choline	5 mM	24 h	*Triticum aestivum*	^	Increased glycine betaine accumulation and maintenance of osmotic potential	Salama et al. (2011)

Physical priming	UV-C	0.85 KJ m^{-2}	1–4 min	*Lactuca sativa*	^	Increased levels of phenolics and flavonoids which efficiently scavenged the diphenyl-picrylhydrazyl radicals	Ouhibi et al. (2014)
Hormone priming	Salicylic acid (SA) + fish flour	0.1 mM SA	Overnight	*Triticum durum*	^	Enhanced activities of phenylalanine ammonia lyase, peroxidase; accumulation of phenolics and flavonoids	Karadag and Yucel (2017)
	Methyl jasmonate	25 μmol L^{-1}	–	*Brassica oleracea*	^	The contents of indolic glucosinolates, glucobrassicin, neo-glucobrassicin, anthocyanins and chlorogenic acid derivatives increased	Hassini et al. (2017)
	Salicylic acid	0.5 mM	Overnight shaking at 150 rpm	*Triticum durum*	^	Accumulation of total phenols, flavonoids and carotenoids; increased activities of phenylalanine ammonia lyase and ascorbic acid oxidase	Yucel and Heybet (2016)
	Melatonin	–	–	*Vicia faba*	^	Increased photosynthetic efficiency, total carbohydrates, total phenols, indole acetic acid, K^+ and Ca^{2+} levels	Dawood and El-Awadi (2015)
	ABA	30 ppm	8 h	*Phaseolus vulgaris*	^	Increased phosphatidylcholine/ phosphatidylethanolamine ratio and confers root cell membrane protection	Salama et al. (2015)
	Melatonin	1 μM	24 h	*Cucumis sativus*	^	ABA catabolism and gibberellic acid biosynthesis were promoted; the antioxidant machinery recharged	Zhang et al. (2014)
	Gibberellic acid (GA_3)	4.5 mM	12 h	*Lactuca sativa*	^	ABA biosynthesis is suppressed which triggers efficient germination	Hela et al. (2012)
	Gibberellic acid (GA_3)	20 mg L^{-1}	24 h	*Foeniculum vulgare*	^	ABA biosynthesis is suppressed which triggers efficient germination	Sedghi et al. (2010)

^ Enhanced
˅ Decreased

Acknowledgements The financial support from Council of Scientific and Industrial Research (CSIR), Government of India, through the Project [38(1387)/14/EMR-II] to Dr. Aryadeep Roychoudhury is gratefully acknowledged.

References

Aloui H, Souguir M, Latique S, Hannachi C (2014) Germination and growth in control and primed seeds of pepper as affected by salt stress. Cercet Agronomice Moldova 47:83–95

Anjum NA, Sofo A, Scopa A, Roychoudhury A, Gill SS et al (2015) Lipids and proteins-major targets of oxidative modifications in abiotic stressed plants. Environ Sci Pollut Res 22:4099–4121

Ashraf M, Iram A (2002) Optimization and influence of seed priming with salts of potassium or calcium in two spring wheat cultivars differing in salt tolerance at the initial growth stages. Agro Chim 46:47–55

Azeem M, Iqbal N, Kausar S, Javed MT, Akram MS, Sajid MA (2015) Efficacy of silicon priming and fertigation to modulate seedling's vigor and ion homeostasis of wheat (*Triticum aestivum* L.) under saline environment. Environ Sci Pollut Res Int 22:14367–14371

Bajehbaj AA (2010) The effects of NaCl priming on salt tolerance in sunflower germination and seedling grown under salinity conditions. Afr J Biotechnol 9:1764–1770

Bakht J, Shafi M, Jamal Y, Sher H (2011) Response of maize (*Zea mays* L.) to seed priming with NaCl and salinity stress. Span J Agric Res 9:252–261

Banerjee A, Roychoudhury A (2016a) Group II late embryogenesis abundant (LEA) proteins: structural and functional aspects in plant abiotic stress. Plant Growth Regul 79:1–17

Banerjee A, Roychoudhury A (2016b) Plant responses to light stress: oxidative damages, photoprotection and role of phytohormones. In: Ahammed GJ, Yu J-Q (eds) Plant hormones under challenging environmental factors. Springer Nature, Dordrecht, pp 181–213

Banerjee A, Roychoudhury A (2017a) Melatonin as a regulator of abiotic stress tolerance in plants. In: Singh VP, Singh S, Mohan Prasad S (eds) Mechanisms behind phytohormonal signalling and crop abiotic stress tolerance. Nova Science Publishers, New York, pp 47–60

Banerjee A, Roychoudhury A (2017b) Abscisic-acid-dependent basic leucine zipper (bZIP) transcription factors in plant abiotic stress. Protoplasma 254:3–16

Banerjee A, Roychoudhury A (2017c) Epigenetic regulation during salinity and drought stress in plants: Histone modifications and DNA methylation. Plant Gene 11:199–204

Banerjee A, Roychoudhury A, Krishnamoorthi S (2016) Emerging techniques to decipher microRNAs (miRNAs) and their regulatory role in conferring abiotic stress tolerance of plants. Plant Biotechnol Rep 10:185–205

Basu S, Roychoudhury A (2014) Expression profiling of abiotic stress-inducible genes in response to multiple stresses in rice (*Oryza sativa* L.) varieties with contrasting level of stress tolerance. BioMed Res Int Article ID: 706890

Bernstein L, Hayward HE (1958) Physiology of salt tolerance. Ann Rev Plant Physiol 9:25–46

Bewley JD, Bradford KJ, Hilhorst HWM, Nonogaki H (2013) Seeds physiology of development. In: Germination and dormancy, 3rd ed. Springer, New York

Bruce TJA, Matthes MC, Napier JA, Pickett JA (2007) Stressful memories of plants: evidence and possible mechanisms. Plant Sci 173:603–608

Bruggink GT, Ooms JJJ, van der Toorn P (1999) Induction of longevity in primed seeds. Seed Sci Res 9:49–53

Chang-Zheng H, Jin H, Zhi-Yu Z, Song-Lin R, Wen-Jian S (2002) Effect of seed priming with mixed-salt solution on germination and physiological characteristics of seedling in rice (*Oryza sativa* L.) under stress conditions. J Zhejiang Univ (Agric Life Sci) 28:175–178

Chen K, Arora R (2011) Dynamics of the antioxidant system during seed osmopriming, post-priming germination, and seedling establishment in spinach (*Spinacia oleracea*). Plant Sci 180:212–220

Das K, Roychoudhury A (2014) Reactive oxygen species (ROS) and response of antioxidants as ROS-scavengers during environmental stress in plants. Front Environ Sci 2:53

Daszkowska-Golec A (2011) *Arabidopsis* seed germination under abiotic stress as a concert of action of phytohormones. OMICS 15:763–774

Dawood MG, EL-Awadi ME (2015) Alleviation of salinity stress on *Vicia faba* L. plants via seed priming with melatonin. Acta Boil Colomb 20:223–235

Ellouzi H, Sghayar S, Abdelly C (2017) H_2O_2 seed priming improves tolerance to salinity; drought and their combined effect more than mannitol in *Cakile maritima* when compared to *Eutrema salsugineum*. J Plant Physiol 210:38–50

Esechie HA (1995) Partitioning of chloride ion in the germinating seed of two forage legumes under varied salinity and temperature regimes. Comm Soil Sci Plant Anal 26:3357–3370

FAO (2011) The state of the world's land and water resources for food and agriculture (SOLAW)-managing systems at risk. Food and Agriculture. Organization of the United Nations, Rome and Earthscan, London

Farhoudi R, Saeedipour S, Mohammadreza D (2011) The effect of NaCl seed priming on salt tolerance, antioxidant enzyme activity, proline and carbohydrate accumulation of muskmelon (*Cucumis melo* L.) under saline condition. Afr J Agric Res 6:1363–1370

Farooq M, Basra SMA, Wahid A, Ahmad N, Saleem BA (2009) Improving the drought tolerance in rice (*Oryza sativa* L.) by exogenous application of salicylic acid. J Agron Crop Sci 195:237–246

Fazlali R, Asli DE, Moradi P (2013) The effect of seed priming by ascorbic acid on bioactive compounds of naked seed pumpkin (*Cucurbita pepo* var. *styriaca*) under salinity stress. Int J Farm Alli Sci 2:587–590

Fercha A, Capriotti AL, Caruso G, Cavaliere C, Samperi R, Stampachiacchiere S, Laganà A (2014) Comparative analysis of metabolic proteome variation in ascorbate-primed and unprimed wheat seeds during germination under salt stress. J Proteome 108:238–257

Gadelha CG, Miranda RS, Alencar NLM, Costa JH, Prisco JT, Gomes-Filho E (2017) Exogenous nitric oxide improves salt tolerance during establishment of *Jatropha curcas* seedlings by ameliorating oxidative damage and toxic ion accumulation. J Plant Physiol 212:69–79

Garg R, Chevala VVSN, Shankar R, Jain M (2015) Divergent DNA methylation patterns associated with gene expression in rice cultivars with contrasting drought and salinity stress response. Sci Rep 5:14922

Gobinathan P, Sankar B, Murali PV, Panneerselvam RN (2009) Effect of calcium chloride on salinity – induced oxidative stress in *Pennisetum typoidies*. Bot Res Int 2:143–148

Gurusinghe SH, Bradford KJ (2001) Galactosyl-sucrose oligosaccharides and potential longevity of primed seeds. Seed Sci Res 11:121–133

Hassini I, Baenas N, Moreno DA, Carvajal M, Boughanmi N, Martinez Ballesta MDC (2017) Effects of seed priming, salinity and methyl jasmonate treatment on bioactive composition of *Brassica oleracea* var. capitata (white and red varieties) sprouts. J Sci Food Agric 97:2291–2299

Hela M, Zargouni H, Tarchoune I, Baatour O, Nasri N, Ben Massoud R et al (2012) Combined effect of hormonal priming and salt treatments on germination percentage and antioxidant activities in lettuce seedlings. Afr J Biotechnol 11:10373–10380

Hubbard M, Germida J, Vujanovic V (2012) Fungal endophytes improve wheat seed germination under heat and drought stress. Botany 90:137–149

Ibrahim EA (2016) Seed priming to alleviate salinity stress in germinating seeds. J Plant Physiol 192:38–46

Jisha KC, Puthur JT (2016) Seed priming with BABA (β-amino butyric acid): a cost-effective method of abiotic stress tolerance in *Vigna radiata* (L.) Wilczek. Protoplasma 253:277–289

Jisha KC, Vijayakumari K, Puthur JT (2013) Seed priming for abiotic stress tolerance: an overview. Acta Physiol Plant 35:1381–1396

Joshi N, Jain A, Arya K (2013) Alleviation of salt stress in *Cucumis sativus* L. through seed priming with calcium chloride. Indian J Appl Res 11:22–25

Jyotsna V, Srivastava AK (1998) Physiological basis of salt stress resistance in pigeonpea (*Cajanus cajan* L.)-II. Pre-sowing seed soaking treatment in regulating early seedling metabolism during seed germination. Plant Physiol Biochem 25:89–94

Karadag B, Yucel NC (2017) Salicylic acid and fish flour pre-treatments affect wheat phenolic and flavonoid compounds, lipid peroxidation levels under salt stress. Cereal Res Commun 45: 192–201

Khan MA, Weber DJ (2008) Ecophysiology of high salinity tolerant plants (tasks for vegetation science), 1st edn. Springer, Amsterdam

Khan HA, Ayub CM, Pervez MA, Bilal RM, Shahid MA, Ziaf K (2009) Effect of seed priming with NaCl on salinity tolerance of hot pepper (*Capsicum annuum* L.) at seedling stage. Soil Environ 28:81–87

Korkmaz A, Şirikçi R (2011) Improving salinity tolerance of germinating seeds by exogenous application of glycine betaine in pepper. Seed Sci Technol 39:377–388

Kubala S, Garnczarska M, Wojtyla Ł, Clippe A, Kosmala A et al (2015) Deciphering priming induced improvement of rapeseed (*Brassica napus* L.) germination through an integrated transcriptomic and proteomic approach. Plant Sci 231:94–113

Maiti R, Pramanik K (2013) Vegetable seed priming: a low cost, simple and powerful techniques for farmers' livelihood. Int J Bio-Resour Stress Manag 4:475–481

Mostofa MG, Hossain MA, Fujita M (2015) Trehalose pretreatment induces salt tolerance in rice (*Oryza sativa* L.) seedlings: oxidative damage and co-induction of antioxidant defense and glyoxalase systems. Protoplasma 252:461–475

Munns R, Tester M (2008) Mechanisms of salinity tolerance. Annu Rev Plant Biol 59:651–681

Nakaune M, Hanada A, Yin YG, Matsukura C, Yamaguchi S (2012) Molecular and physiological dissection of enhanced seed germination using short-term low-concentration salt seed priming in tomato. Plant Physiol Biochem 52:28–37

Nawaz A, Amjad M, Jahangir MM, Khan SM, Cui H, Hu J (2012) Induction of salt tolerance in tomato (*Lycopersicon esculentum* Mill.) seeds through sand priming. Aust J Crop Sci 6:1199–1203

Netondo GW, Onyango JC, Beck E (2004) Sorghum and salinity: II. Gas exchange and chlorophyll fluorescence of sorghum under salt stress. Crop Sci 44:806–811

Ouhibi C, Attia H, Rebah F, Msilini N, Chebbi M et al (2014) Salt stress mitigation by seed priming with UV-C in lettuce plants: growth, antioxidant activity and phenolic compounds. Plant Physiol Biochem 83:126–133

Paparella S, Araújo SS, Rossi G, Wijayasinghe M, Carbonera D, Balestrazzi A (2015) Seed priming: state of the art and new perspectives. Plant Cell Rep 34:1281–1293

Pastor V, Luna E, Mauch-Mani B, Ton J, Flors V (2013) Primed plants do not forget. Environ Exp Bot 94:46–56

Paul S, Roychoudhury A (2016) Seed priming with spermine ameliorates salinity stress in the germinated seedlings of two rice cultivars differing in their level of salt tolerance. Trop Plant Res 3:616–633

Paul S, Roychoudhury A (2017) Effect of seed priming with spermine/spermidine on transcriptional regulation of stress-responsive genes in salt-stressed seedlings of an aromatic rice cultivar. Plant Gene 11:133–142

Paul S, Roychoudhury A, Banerjee A, Chaudhuri N, Ghosh P (2017) Seed pre-treatment with spermidine alleviates oxidative damages to different extent in the salt (NaCl)-stressed seedlings of three indica rice cultivars with contrasting level of salt tolerance. Plant Gene 11:112–123

Rajjou L, Duval M, Gallardo K, Catusse J, Bally J, Job C, Job D (2012) Seed germination and vigor. Annu Rev Plant Biol 63:507–533

Ratikanta KM (2011) Seed priming: an efficient farmers' technology to improve seedling vigour, seedling establishment and crop productivity. Int J Bio-Resour Stress Manag 2:297

Roychoudhury A, Banerjee A (2016) Endogenous glycine betaine accumulation mediates abiotic stress tolerance in plants. Trop Plant Res 3:105–111

Roychoudhury A, Chakraborty M (2013) Biochemical and molecular basis of varietal difference in plant salt tolerance. Ann Rev Res Biol 3:422–454

Roychoudhury A, Roy C, Sengupta DN (2007) Transgenic tobacco plants overexpressing the heterologous *lea* gene *Rab16A* from rice during high salt and water deficit display enhanced tolerance to salinity stress. Plant Cell Rep 26:1839–1859

Roychoudhury A, Basu S, Sarkar SN, Sengupta DN (2008) Comparative physiological and molecular responses of a common aromatic indica rice cultivar to high salinity with non-aromatic indica rice cultivars. Plant Cell Rep 27:1395–1410

Roychoudhury A, Basu S, Sengupta DN (2009) Effects of exogenous abscisic acid on some physiological responses in a popular aromatic indica rice compared with those from two traditional non-aromatic indica rice cultivars. Acta Physiol Plant 31:915–926

Roychoudhury A, Basu S, Sengupta DN (2012) Antioxidants and stress-related metabolites in the seedlings of two indica rice varieties exposed to cadmium chloride toxicity. Acta Physiol Plant 34:835–847

Roychoudhury A, Paul S, Basu S (2013) Cross-talk between abscisic acid-dependent and abscisic acid-independent pathways during abiotic stress. Plant Cell Rep 32:985–1006

Sadeghi H, Khazaei F, Yari L, Sheidaei S (2011) Effect of seed osmopriming on seed germination behaviour and vigor of soybean (*Glycine max* L.). J Agric Biol Sci 6:39–43

Salama KHA, Mansour MMF, Hassan NS (2011) Choline priming improves salt tolerance in wheat (*Triticum aestivum* L.). Aust J Basic Appl Sci 5:126–132

Salama KHA, Ahmed HFS, El-Araby MMA (2015) Interaction of exogenous abscisic acid and salinity on the lipid root plasma membrane of *Phaseolus vulgaris* L. Egypt J Exp Biol (Bot) 11:189–196

Sedghi M, Nemati A, Esmaielpour B (2010) Effect of seed priming on germination and seedling growth of two medicinal plants under salinity. Emir J Food Agric 22:130–139

Sharma AD, Rathore SVS, Srinivasan K, Tyagi RK (2014) Comparison of various seed priming methods for seed germination, seedling vigour and fruit yield in okra (*Abelmoschus esculentus* L. Moench). Sci Hortic 165:75–81

Sharma KK, Singh US, Sharma P, Kumar A, Sharma L (2015) Seed treatments for sustainable agriculture – a review. J Appl Nat Sci 7:521–539

Siri B, Vichitphan K, Kaewnaree P, Vichitphan S, Klanrit P (2013) Improvement of quality, membrane integrity and antioxidant systems in sweet pepper (*Capsicum annuum* Linn.) seeds affected by osmopriming. Aust J Crop Sci 7:2068–2073

Summart J, Thanonkeo P, Panichajakul S, Prathepha P, McManus MT (2010) Effect of salt stress on growth, inorganic ion and proline accumulation in Thai aromatic rice, Khao Dawk Mail 105, callus culture. Afr J Biotechnol 9:145–152

Tanou G, Fotopoulos V, Molassiotis A (2012) Priming against environmental challenges and proteomics in plants: update and agricultural perspectives. Front Plant Sci 3:216

Thiam M, Champion A, Diouf D, Mame Ourèye SY (2013) NaCl effects on *in vitro* germination and growth of some Senegalese cowpea (*Vigna unguiculata* (L.) Walp.) Cultivars. ISRN Biotech 11

Varier A, Vari AK, Dadlani M (2010) The sub cellular basis of seed priming. Cur Sci 99:450–456

Xiao-Fang S, Qing Song Z, You Liang L (2000) Regulations of salt tolerance of cotton plants at seedling emergence stage by soaking seeds in Pix (DPC) and $CaCl_2$ solutions. Jiangsu J Agric Sci 16:204–207

Younesi O, Moradi A (2015) Effect of priming of seeds of *Medicago sativa* 'bami' with gibberellic acid on germination, seedlings growth and antioxidant enzymes activity under salinity stress. J Hortic Res 22:167–174

Yucel NC, Heybet EH (2016) Salicylic acid and calcium treatments improves wheat vigor, lipids and phenolics under high salinity. Acta Chim Slov 63:738–746

Zavariyan A, Rad M, Asghari M (2015) Effect of seed priming by potassium nitrate on germination and biochemical indices in *Silybum marianum* L. under salinity stress. Int J Life Sci 9:23–29

Zhang HJ, Zhang N, Yang RC, Wang L, Sun QQ et al (2014) Melatonin promotes seed germination under high salinity by regulating antioxidant systems, ABA and GA_4 interaction in cucumber (*Cucumis sativus* L.). J Pineal Res 57:269–279

6 Seed Priming with Plant Growth Regulators to Improve Crop Abiotic Stress Tolerance

Elouaer Mohamed Aymen

Abstract

Plants are frequently subjected to abiotic stress such as drought, salinity, heat, and cold which constitutes a major limitation to agricultural production worldwide. The unfavorable environmental conditions that plants encounter in vegetative cycle perturb their metabolic reactions and negatively affect growth at cellular and biochemical plant levels. Preventing crop losses and generating more food to meet the demands of growing human populations have gained importance. Identifying plant mechanisms to neutralize abiotic stresses and sustain their growth and survival under unfavorable conditions holds huge importance. Research studies have revealed that plant growth regulators (PGR) confirm their significance as metabolic engineering targets for producing abiotic stress-tolerant crop plants. In addition, seed priming has shown its importance as a powerful technique to improve germination, growth, and yield of crops under unfavorable environment conditions. The combination of the two effects, seed priming with PGR, could have very prevailing results. In this context, during this chapter, we evaluate the effect of seed priming with PGR in plant growth development and abiotic stress tolerance.

Keywords

Abiotic stress · Seed priming · Plant growth regulators · Growth · Germination

E. M. Aymen (✉)
High Institute of Agronomy, Chott Mariem, 4041, Sousse, Tunisia

© Springer Nature Singapore Pte Ltd. 2018

A. Rakshit, H. B. Singh (eds.), *Advances in Seed Priming*,
https://doi.org/10.1007/978-981-13-0032-5_6

6.1 Introduction

The human population is growing rapidly and requires a considerable increase in agricultural productivity worldwide. However, diverse abiotic stresses are limiting crop productivity (Wani and Sah 2014). In order to nourish the rising world population, crop productivity must be amplified in the near future. That's why plant breeders and biotechnologists should improve crop tolerance to abiotic stress by recognizing these mechanisms. However, plant machineries leading to environmental stress response and tolerance are complex (Qin et al. 2011). Since the complication of stress tolerance, conventional breeding trials have met little success. That's why original and effective approaches should be developed. Seed priming is a revolutionary technique used to improve seed germination and seedling growth in abiotic stress conditions. During this process, a series of physiological and biochemical processes are triggered leading to improving plant growth under stress conditions (Eisvand et al. 2010). Seeds can be soaked in solutions containing exogenous molecules such as salts (Khan et al. 2009a, b) or plant growth regulators (PGR) (Nakaune et al. 2012). Seed priming with PGR pretreatment is a commonly used strategy to improve seed germination and seedling growth in unfavorable conditions (Masood et al. 2012; Hu et al. 2013). Seeds presoaked with optimal concentration of PGR enhance germination, growth, and yield of crops under stress condition by rising nutrient reserves through improved physiological activities and root profusion (Afzal et al. 2002; Akbari et al. 2007). PGR are organic compounds produced in extremely small amount and play a vital function in growth, expansion, and yield of crops. They regulate, as chemical messengers, a range of cellular processes in higher plants and coordinate diverse signal transduction pathways during abiotic stress response (Vob et al. 2014; Kazan 2015). For instance, seeds of rye (*Secale montanum*) pretreated with gibberellic acid (GA3) increased germination under water deficit (Ansari et al. 2013). Khan et al. (2009a, b) confirmed that pepper seeds (*Capscum annum* L.) pretreated with salicylic acid resulted in better germination and seedling growth under salt stress. Furthermore, ethylene reduces high temperature effect on seed germination of lettuce (Nascimento et al. 2004).

6.2 Abiotic Stresses: World Agricultural Challenge

Exploring how abiotic stresses can influence plant growth at the physiological, biochemical, and molecular levels is decisive to advance crop production, since stresses cause crop losses (Kazan 2015). Abiotic stresses, including drought, salinity, chilling, freezing, heat, and UV radiation, are the main environmental factors restraining crop production. They negatively influence growth, biomass production, and yields of food crops threatening consequently food security worldwide (Kaur et al. 2008; Thakur et al. 2010). Among these stresses, drought, salinity, and temperature severity are the most common abiotic stresses limiting crop productivity in the world (Jaleel et al. 2009; Thakur et al. 2010). They affect plant survival, pigment content, membrane integrity, water relations, osmotic adjustments, and photosynthetic activity (Sanghera et al. 2011; Pathak et al. 2014). Drought and salinity affect together more than 55% of the world's agricultural land (Dos Reis et al. 2012). Since abiotic stress tolerance is multigenic in nature (Collins et al. 2008), an enormous challenge was undertaken to comprehend key mechanisms to go forward in selective breeding purposes. Understanding the machinery of plants' environmental stress tolerance is of critical importance for the development of stress-tolerant and high-yielding food crop cultivars.

6.2.1 Plant Drought Tolerance

Drought tolerance is the ability to survive and produce stable yields under water scarcity. Drought stress decreases plants' photosynthetic rate, decreasing consequently the amount of assimilates available for export to the sink organs (Kim et al. 2000). Abscisic acid (ABA) is a key plant growth regulator in the response and adaptation of plants to water scarcity. It is involved in stomatal closure, accumulation of osmoprotectants, and changes in gene expression (Umezawa et al. 2010). The biosynthesis of osmoprotectants such as amino acid, amines, and carbohydrates is another indispensable strategy for plant resistance to water stress. The most common osmoprotectants are proline, glycine betaine, fructans, starch, and mono- and disaccharides.

6.2.2 Plant Heat Tolerance

Heat stress perturbs cellular homeostasis and causes protein denaturation and dysfunction in plant cells, leading to brutal growth retardation. During this stress, electron transport is altered affecting electron flow from oxygen-evolving complex (OEC) toward the acceptor side of photosystem II (PSII). These alterations affect the generation of ATP, Rubisco for carbon fixation, and starch (Asthir 2015). Drought stress generates the accumulation of ROS leading to a severe damage in DNA and peroxidation of membrane lipids and pigments. Other changes include a decrease in photosynthetic pigment ratio and inhibitions of stomatal conductance and photosynthesis rate. These alterations ultimately reduce the partitioning of photosynthates, which manifest by reduced growth and economic yield. Other morphological damages associated with heat stress comprise scorching of leaves, branches, and stems, leaf senescence, fruit discoloration, and damage (Hasanuzzaman et al. 2013).

6.2.3 Plant Cold Tolerance

Cold stress occurs at temperatures less than 20 °C. Chilling (<20 °C) or freezing (<0 °C) temperatures can trigger the formation of ice in plant tissues, cause cellular dehydration and leakage of intracellular solutes, and reduce plasma membrane integrity (Chinnusamy et al. 2007). Consequently, cold stress severely affects plant growth and leads to substantial crop losses (Sanghera et al. 2011). To cope with this unfavorable condition, plants adopt several strategies such as activating primary metabolisms, raising the level of antioxidants, and maintaining osmotic balance (Miura and Furumoto 2013). During cold stress, membrane rigidification occurs as opposed to heat stress. This process is the upstream trigger for the induction of cytosolic Ca^{2+} signatures leading to a transient increase in cytosolic Ca^{2+} levels (Knight et al. 1991).

6.3 Plant Growth Regulators: Key Mediators of Plant Responses to Environmental Stresses

Plants have to adjust their development to respond to various abiotic stresses. Plant growth regulators (PGR) are cells signaling molecules acting in very small quantities that mediate these responses. Their crucial functions are advancing plant adaptation to an altering environment by mediating growth, development, and nutrient allocation (Fahad et al. 2015a, b). PGR are endogenous substances responsible in adjusting physiological and molecular responses for plant survival. They include gibberellins (GAs), salicylic acid (SA), auxin (IAA), ethylene (ET), cytokinins (CKs), brassinosteroids (BRs), abscisic acid (ABA), and jasmonates (JAs).

6.3.1 Abscisic Acid (ABA), the Abiotic Stress Hormone

Abscisic acid (ABA) is the most studied PGR. It plays an important role throughout numerous plant physiological processes and developmental stages including stomatal closure, embryo morphogenesis, seed dormancy, and synthesis of storage proteins and lipids (Sreenivasulu et al. 2010). ABA is a vital messenger in the adaptive response of plants to abiotic stress. During this response, endogenous ABA levels increase rapidly, activating specific signaling pathways and altering gene expression levels (O'Brien and Benkova 2013). Zhang et al. (2006) and Hossain et al. (2010) stated a substantial increase in ABA concentration upon exposure of plants to salinity and drought. It regulates the expression of different stress-responsive genes implicated in the accumulation of compatible osmolytes and the synthesis of proteins and antioxidant enzymes (Chaves et al. 2003; Verslues et al. 2006).

6.3.2 Auxins (IAA)

IAA (indole-3-acetic acid) is a multifunctional PGR and is vital for plant growth under stress conditions (Kazan 2013). IAA boosts plant root and shoot growth and plays consequently a fundamental part in plant adaptation to salt stress (Egamberdieva 2009; Iqbal et al. 2014; Fahad et al. 2015a, b). Auxin stimulates the transcription of primary auxin response genes identified and characterized in several plant species (Javid et al. 2011).

6.3.3 Gibberellins (GAs)

GAs improve seed germination, leaf expansion, stem elongation, fruit development, and abiotic stress response and adaptation (Yamaguchi 2008; Colebrook et al. 2014). It interacts with other PGR in many stimulus-response processes (Munteanu et al. 2014).

6.3.4 Salicylic Acid (SA)

SA plays a vital role in the regulation of plant growth, ripening, and responses to abiotic stresses (Khodary 2004; Miura et al. 2013; Miura and Tada 2014). Gharib and Hegazi (2010) showed that SA stimulated growth of bean seedlings and reduced the adverse effect of cold and chilling stresses.

6.3.5 Cytokinins (CKs)

CKs are involved in many plant growth processes and abiotic stress tolerance (Nishiyama et al. 2011; Kang et al. 2012; O'Brien and Benkova 2013). CKs are often considered ABA antagonists (Pospíšilová 2003). It has been linked to different abiotic stress tolerance like cold stress and freezing stress (Jeon et al. 2010). Salinity or osmotic stress shows an effect in the expression levels of CK receptors and metabolism, respectively, in *Arabidopsis* and maize (Zalabák et al. 2013).

6.3.6 Jasmonates (JAs)

Jasmonates are involved in plant development including reproductive processes, secondary metabolism, and plant responses to environmental stresses (Pauwels et al. 2009; Seo et al. 2001; Fahad et al. 2015a, b). Exogenous application of JA significantly reduced salinity and heavy metal stress symptoms in plants by activating the antioxidant machinery (Yoon et al. 2009; Yan et al. 2013). Wang et al. (2010) have reported a significant increase in endogenous levels of JA in rice roots under salinity stress. In addition, JA confers tolerance to metal stress in plants via the accumulation of phytochelatins (Maksymiec et al. 2007).

6.3.7 Ethylene (ET)

ET, a gaseous PGR, is involved in plant growth and development, notably fruit ripening, flower senescence, leaf and petal abscission, and stress response regulation (Gamalero and Glick 2012; Groen and Whiteman 2014). Enhanced abiotic tolerance was achieved with higher endogenous ET concentrations in plants (Shi et al. 2012; Groen and Whiteman 2014). ET also induces plants' defense response to heat stress (Larkindale et al. 2005). Yin et al. (2015) have shown that ET and ABA act in synergy or in antagonism to control plant growth.

6.4 Seed Priming as a Strategy to Improve Abiotic Stress Tolerance

Recently, diverse strategies have been employed to induce abiotic stress tolerance in plants. Seed priming is an effective, practical, and low-cost technique to obtain rapid emergence, high seedling vigor, and better crop yields under unfavorable environmental conditions (Jisha et al. 2013; Paparella et al. 2015). It is a controlled hydration technique triggering metabolic processes during early phase of germination before radicle protrusion (Hussain et al. 2015). Higher and synchronized germination of primed seeds is due to reduction in the lag time of imbibition (Brocklehurst and Dearman 2008), enzyme activation (Lee and Kim 2000), buildup of germination-enhancing metabolites (Hussain et al. 2015), metabolic repair during imbibition (Farooq et al. 2006), and osmotic adjustment (Bradford 1986). Primed plants exhibit activation of cellular defense responses, which imparts abiotic stress tolerance (Jisha et al. 2013). Various seed priming techniques have been employed under different environmental stresses including hydropriming, osmopriming, chemical priming, nutrient priming, and hormonal priming (Jisha et al. 2013; Paparella et al. 2015). During seed priming, germination process is induced by soaking seeds in solutions containing exogenous molecules such as salts (Khan et al. 2009a, b), metals (Mirshekari et al. 2012), or hormones (Nakaune et al. 2012). Varier et al. (2010) and Eisvand et al. (2010) suggest that seed priming activates a series of physiological processes that improve plant growth under stressful conditions, including the induction of antioxidant systems.

6.4.1 Seed Priming with Plant Growth Regulators

PGR pretreatment is a commonly used priming approach to improve seed germination under stressful conditions (Jisha et al. 2013; Hu et al. 2013). It can be used to advance germination, seedling growth, and yield under drought, salinity, metal, cold, and heat stresses.

6.4.1.1 Seed Priming with PGR Under Water Deficit

Seeds of rye (*Secale montanum*) primed with gibberellic acid increased germination under water deficit (Ansari et al. 2013). ABA-primed seeds of *Brassica napus* exhibited earlier germination and higher final percent radicle protrusion than non-primed control seeds, under water stress (Gao et al. 2002). Seeds of *Agropyron elongatum* primed with gibberellin and abscisic acid exhibited induced CAT and SOD activities under drought conditions when compared to unprimed seeds (Eisvand et al. 2010). Farooq et al. (2013) have shown that seeds primed with ascorbic acid improve emergence, growth, yield, and water statue of wheat seedlings under water deficit. Priming with ascorbic acid showed significant effects on germination percentage, shoot length, root length, vigor index, and CAT and POX activity in rapeseed (*Brassica napus* L.) plant under drought condition (Razaji et al. 2014).

6.4.1.2 Seed Priming with PGR Under Salt Stress

In pepper (*Capscum annum* L.), Khan et al. (2009a, b) showed that pretreatment with acetylsalicylic acid and salicylic acid resulted in greater uniformity of germination and establishment of seedlings under high salinity. ABA-primed seeds of *Brassica napus* exhibited earlier germination and higher final percent radicle protrusion than non-primed control seeds, under salt stress (100 mM NaCl) (Gao et al. 2002). In wheat seed germination, auxin pretreatments increased the hypocotyl length, seedling fresh and dry weight, and hypocotyl dry weight under saline conditions (Akbari et al. 2007). Salicylic acid priming in fennel seeds also showed better germination under salt stress (Farahbakhsh 2012). Iqbal et al. (2011) have reported that seed priming with gibberellic acid induced an increase in grain yield of wheat plants, modulation of ion uptake and partitioning, and hormone homeostasis under saline conditions.

6.4.1.3 Seed Priming with PGR Under Heat Stress

Additionally, ethylene was used to minimize the effect of high temperatures on seed germination of lettuce (*Lactuca sativa* L.) (Nascimento et al. 2004). Rehman et al. (2012) have shown that seed priming with salicylic acid improved temperature stress resistance in spring maize through an earlier emergence, increased seedling dry weight and tissue water status, and improved membrane stability. Seed priming with salicylic acid or jasmonic acid improves growth, carbohydrate content, and chilling resistance in sunflower (*Helianthus annuus* L.) (Gornik and Lahuta 2017). Singh and Singh (2016) have shown that seed priming with three levels of salicylic acid (0.25 mM, 0.5 mM, and 0.75 mM) improves growth, flowering, yield, and fruit quality under high-temperature stress conditions.

6.4.1.4 Seed Priming with PGR Under Cold Stress

The incorporation of methyl jasmonate (3 μM) into the priming solution on low temperature improves germination and emergence performance of watermelon (*Citrullus lanatus*) cv. Crimson Sweet (Korkmaz et al. 2004). Gamel et al. (2017) have shown that seed priming with 100 ppm gibberellic acid improves germination, growth, yield, and fruit quality of three tomato cultivars under low temperature. Ansari and Zadeh (2012) have shown that seed priming with gibberellic acid (25 ppm) advances germination and seedling growth of mountain rye (*Secale montanum*) under cold stress.

6.4.1.5 Seed Priming with PGR Under Metal Stress

Seed priming with ethylene (100 μM) improves germination parameters of pigeon pea under cadmium stress (Sneideris et al. 2014). PGR priming using auxin, cytokinin, and gibberellic acid at concentration of 10–100 μM was the most appropriate priming treatment for soybean (*Glycine max*) seeds grown under lead (Pb) stress conditions (Abu-Muriefah 2017). Seed priming with jasmonate advances growth and activity of SOD and POD and increases significantly the accumulation of chlorophyll and carotenoid and neutralizes the toxic effect of Cu^{2+} on *Cajanus cajan* seedlings under copper stress (Poonam et al. 2013).

6.5 Conclusion and Future Perspectives

It can be concluded that seed priming with PGR has the potential for improving crop abiotic stress tolerance which provides new opportunities to maintain sustainable crop production to feed the growing population under changing environmental conditions. Even though, with rapid development of genomic technology, significant attempts have been done on the way to decoding the plant abiotic stress responses, many challenges still lie ahead to uncover the complexity of stress signal transduction pathways. More hard work will be required at the genetic level of PGR biosynthetic pathway. The success in elucidating roles of PGR in stress tolerance at molecular levels will help in showing positive effects of seed priming with PGR and their substitutes in improving stress tolerance in a wide range of crop species. However, more research will be needed in unraveling the mechanism of PGR, especially with stress-responsive genes.

References

Abu-Muriefah SS (2017) Phytohormonal priming improves germination and antioxidant enzymes of soybean (Glycine max) seeds under lead (Pb) stress. Biosci Res 14(1):42–56

Afzal I, Basra SMA, Ahmad N, Cheema MA, Warriach EA, Khaliq A (2002) Effect of priming and growth regulator treatment on emergence. Int J Agric Biol 4:306–306

Akbari G, Sanavy SA, Yousefzadeh S (2007) Effect of auxin and salt stress (NaCl) on seed germination of wheat cultivars (Triticum aestivum L.). Pak J Biol Sci 10:2557–2561

Ansari O, Zadeh FS (2012) Does gibberellic acid (GA), salicylic acid (SA) and ascorbic acid (ASC) improve mountain rye (Secale montanum) seeds germination and seedlings growth under cold stress? Int Res J Appl Basic Sci 3(8):1651–1657

Ansari O, Azadi MS, Sharif-Zadeh F, Younesi E (2013) Effect of hormone priming on germination characteristics and enzyme activity of mountain rye (*Secale montanum*) seeds under drought stress conditions. J Stress Physiol Biochem 9:61–71

Asthir B (2015) Protective mechanisms of heat tolerance in crop plants. J Plant Interact 10:202–210

Bradford KJ (1986) Manipulation of seed water relations via osmotic priming to improve germination under stress conditions. Horticult Sci 21:1105–1112

Brocklehurst PA, Dearman J (2008) Interaction between seed priming treatments and nine seed lots of carrot, celery and onion II. Seedling emergence and plant growth. Ann Appl Biol 102:583–593

Chaves MM, Maroco JP, Pereira JS (2003) Understanding plant responses to drought: from genes to the whole plant. Funct Plant Biol 30:239–264

Chinnusamy V, Zhu J, Zhu JK (2007) Cold stress regulation of gene expression in plants. Trends Plant Sci 12:444–445

Colebrook EH, Thomas SG, Phillips AL, Hedden P (2014) The role of gibberellins signaling in plant responses to abiotic stress. J Exp Biol 217:67–75

Collins NC, Tardieu F, Tuberosa R (2008) Quantitative trait loci and crop performance under abiotic stress: where do we stand? Plant Physiol 147:469–486

Dos Reis SP, Lima AM, de Souza CRB (2012) Recent molecular advances on downstream plant responses to abiotic stress. Int J Mol Sci 13:8628–8647

Egamberdieva D (2009) Alleviation of salt stress by plant growth regulators and IAA producing bacteria in wheat. Acta Physiol Plant 31:861–864

Eisvand HR, Tavakkol-Afshari R, Sharifzadeh F, Maddah Arefi H, Hesamzadeh Hejazi SM (2010) Effects of hormonal priming and drought stress on activity and isozyme profiles of antioxidant enzymes in deteriorated seed of tall wheatgrass (*Agropyron elongatum* host). Seed Sci Technol 38:280–297

Fahad S, Hussain S, Bano A, Saud S, Hassan S, Shan D, Khan FA, Khan F, Chen YT, Wu C, Tabassum MA, Chun MX, Afzal M, Jan A, Jan MT, Huang JL (2015a) Potential role of phytohormones and plant growth-promoting rhizobacteria in abiotic stresses: consequences for changing environment. Environ Sci Pollut Res 22:4907–4921

Fahad S, Nie L, Chen Y, Wu C, Xiong D, Saud S, Hongyan L, Cui K, Huang J (2015b) Crop plant hormones and environmental stress. Sustain Agric Rev 15:371–400

Farahbakhsh H (2012) Germination and seedling growth in unprimed and primed seeds of fennel as affected by reduced water potential induced by NaCl. Int Res J Appl Basic Sci 3:737–744

Farooq M, Basra SMA, Hafeez K (2006) Seed invigoration by osmohardening in coarse and fine rice. Seed Sci Technol 34:181–187

Farooq M, Irfan M, Aziz T, Ahmad I, Cheema SA (2013) Seed priming with ascorbic acid improves drought resistance of wheat. J Agron Crop Sci 199(1):12–22

Gamalero E, Glick BR (2012) Ethylene and abiotic stress tolerance in plants. In: Ahmed P, Prasad MNV (eds) Environmental adaptations and stress tolerance of plants in the era of climate change. Springer, New York, pp 395–412

Gamel RE, Elsayed A, Bashasha J, Haroun S (2017) Priming tomato cultivars in β-sitosterol or gibberellic acid improves tolerance for temperature stress. Int J Bot 13:1–14

Gao YP, Bonham-Smith PC, Gusta LV (2002) The role of peroxiredoxin antioxidant and calmodulin in ABA-primed seeds of *Brassica napus* exposed to abiotic stresses during germination. J Plant Physiol 159:951–958

Gharib FA, Hegazi AZ (2010) Salicylic acid ameliorates germination, seedling growth, phytohormone and enzymes activity in bean (*Phaseolus vulgaris* L.) under cold stress. J Am Sci 6:675–683

Gornik K, Lahuta BL (2017) Application of phytohormones during seed hydropriming and heat shock treatment on sunflower (*Helianthus annuus* L.) chilling resistance and changes in soluble carbohydrates. Acta Physiol Plant 39:118

Groen SC, Whiteman NK (2014) The evolution of ethylene signaling in plant chemical ecology. J Chem Ecol 40:700–716

Hasanuzzaman M, Nahar K, Alam MM, Roychowdhury R, Fujita M (2013) Physiological, biochemical, and molecular mechanisms of heat stress tolerance in plants. Int J Mol Sci 14(5):9643–9684

Hossain MA, Cho JI, Han M, Ahn CH, Jeon JS, An G, Park PB (2010) The ABRE-binding bZIP transcription factor OsABF2 is a positive regulator of abiotic stress and ABA signaling in rice. J Plant Physiol 167(17):1512–1520

Hu YF, Zhou G, Na XF, Yang L, Nan WB, Zhang YO, Li JL, Bi YR (2013) Cadmium interferes with maintenance of auxin homeostasis in Arabidopsis seedlings. J Plant Physiol 170:965–975

Hussain S, Zheng M, Khan F, Khaliq A, Fahad S, Peng S (2015) Benefits of rice seed priming are offset permanently by prolonged storage and the storage conditions. Sci Rep 5:8101

Iqbal N, Nazar R, Iqbal MRK, Masood A, Nafees AK (2011) Role of gibberellins in regulation of source sink relations under optimal and limiting environmental conditions. Curr Sci 100:998–1007

Iqbal N, Umar S, Khan NA, Khan MIR (2014) A new perspective of phytohormones in salinity tolerance: regulation of proline metabolism. Environ Exp Bot 100:34–42

Jaleel CA, Manivannan P, Wahid A, Farooq M, Somasundaram R, Panneerselvam R (2009) Drought stress in plants: a review on morphological characteristics and pigments composition. Int J Agric Biol 11:100–105

Javid MG, Sorooshzadeh A, Moradi F, Sanavy SAMM, Allahdadi I (2011) The role of phytohormones in alleviating salt stress in crop plants. Aust J Crop Sci 5:726–734

Jeon J, Kim NY, Kim S, Kang NY, Novák O, Ku SJ (2010) A subset of cytokinin two component signaling system plays a role in cold temperature stress response in Arabidopsis. J Biol Chem 285:23371–23386

Jisha KC, Vijayakumari K, Puthur JT (2013) Seed priming for abiotic stress tolerance: an overview. Acta Physiol Plant 35:1381–1396

Kang NY, Cho C, Kim NY, Kim J (2012) Cytokinin receptor-dependent and receptor-independent path ways in the dehydration response of Arabidopsis thaliana. J Plant Physiol 169:1382–1391

Kaur G, Kumar S, Nayyar H, Upadhyaya HD (2008) Cold stress injury during the pod-filling phase in chickpea (*Cicer arietinum* L.): effects on quantitative and qualitative components of seeds. J Agron Crop Sci 194(6):457–464

Kazan K (2013) Auxin and the integration of environmental signals into plant root development. Ann Bot 112:1655–1665

Kazan K (2015) Diverse roles of jasmonates and ethylene in abiotic stress tolerance. Trends Plant Sci 20:219–229

Khan HA, Ayub CM, Pervez MA, Bilal RM, Shahid MA, Ziaf K (2009a) Effect of seed priming with NaCl on salinity tolerance of hot pepper (*Capsicum annuum* L.) at seedling stage. Soil Environ 28:81–87

Khan HA, Pervez MA, Ayub CM, Ziaf K, Bilal RM, Shahid MA, Akhtar N (2009b) Hormonal priming alleviates salt stress in hot pepper (*Capsicum annuum* L.). Soil Environ 28:130–135

Khodary SEA (2004) Effect of salicylic acid on growth, photosynthesis and carbohydrate metabolism in salt stressed maize plants. Int J Agric Biol 6:5–8

Kim JY, Mahé AB, Rangeon J, Prioul JL (2000) A maize vacuolar invertase, IVR2, is induced by water stress. Organ/tissue specificity and diurnal modulation of expression. Plant Physiol 124:71–84

Knight MR, Campbell AK, Smith SM, Trewavas AJ (1991) Transgenic plant aequorin reports the effects of touch and cold-shock and elicitors on cytoplasmic calcium. Nature 352(1991):524–526

Korkmaz A, Tiryaki I, Nuri Nas M, Ozbay N (2004) L'inclusion d'un régulateur de croissance à la solution de pré-germination améliore la germination et la levée des graines de pastèque à basse température. Can J Plant Sci 84:1161–1165

Larkindale J, Hall DJ, Knight MR, Vierling E (2005) Heat stress phenotypes of Arabidopsis mutants implicate multiple signaling pathways in the acquisition of thermo-tolerance. Plant Physiol 138:882–897

Lee SS, Kim JH (2000) Total sugars, α-amylase activity and emergence after priming of normal and aged rice seeds. Kor J Crop Sci 45:108–111

Maksymiec W, Wojcik M, Krupa Z (2007) Variation in oxidative stress and photochemical activity in *Arabidopsis thaliana* leaves subjected to cadmium and excess copper in the presence or absence of jasmonate and ascorbate. Chemosphere 66:421–427

Masood A, Iqbal N, Khan NA (2012) Role of ethylene in alleviation of cadmium-induced capacity inhibition by sulphur in mustard. Plant Cell Environ 35:524–533

Mirshekari B, Baser S, Allahyari S, Hamendanlu N (2012) 'On-farm' seed priming with Zn plus Mn is an effective way to improve germination and yield of marigold. Afr J Microbiol Res 6:5796–5800

Miura K, Furumoto T (2013) Cold signaling and cold response in plants. Int J Mol Sci 14(3):5312–5337

Miura K, Tada Y (2014) Regulation of water, salinity, and cold stress responses by salicylic acid. Front Plant Sci 5:4

Miura K, Okamoto H, Okuma E, Shiba H, Kamada H, Hasegawa PM, Murata Y (2013) SIZ1 deficiency causes reduced stomatal aperture and enhanced drought tolerance via controlling salicylic acid-induced accumulation of reactive oxygen species in Arabidopsis. Plant J 49:79–90

Munteanu V, Gordeev V, Martea R, Duca M (2014) Effect of gibberellins cross talk with other phytohormones on cellular growth and mitosis to endoreduplication transition. Int J Adv Res Biol Sci 1(6):136–153

Nakaune M, Hanada A, Yin Y, Matsukura C, Yamaguchi S, Ezura H (2012) Molecular and physiological dissection of enhanced seed germination using short-term low-concentration salt seed priming in tomato. Plant Physiol Biochem 52:28–37

Nascimento MM, Lemos JAC, Abranches J, Gonçalves RB, Burne RA (2004) Adaptive acid tolerance response of Streptococcus sobrinus. J Bacteriol 186:6383–6390

Nishiyama R, Watanabe Y, Fujita Y, Tien LD, Kojima M, Werner T, Vankova R, Yamaguchi-Shinozaki K, Shinozaki K, Kakimoto T, Sakakibara H, Schmuelling T, Lam-Son PT (2011) Analysis of cytokinin mutants and regulation of cytokinin metabolic genes reveals important regulatory roles of cytokinins in drought, salt and abscisic acid responses, and abscisic acid biosynthesis. Plant Cell 23:2169–2183

O'Brien JA, Benkova E (2013) Cytokinin cross-talking during biotic and abiotic stress responses. Front Plant Sci 4:451

Paparella S, Araújo SS, Rossi G, Wijayasinghe M, Carbonera D, Balestrazzi A (2015) Seed priming: state of the art and new perspectives. Plant Cell Rep 34:1281–1293

Pathak MR, Da Silva Teixeira JA, Wani SH (2014) Polyamines in response to abiotic stress tolerance through transgenic approaches. GM Crops Food 5:87–96

Pauwels L, Inze D, Goossens A (2009) Jasmonate-inducible gene: what does it mean? Trends Plant Sci 14:87–91

Poonam S, Kaur H, Geetika S (2013) Effect of Jasmonic acid on photosynthetic pigments and stress markers in *Cajanus cajan* (L.) Millsp. seedlings under copper stress. Am J Plant Sci 4:817–823

Pospíšilová J (2003) Participation of phytohormones in the stomatal regulation of gas exchange during water stress. Biol Plant 46:491–506

Qin F, Shinozaki K, Yamaguchi-Shinozaki K (2011) Achievements and challenges in understanding plant abiotic stress responses and tolerance. Plant Cell Physiol 52:1569–1582

Razaji A, Farzanian M, Sayfzadeh S (2014) The effects of seed priming by ascorbic acid on some morphological and biochemical aspects of rapeseed (Brassica napus L.) under drought stress condition. Int J Biosci 4(1):432–442

Rehman H, Afzal I, Farooq M, Aziz T, Ahmad SM (2012) Improving temperature stress resistance in spring maize by seed priming. Proceedings of 3rd International Conference 'Frontiers in Agriculture'. Dankook International Cooperation on Agriculture, Dankook University, Cheonansi, Republic of Korea

Sanghera GS, Wani SH, Hussain W, Singh NB (2011) Engineering cold stress tolerance in crop plants. Curr Genomics 14:30–43

Seo HS, Song JT, Cheong JJ, Lee YH, Lee YH, Hwang I, Lee JS, Yang DC (2001) Jasmonic acid carboxyl methyl transferase: a key enzyme for jasmonate regulated plant responses. Proc Natl Acad Sci U S A 98:4788–4793

Shi Y, Tian S, Hou L, Huang X, Zhang X, Guo H, Yang S (2012) Ethylene signaling negatively regulates freezing tolerance by repressing expression of CBF and type-A ARR genes in Arabidopsis. Plant Cell 24:2578–2595

Singh SK, Singh PK (2016) Effect of seed priming of tomato with salicylic acid on growth, flowering, yield and fruit quality under high temperature stress conditions. Int J Adv Res 4(2):723–727

Sneideris LC, Gavassi MA, Campos ML, Damião VD, Carvalho RF (2014) Effects of hormonal priming on seed germination of pigeon pea under cadmium stress. An Acad Bras Cienc 87(3):1847–1852

Sreenivasulu N, Radchuk V, Alawady A, Borisjuk L, Weier D, Staroske N (2010) De-regulation of abscisic acid contents causes abnormal endosperm development in the barley mutant seg8. Plant J. 64:589–603

Thakur P, Kumar S, Malik JA, Berger JD, Nayyar H (2010) Cold stress effects on reproductive development in grain crops: an overview. Environ Exp Bot 67(3):429–443

Umezawa T, Nakashima K, Miyakawa T, Kuromori T, Tanokura M, Shinozaki K, Yamaguchi-Shinozaki K (2010) Molecular basis of the core regulatory network in ABA responses: sensing, signaling and transport. Plant Cell Physiol 51:1821–1839
Varier A, Vari AK, Dadlani M (2010) The subcellular basis of seed priming. Curr Sci India 99:451–456
Verslues PE, Agarwal M, Katiyar-Agarwal S, Zhu J, Zhu JK (2006) Methods and concepts in quantifying resistance to drought, salt and freezing, abiotic stresses that affect plant water status. Plant J 45:523–539
Vob U, Bishopp A, Farcot E, Bennett MJ (2014) Modelling hormonal response and development. Trends Plant Sci 19:311–319
Wani SH, Sah SK (2014) Biotechnology and abiotic stress tolerance in rice. J Rice Res 2:e105
Wang W, Mao J, Zhang Z, Tang Q, Xie Y, Zhu J, Zhang L, Liu Z, Shi Y, Goodfellow M (2010) Deinococcus wulumuqiensis sp. nov., and Deinococcus xibeiensis sp. nov., isolated from radiation polluted soil. Int J Syst Evol Microbiol 60:2006–2010
Yamaguchi S (2008) Gibberellin metabolism and its regulation. Annu Rev Plant Physiol 59:225–251
Yan Z, Chen J, Li X (2013) Methyl jasmonate as modulator of Cd toxicity in *Capsicum frutescens* var. fasciculatum seedlings. Ecotoxicol Environ Saf 98:203–209
Yin CC, Ma B, Collinge DP, Pogson BJ, He SJ, Xiong Q, Zhang JS (2015) Ethylene responses in rice roots and coleoptiles are differentially regulated by a carotenoid isomerase-mediated abscisic acid pathway. Plant Cell 27:1061–1081
Yoon JY, Hamayun M, Lee SK, Lee IJ (2009) Methyl jasmonate alleviated salinity stress in soybean. J Crop Sci Biotechnol 12:63–68
Zalabák D, Pospíšilová H, Šmehilová M, Mrízová K, Frébort I, Galuszka P (2013) Genetic engineering of cytokinin metabolism: prospective way to improve agricultural traits of crop plants. Biotech Adv 31:97–117
Zhang J, Jia W, Yang J, Ismail AM (2006) Role of ABA in integrating plant responses to drought and salt stresses. Field Crops Res 97:111–119

Addressing Stresses in Agriculture Through Bio-priming Intervention

7

Deepranjan Sarkar, Sumita Pal, Ms. Mehjabeen, Vivek Singh, Sonam Singh, Subhadip Pul, Jancy Garg, Amitava Rakshit, and H. B. Singh

Abstract

Concurrent occurrences of different stresses, i.e. biotic and abiotic, are very common in the environment of plants which consequently reduce yield. As cost-effective options are very limited, bio-priming is a suitable tool to address the numerous challenges associated with agriculture. Plant growth benefits are easily attainable through this technique while managing the natural resources and enhancing the environmental sustainability.

Keywords

Bio-priming · Alleviation · Abiotic stress · Biotic stress

7.1 Introduction

Various stress factors are responsible for lowering of the agricultural production, which create the need of identifying these stresses, note the changes occurring in plant system and develop suitable tool which can help plant to easily fight against them. Abiotic stresses mainly include extreme temperatures (heat, cold), drought (limited precipitation, drying winds), heavy metals and salinity (Krasensky and

D. Sarkar · S. Pal (✉) · M. Mehjabeen · S. Singh · S. Pul · J. Garg · A. Rakshit
Department of Soil Science and Agricultural Chemistry, Institute of Agricultural Sciences, BHU, Varanasi, India

V. Singh
National Bureau of Agriculturally Important Microorganisms (NBAIM), Mau Nath Bhanjan, Uttar Pradesh, India

H. B. Singh
Department of Mycology and Plant Pathology, Institute of Agricultural Sciences, BHU Varanasi, Uttar Pradesh, India

© Springer Nature Singapore Pte Ltd. 2018

A. Rakshit, H. B. Singh (eds.), *Advances in Seed Priming*,
https://doi.org/10.1007/978-981-13-0032-5_7

Jonak 2012), but the notable biotic stresses are viruses, fungi, bacteria, weeds, insects and other pests and pathogens (Suzuki et al. 2014). Wang et al. (2003) reported that abiotic stresses can reduce average yields by >50% for most major crop plants. Increasing abiotic stress like increase in temperature can further facilitate pathogen spread and weaken the defence mechanisms of plants (Suzuki et al. 2014). Genetic manipulation to generate stress-tolerant crops is one area where many research works are being carried out with the help of breeding programmes. But these practices are time-taking and involve huge investments. Cost-effective options are very limited, indicating the urgent need of solution which is agriculturally economical for the farmers to easily adapt it. Bio-priming is a seed treatment process using microbial agents to accelerate germination and improve seedling establishment under adverse conditions (Bhatt et al. 2015; Singh et al. 2016b).

7.2 How Is Bio-priming Effective?

The beneficial effects of microbe-plant interactions include plant growth promotion and induced disease resistance and tolerance to abiotic stresses (Table 7.1, Fig. 7.1). Bio-primed plants adapt themselves to environmental stresses by metabolic adjustments in their system, leading to the accumulation of several compatible organic solutes like sugars, polyamines, betaines, quaternary ammonium compounds, polyhydric alcohols, proline and other amino acids (Sandhya et al. 2010; Ray et al. 2016). Development of induced resistance in plants treated with microbes help to overcome pathogen infection and control the plant diseases (El-Mohamedy et al. 2015). Microorganisms like *Trichoderma* augment deep root growth, helping the plant in water acquisition and nutrient uptake during adverse conditions (Rawat et al. 2012; Keswani et al. 2016).

7.3 Precautions Needed

Region-specific bio-inoculants must be screened and isolated according to their traits linked with the environment in which they can survive, tolerate and effectively function in adverse conditions (Bisen et al. 2015; Rakshit et al. 2015). Successful inoculation is dependent upon type of soil, optimum temperature, concentration of nutrients and salt, moisture level, etc. Modern screening methods focussing on the microbial traits, including their culturing, storage, transportation and application, should be given special attention. Mechanisms underlying the actions of microbes (root colonization, parasitism, antibiosis, antagonism, etc.) must be studied (du Jardin 2015).

7.4 Present Status

Numerous strains have been identified as seed dressers, potential rhizosphere colonizers and stimulants for plant growth and health beyond the germination and seedling mergence stage of different crop species (Singh et al. 2016a, b; Singh 2016).

Table 7.1 Effect of bio-priming in alleviation of various stresses

	Crop	Stress	Microbial agents	Betterment	References
Pulses	Lentil	Strain stress (*Pythium* spp.)	*Rhizobium leguminosarum* bv. *viceae*	Good seedling emergence and vigour. Yield improvement by 43.11%	Huang and Erickson (2007)
	Pea	Strain stress (*Pythium* spp.)	*Rhizobium leguminosarum* bv. *viceae*	Yield improvement by 41.35%	Huang and Erickson (2006)
	Mung bean	Strain stress (root rot) *Macrophomina* spp.	*T. harzianum*– RA	Root rot incidence decrement by 12.45%	Raguchander et al. (1997)
	Chickpea	Biotic stress (*Sclerotium rolfsii*)	*Mesorhizobium* (RL091), fluorescent *Pseudomonas* (PHU094), *Trichoderma* (THU0816)	Plant mortality (%) decreased by 76 ± 4%	Singh et al. (2014)
Cereals	Wheat	Salinity stress	*Trichoderma harzianum* (Th-19)	Increase in germination % by 0.87 during the salinity stress	Rawat et al. (2011)
	Wheat	Salinity stress	*Trichoderma longibrachiatum*	Superoxide dismutase, catalase, peroxidase increased by 29%, 39% and 19% respectively	Zhang et al. (2016)
	Rice	Nutrient stress	*Azospirillum amazonense*	NUE increased up to 18.5%	Rodrigues et al. (2008)
	Maize	Biotic stress (*Fusarium* spp.)	*Burkholderia cepacia*	Inhibited *F. solani* spp. by 43%, *F. moniliforme* 58%, *F. proliferatum* spp. 66%, *F. subglutinans* ISPV 517 74% and *F. graminearum* spp. 74%	Bevivino et al. (1998)
	Wheat and barley	Biotic stress (*Fusarium culmorum*)	*P. fluorescens* strain MKB 249, *Pseudomonas* sp. strain MKB 158	Level of disease symptoms decreased by 53–91%	Khan et al. (2005)
	Maize	Drought	*Pseudomonas* spp.	Improved plant biomass, relative water content, leaf water potential, root adhering soil/root tissue ratio, aggregate stability and mean weight diameter, decreased leaf water loss	Sandhya et al. (2010)
	Rice	Salinity	*Trichoderma harzianum*	Alleviation of oxidative damage and significant increase in length and fresh weight of shoot and root, number of leaves, leaf area, photosynthetic rate, chlorophyll fluorescence, chlorophyll content	Rawat et al. (2012)
	Baby corn	Nutrient stress with reference to N and P	*Trichoderma harzianum*	Improved nitrogen acquisition efficiency	Meena et al. (2016)

(continued)

Table 7.1 (continued)

	Crop	Stress	Microbial agents	Betterment	References
Fruits and vegetables	Cucumber	Nutrient stress	*T. asperellum* strain T 34	Increase in micronutrient use efficiency	Santiago et al. (2012)
	Tomato	Nutrient stress	*T. harzianum* T969	PUE increase by 65.8%	Azarmi et al. (2011)
	Melon	Strain stress (*Fusarium oxysporum* f.sp. *melonis*)	*Glomus intraradices* and *T. harzianum*	Disease incidence decreased by 50% (*T. harzianum*) and significantly by *Glomus intraradices*	Martínez-Medina et al. (2010)
	Broccoli	Nutrient stress	*P. fluorescens*, AM fungi	Increase in NUE up to 235.42% and PUE up to 163.33%	Tanwar et al. (2013)
	Green bean	Pathogens: *Rhizoctonia solani* and *Fusarium solani*	*Trichoderma harzianum*	Improved plant growth, total yield and its components, green pod nutritional value parameters	El-Mohamedy et al. (2015)
	Tomato	Osmotic	*Enterobacter* spp.	Improved germination and enhanced seedling growth	Bhatt et al. (2015)
	Tomato	Biotic stress (root-knot nematode)	*Meria coniospora*	Galls/root system decreased from 9.5 ± 4 (control) to 1.5 ±	Jansson et al. (1985)
Oilseeds	Soybean	Nutrient stress (Fe)	*Trichoderma virens* As19–1	Fe uptake increased to 77%	Entesari et al. (2013)
	Groundnut	Biotic stress stem rot disease (*Sclerotium rolfsii*)	*Trichoderma harzianum* (ITCC - 4572)	% disease inhibition by 57%	Ganesan et al. (2007)
	Safflower	Biotic stress *Macrophomina phaseolina* (root rot disease)	*Trichoderma harizianum, Pseudomonas fluorescens* and *Bacillus subtilis*	Disease incidence due to *Bacillus subtilis* decreased by 16%and *Trichoderma harizianum* and *Pseudomonas fluorescens* by 18%	Govindappa et al. (2010)
Plantation crops	Sugarcane	Nutrient stress	*P. fluorescens* strain R62+R81	PUE increased by 0.719	Yadav et al. (2013)
	Tea	Nutrient stress	*Azospirillum brasilense*	Increase in NUE up to 65%, PUE up to 25% and KUE by 16%	Thomas et al. (2010)

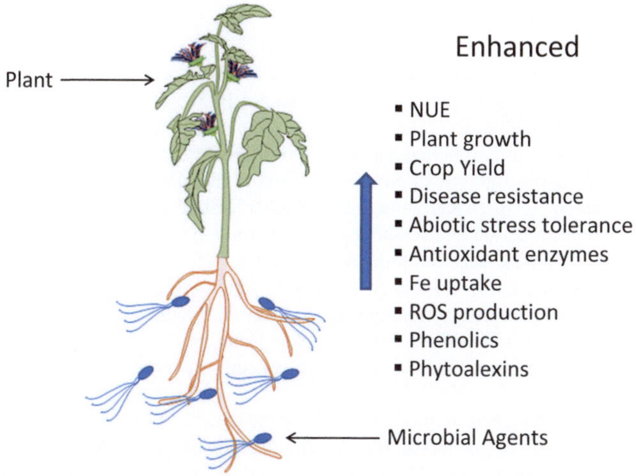

Fig. 7.1 Systemic representation of microbial-mediated growth promotion and defence responses in plants

Bio-priming is an important means of managing many soil and seed-borne diseases (Rao et al. 2009). The technique has enormous potential to have multifunctional impact on soil-plant system which can improve NUE, nutrient uptake, plant growth and plant tolerance to multi-stress by reducing the dependency on chemical inputs and thereby lowering the environmental pollution and increasing the agricultural sustainability (Rakshit et al. 2014).

7.5 Conclusion

Exploration of microbes in bio-priming sector can contribute a lot towards curbing the stress factors that the plants face in their environment. Present understanding of the microbial diversity and their effective functioning along with the mechanisms involved in their actions has contributed in the advancement of technology and commercialisation of inoculants.

7.6 Future Thrust

Bio-priming has a huge scope in practising modern agriculture. Isolation of specific inoculum or development of consortia (crop based) is quite challenging for the researchers. Screening and identification of microbes as a particular stress tolerant and those which can acclimatize themselves in varied adverse conditions with longer shelf lives are of utmost necessary. Further extensive research can help in sustainable management of resources.

References

Azarmi R, Hajieghrari B, Giglou A (2011) Effect of *Trichoderma* isolates on tomato seedling growth response and nutrient uptake. Afr J Biotechnol 10:5850–5855

Bevivino A, Sarrocco S, Dalmastri C, Tabacchioni S, Cantale C, Chiarini L (1998) Characterization of a free-living maize-rhizosphere population of *Burkholderia cepacia*: effect of seed treatment on disease suppression and growth promotion of maize. FEMS Microbiol Ecol 27:225–237

Bhatt RM, Selvakumar G, Upreti KK, Boregowda PC (2015) Effect of biopriming with Enterobacter strains on seed germination and seedling growth of tomato (*Solanum lycopersicum* L.) under osmotic stress. Proc Natl Acad Sci India Sect B Biol Sci 85:63–69

Bisen K, Keswani C, Mishra S, Saxena A, Rakshit A, Singh HB (2015) Unrealized potential of seed biopriming for versatile agriculture. In: Rakshit A, Singh HB, Sen A (eds) Nutrient use efficiency: from basics to advances. Springer, New Delhi, pp 193–206

du Jardin P (2015) Plant biostimulants: definition, concept, main categories and regulation. Sci Hortic 196:3–14

El-Mohamedy RS, Shafeek MR, Fatma AR (2015) Management of root rot diseases and improvement growth and yield of green bean plants using plant resistance inducers and biological seed treatments. J Agric Technol 11:1219–1234

Entesari M, Sharifzadeh F, Ahmadzadeh M, Farhangfar M (2013) Seed biopriming with *Trichoderma* species and *Pseudomonas fluorescens* on growth parameters, enzymes activity and nutritional status of soybean. Intl J Agron Plant Prod 4:610–619

Ganesan S, Ganesh RK, Sekar R (2007) Integrated management of stem rot disease (*Sclerotium rolfsii*) of groundnut (*Arachis hypogaea* L.) using *Rhizobium* and *Trichoderma harzianum* (ITCC – 4572). Turk J Agric For 31:103–108

Govindappa M, Lokesh S, Rai VR, Naik VR, Raju SG (2010) Induction of systemic resistance and management of safflower *Macrophomina phaseolina* root-rot disease by biocontrol agents. Arch Phytopathol Plant Protect 43:26–40

Huang HC, Erickson RS (2007) Effect of Seed Treatment with *Rhizobium leguminosarum* on *Pythium* damping-off, seedling height, root nodulation, root biomass, shoot biomass, and seed yield of Pea and Lentil. J Phytopathol 155:31–37

Jansson HB, Jeyaprakash A, Zuckerman BM (1985) Control of root-knot nematodes on tomato by the Endoparasitic fungus *Meria coniospora*. J Nematol 17:327–329

Keswani C, Bisen K, Singh V, Sarma BK, Singh HB (2016) Formulation technology of biocontrol agents: present status and future prospects. In: Arora NK, Mehnaz S, Balestrini R (eds) Bioformulations: for sustainable agriculture. Springer, India, pp 35–52

Khan MR, Fischer S, Egan D, Doohan FM (2005) Biological control of Fusarium seedling blight disease of wheat and barley. Phytopathology 96:386–394

Krasensky J, Jonak C (2012) Drought, salt, and temperature stress-induced metabolic rearrangements and regulatory networks. J Exp Bot 6:1593–1608

Martínez-Medina A, Pascua JA, Pérez-Alfocea F, Albacete A, Roldán A (2010) *Trichoderma harzianum* and *Glomus intraradices* modify the hormone disruption induced by *Fusarium oxysporum* infection in melon plants. Phytopathology 100:682–688

Meena SK, Rakshit A, Meena VS (2016) Effect of seed bio-priming and N doses under varied soil type on nitrogen use efficiency (NUE) of wheat (*Triticum aestivum* L.) under greenhouse conditions. Biocat Agric Biotech 6:68–75

Raguchander T, Rajappan K, Samiappan R (1997) Evaluating methods of application of biocontrol agent in the control of mungbean root rot. Indian Phytopathol 50:229–234

Rakshit A, Pal S, Meena S, Manjhee B, Preetipriya RS, Rai A, Bhowmick MK, Singh HB (2014) Seed bio-priming: a potential tool in integrated resource management. SATSA Mukhaptra Annu Tech Issue 18:94–103

Rakshit A, Sunita K, Pal S, Singh A, Singh HB (2015) Bio-priming mediated nutrient use efficiency of crop species. In: Rakshit A, Singh HB, Sen A (eds) Nutrient use efficiency: from basics to advances. Springer, New Delhi, pp 181–191

Rao MSL, Kulkarni S, Lingaraju S, Nadaf HL (2009) Bio-priming of seeds: a potential tool in the integrated management of Alternaria blight of sunflower. Helia 32:107–114

Rawat L, SinghY SN, Kumar J (2011) Alleviation of the adverse effects of salinity stress in wheat (*Triticumaestivum* L.) by seed biopriming with salinity tolerant isolates of *Trichoderma harzianum*. Plant Soil 347:387–400

Rawat L, Singh Y, Shukla N, Kumar J (2012) Seed biopriming with salinity tolerant isolates of *Trichoderma harzianum* alleviates salt stress in rice: growth, physiological and biochemical characteristics. J Plant Pathol 94:353–365

Ray S, Singh V, Singh S, Sarma BK, Singh HB (2016) Biochemical and histochemical analyses revealing endophytic *Alcaligenes faecalis* mediated suppression of oxidative stress in *Abelmoschus esculentus* challenged with *Sclerotium rolfsii*. Plant Physiol Biochem 109:430–441

Rodrigues EP, Rodrigues LS, de Oliveira ALM, Baldani VLD, dos Santos Teixeira KR, Urquiaga S, Reis VM (2008) *Azospirillum amazonense* inoculation: effects on growth, yield and N2 fixation of rice (*Oryza sativa* L.). Plant Soil 302:249–261

Sandhya V, Ali SZ, Grover M, Reddy G, Venkateswarlu B (2010) Effect of plant growth promoting *Pseudomonas* spp. on compatible solutes, antioxidant status and plant growth of maize under drought stress. Plant Growth Regul 62:21–30

Santiago AD, García-López AM, Quintero JM, Avilés M, Delgado A (2012) Effect of *Trichoderma asperellum* strain T34 and glucose addition on iron nutrition in cucumber grown on calcareous soils. Soil Biol Biochem 57:598–605

Singh HB (2016) Seed biopriming: a comprehensive approach towards agricultural sustainability. Indian Phytopathol 69:203–209

Singh A, Jain A, Sarma BK, Upadhyaya RS, Singh HB (2014) Rhizosphere competent microbial consortium mediates rapid changes in phenolic profiles in chickpea during *Sclerotium rolfsii* infection. Microbiol Res 169:353–360

Singh V, Upadhyay RS, Sarma BK, Singh HB (2016a) Seed bio-priming with *Trichoderma asperellum* effectively modulate plant growth promotion in pea. Int J Agric Environ Biotechnol 9:361–365

Singh V, Upadhyay RS, Sarma BK, Singh HB (2016b) Trichoderma asperellum spore dose depended modulation of plant growth in vegetable crops. Microbiol Res 193:74–86

Suzuki N, Rivero RM, Shulaev V, Blumwald E, Mittler R (2014) Abiotic and biotic stress combinations. New Phytol 203:32–43

Tanwar A, Aggarwal A, Kaushish S, Chauhan S (2013) Interactive effect of AM Fungi with *Trichoderma viride* and *Pseudomonas fluorescens* on growth and yield of broccoli. Protect Sci 49:140–148

Thomas J, Ajay D, Raj Kumar R, Mandal A (2010) Influence of beneficial microorganisms during in vivo acclimatization of in vitro-derived tea (*Camellia sinensis*) plants. Plant Cell Tiss Organ Cult 101:365–370

Wang WX, Vinocur B, Altman A (2003) Plant responses to drought, salinity and extreme temperatures: towards genetic engineering for stress tolerance. Planta 218:1–14

Yadav SK, Dave A, Sarkar A, Singh HB, Sarma BK (2013) Co-inoculated biopriming with Trichoderma, Pseudomonas and Rhizobium improves crop growth in Cicer arietinum and Phaseolus vulgaris. Intl J Agric Environ Biotechnol 6(2):255–259

Zhang S, Gan Y, Xu B (2016) Application of plant-growth-promoting fungi *Trichoderma longibrachiatum* T6 enhances tolerance of wheat to salt stress through improvement of antioxidative defense system and gene expression. Front Plant Sci 7:1405

Role of Microbial Seed Priming and Microbial Phytohormone in Modulating Growth Promotion and Defense Responses in Plants

8

Vivek Singh, Anupam Maharshi, Dhananjaya P. Singh, Ram Sanmukh Upadhyay, Birinchi Kumar Sarma, and Harikesh Bahadur Singh

Abstract

Plant growth and development are greatly affected by various biotic and abiotic stresses. Various strategies are utilized to minimize stresses in plants. Seed priming with beneficial microorganisms is one of the most beneficial methods to improve plant growth and development and induce systemic tolerance in plants towards biotic as well as abiotic stresses. Seed priming is a method of conditioning the seeds by plant growth-promoting microbes which provide better abilities to the plant to withstand various environmental challenges beginning from seed germination. Seed priming with beneficial microorganism is also known as bio-priming that enhances seed germination, protects germinating seed from different phytopathogens, and provides suitable conditions for establishment of the plant. Bio-priming has several mechanisms to stimulate morphogenesis and plant immunity, viz., production of phytohormones, induced expression of plant growth-promoting genes, increased nutrient status into the plant, mycoparasitism, antibiosis, induced phenolic production, activation of antioxidant production, and systemic defense activation. Some important microorganisms that synthesize

V. Singh
ICAR-National Bureau of Agriculturally Important Microorganisms, Maunath Bhanjan, India

Department of Botany, Institute of Science, Banaras Hindu University, Varanasi, India

A. Maharshi · B. K. Sarma · H. B. Singh (✉)
Department of Mycology and Plant Pathology, Institute of Agricultural Sciences, BHU, Varanasi, Uttar Pradesh, India

D. P. Singh
ICAR-National Bureau of Agriculturally Important Microorganisms, Maunath Bhanjan, India

R. S. Upadhyay
Department of Botany, Institute of Science, Banaras Hindu University, Varanasi, India

© Springer Nature Singapore Pte Ltd. 2018

A. Rakshit, H. B. Singh (eds.), *Advances in Seed Priming*,
https://doi.org/10.1007/978-981-13-0032-5_8

phytohormones include *Azotobacter* spp., *Pantoea agglomerans*, *Rhodospirillum rubrum*, *Rhizobium* spp., *Bacillus subtilis*, *Pseudomonas fluorescens*, *Paenibacillus polymyxa*, *Trichoderma* spp., *Pseudomonas putida*, *Rhizobium phaseoli*, *Bacillus cereus*, and *Acinetobacter calcoaceticus*. The main objective of this chapter is to enlighten the importance of seed priming with microorganisms and the role of different phytohormones produced by them in modulating growth, development, and defense activation in the host plant.

Keywords
Seed biopriming · Microbial phytohormone · Plant growth promotion · Defense responses

8.1 Introduction

Seed bio-priming is the technique of treating seeds with plant growth-promoting microbes (PGPM) that also enhances the basal level of plant defense by inducing the plant immunological memory. It is also defined as priming of seeds with beneficial microorganism followed by seed hydration (Singh et al. 2016a, b; Singh 2016; Rakshit et al. 2014; Rakshit et al. 2015a, b). During bio-priming, the stimulus from beneficial microbes is perceived and leads changes in the physiological, transcriptional, metabolic, and epigenetic behavior of the plant (Meena et al. 2016; Meena et al. 2017). These altered behaviors show higher defense responses resulting in enhanced resistance and stress tolerance (Mauch-Mani et al. 2017). Seed bio-priming process comprises different priming methods such as incubating under moist conditions in a plastic bag (Callan et al. 1991) or in a moistened finely ground coal or lignite materials (solid matrix priming) (Harman and Taylor 1988) and drum priming (Bennett and Whipps 2008). During bio-priming, there is an increase in microbial population on the seed surface. Callan et al. (1990) reported that during bio-priming, the bacterial population increases from 10- to over 10,000-fold depending on the initial level of inoculum. However, it is important to sterilize the seeds to remove or reduce the undesired microflora before priming. If the seeds were infected with the pathogen, their population will also increase during the priming process causing adverse and undesirable effect on emerging plants (Reddy 2013; Singh et al. 2013a, b; Jain et al. 2015). On the other hand, multiplication of undesired indigenous microorganisms may reduce the survivability of beneficial microbes used for seed bio-priming process (Wright et al. 2003). Seed bio-priming ensures better survival of the plants right from seed germination to maturity stage and also plays a role in enhancing nutritional value of the edible parts of the crop plants.

8.2 Seed Priming Regulate Microbe-Plant Interaction

Seed bio-priming provides perfect conditions for the microbial inoculum to colonize the seed surface (McQuilken et al. 1998). It helps the microbes to be established in the rhizosphere zone of the plant and provides long-term protection to plants from various biotic stresses along with promotion in plant growth (Yadav et al. 2015) (Figs. 8.1 and 8.2). The interaction of beneficial microbes with plant leads to establishment of various types of relationships, in which microbes benefit through utilization of carbon compounds released from the colonized plants. On the other hand, organic compounds from plant root exudates also enhance the microbial biomass and their functional activity in the rhizosphere zone (Ortíz-Castro et al. 2009). Free-living microbes such as saprophytic fungal species of the genus *Trichoderma* and important plant growth-promoting rhizobacteria (PGPR) are able to enhance plant growth and inhibit the soil-borne plant pathogens (Ortíz-Castro et al. 2009; Keswani et al. 2016) (Fig. 8.1). These beneficial microbes overcome the challenges of pathogens by various mechanisms such as mycoparasitism, production of phytohormones, competition for nutrients and space, and several others (Keswani et al. 2014; Bisen et al. 2015). Further, application of microorganisms also helps in decomposition and mineralization of organic matter and increase the availability of important nutrients such as iron and phosphorus (Valencia-Cantero et al. 2007; Ortíz-Castro et al. 2009). Plants have evolved a number of the physical and chemical barrier to protect it from pathogens, while some of the inducible responses are activated after pathogen perception (Hammond-Kosack and Jones 1996). This recognition step was achieved by microbe-associated molecular patterns (MAMPs) that are common to many classes of microbes (Ortíz-Castro et al. 2009).

The ultimate role of seed bio-priming is to provide better health and protection to the plant from any harmful environment during initial plant growth stage (et al. 2016b). The presence of microbial population in rhizosphere plays different important roles as mentioned above and their presence is also directly related to the synthesis of phytohormones. A wide range of microorganisms is capable of producing phytohormones that regulate plant growth and development

Fig. 8.1 Effect of seed priming with beneficial microorganism on plant growth promotion in bitter gourd; (**a**) control; (**b**) treatment (primed with beneficial microorganism). Microbial production of phytohormones and their role in plant growth

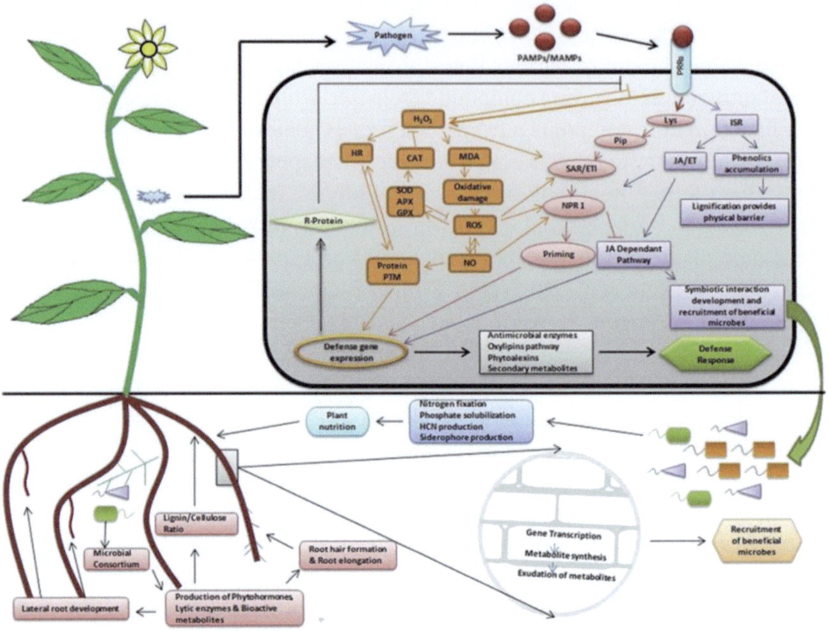

Fig. 8.2 Systemic representation of seed bio-priming with beneficial microbe-mediated growth promotion and defense responses in plants (Sarma et al. 2015)

(Ortíz-Castro et al. 2009; Spaepen and Vanderleyden 2011; Glick 2012). Auxin is one of the most important phytohormones that play a crucial role in plant growth and development. Among various forms of auxins, IAA (indole-3-acetic acid) is the most studied and most common auxin (Spaepen and Vanderleyden 2011; Glick 2012). IAA plays a major role in cell division, extension, and differentiation, enhances xylem and root development, promotes tuber and seed germination. It also helps in initial adventitious and lateral root formation, enhances photosynthesis, pigment formation, and mediate responses to gravity, light, and fluorescence (Tsavkelova et al. 2006; Spaepen and Vanderleyden 2011; Glick 2012). Several bacterial species are reported to synthesize auxins such as indole-3-acetic acid (IAA), indole-3-butyric acid (IBA), or their precursors (Martínez-Morales et al. 2003; Spaepen et al. 2008). Patten and Glick (2002) reported the IAA synthesis by wild-type *Pseudomonas putida* GR12-2. Likewise, fungal species also synthesizes auxins as found in *Arabidopsis thaliana* and maize (*Zea mays*) that was inoculated with *Trichoderma* leading to changes in the root system architecture and enhanced plant growth. The effects were observed as increased root hair growth and lateral root formation (Bjorkman et al. 1998; Harman et al. 2004; Contreras-Cornejo et al. 2009).

Cytokinins stimulate cell division and differentiation processes in meristems both in plant shoots and roots. They are also involved in the processes including primary root growth, shoot formation, and callus formation. Cytokinins help plants

Table 8.1 Phytohormones synthesized by microorganisms

Phytohormones	Microorganisms	References
Auxins	*Azotobacter vinelandii* and *Azotobacter beijerinckii*	Azcon and Barea (1975)
	Trichoderma virens and *Trichoderma atroviride*	Contreras-Cornejo et al. (2009)
	Azospirillum brasilense	Martínez-Morales et al. (2003)
	Pseudomonas putida	Mayak et al. (1999)
	Rhizobium phaseoli	Atzorn et al. (1988)
Cytokinin	*Paenibacillus polymyxa*	Timmusk et al. (1999)
	Pseudomonas fluorescens	García de Salamone et al. (2001)
	Azotobacter vinelandii Azotobacter beijerinckii	Azcón and Barea (1975)
	Bacillus subtilis	Arkhipova et al. (2005)
	Azotobacter chroococcum	Nieto and Frankenberger (1989)
	Azotobacter vinelandii	Taller and Wong (1989)
	Paenibacillus polymyxa	Timmusk et al. (1999)
Gibberellins	*Azotobacter vinelandii* and *Azotobacter beijerinckii*	Azcón and Barea (1975)
	Rhizobium phaseoli	Atzorn et al. (1988)
	Acinetobacter calcoaceticus	Kang et al. (2009)

to maintain the pool of totipotent stem cells in their root and shoot meristems (Howell et al. 2003; Leibfried et al. 2005), while the gibberellins (GAs) stimulate growth and control many plant developmental processes including seed germination, stem growth, sex expression, and fruit formation (Bomke and Tudzynski 2009). Cytokinin production was reported in the cell-free medium of few strains of *Rhizobium* spp., *Pantoea agglomerans*, *Pseudomonas fluorescens*, *Azotobacter* spp., *Bacillus subtilis*, *Rhodospirillum rubrum*, and *Paenibacillus polymyxa* (Glick 2012). Gibberellin production was also reported in the case of *Rhizobium phaseoli*, *Bacillus cereus*, and *Acinetobacter calcoaceticus* (Atzorn et al. 1988; Joo et al. 2005; Kang et al. 2009) (Table 8.1).

8.3 Role of Bio-priming in Nutritional Values of Plant

The present world population of about 7.2 billion is expected to be more than 9.6 billion by the end of 2050 (UNDESA 2013). Continuous increasing demand of food supply leads to the higher use of chemical fertilizers and pesticides which is harmful for the human health and environment also. Therefore researchers are concentrating more on the use of bio-fertilizers and biopesticides as an alternative to chemicals for sustainable agriculture. In this regard several attempts have been made toward nutrient-rich high-quality food production in sustainable agriculture

(Raja 2013; Patel et al. 2015). Seed bio-priming has the potentiality to enhance plant growth, yield, and nutritional values of the end product. It also has the ability to enhance uniform germination of seeds which leads to rapid, uniform, and high establishment of crops, hence improving harvest quality and yield (Mahmood et al. 2016). Several mechanisms are reported to define the role of bio-priming to enhance nutritional value of the plant products, viz., phosphate solubilization, increased N_2 fixation, increased production of plant growth-promoting compounds like phytohormones, production of antibiotics, and decomposition of organic matter (Sinha et al. 2014). Rhizosphere microbes play a very crucial role in enhanced uptake of three essential nutrients N, P, and K (Sarma et al. 2015). It is well elaborated that N, P, and K are the major constituents of several enzymes, hormones, amino acids, and genetic material in plants having role in various physiological processes in the plant (Maathuis 2009; Krouk et al. 2010; Chevalier and Rossignol 2011). Yadav et al. (2017) reported increased nutrient uptake of N, P, K, Na, Ca, and organic matter in the seed, foliage, and pericarp of the chickpea plants bio-primed with *Pseudomonas fluorescens* OKC and *Trichoderma asperellum* T42. Additionally, nutritional qualities of the seed, foliage, and pericarp of the chickpea plants, viz., total phenolic and protein content, carbohydrate content, total flavonoid content, and reducing power, were also found to be increased during seed bio-priming. There was also increment in number of branches, heads per plant, diameter of head, grain number per head, grains per plant, 1000 grain weight, oil content, and grain yield in safflower bio-primed with *Pseudomonas* spp. (Sharif 2012). Recently, Singh et al. (2018) demonstrated that treatment with *Trichoderma asperellum* T42 enhanced nitrogen utilization in tobacco. Treatment with microbial consortium of *Pseudomonas fluorescens* OKC, *Trichoderma asperellum* T42, and *Rhizobium* sp. RH4 was found to have a significant role in enhancement of seed germination and seedling growth in both chickpea and rajma (Yadav et al. 2013). It is also well reported that certain microbes also have the ability to produce vitamins and other nutrients. *Azotobacter vinelandii* strain ATCC 12837 and *A. chroococcum* strain H23 produce the vitamin B group members, viz., niacin, pantothenic acid, thiamine, riboflavin, and biotin, after 72 h of growth in chemically defined media (Revillas et al. 2000). Therefore, it can be claimed that suitable seed bio-priming can improve nutritional value of the plant and plant harvest.

8.4 4Role of Seed Bio-priming in Biotic and Abiotic Stress Management

In natural environment plants continuously face biotic and abiotic challenges, which are also increasing day by day as an outcome of the changing climate. These environmental stresses are becoming continuous threat to plant production as well as productivity (Sarma et al. 2012). The adverse effect of chemical pesticides is well known, and it is the main reason to emphasize the microbe-mediated alleviation of these stresses. Beneficial microbial consortium efficiently protects the plant and boosts the tolerance level of plants under stresses (Ray et al. 2017). Several reports are available in relation to PGPR-mediated stress tolerance and increased yield, viz., in pea (Jain et al. 2013; Patel et al. 2016), lettuce (Kohler et al. 2009), chickpea (Sarkar et al. 2014), wheat (Jaderlund et al. 2008; Chakraborty et al. 2013; Nadeem et al. 2013; Kumar et al. 2014), maize (Rojas-Tapias et al. 2012), rice (Bal et al. 2013; Jha et al. 2013; Lavakush et al. 2014), broad bean (Younesi and Moradi 2014), soybean (Masciarelli et al. 2014), and groundnut (Paulucci et al. 2015). An increase in phenolic contents was found during pathogen infection site, i.e., in the collar region and leaves of chickpea during challenged inoculation with *Sclerotium rolfsii* (Singh et al. 2014). Similarly, a significant increment of total phenolic, flavonoid, ascorbic acid, and protein contents, free radical-scavenging activity, hydroxyl radical-scavenging activity, iron chelation, and reducing power were reported in the seeds and pericarp of pods of pea treated with the microbial consortium comprising *Bacillus subtilis* BHHU100, *Trichoderma harzianum* TNHU27, and *Pseudomonas aeruginosa* PJHU15 (Jain et al. 2014). It is also well recognized that the mixtures of various strains of *Trichoderma* perform better in plant disease management than an individual strain (Kumar et al. 2016). Singh et al. (2013a, b) also showed that high and uniform deposition of lignin in cambial cells of chickpea during challenge with *S. rolfsii* when bio-primed with the triple microbial consortium (*P. aeruginosa* PHU094, *T. harzianum* THU0816, *Mesorhizobium* species RL091) compared to the individual microbial applications.

Systemic resistance is the type of tolerance of plant against a wide range of plant pathogens induced by the interaction with a mild strain of the pathogen. Seed bio-priming provides significant resistance through induced systemic resistance. A wheat endophytic bacterium (*Pseudomonas aeruginosa* PW09) triggers the induced systemic resistance against *Sclerotium rolfsii* infection and increased biomass in cucumber under NaCl stress by enhancing the phenylpropanoid metabolism, antioxidant activities, and proline accumulation (Pandey et al. 2012). Patel et al. (2016) reported the *Pseudomonas fluorescens* OKC and *Trichoderma asperellum* T42 mediated induced expression of *LOX1* and *COI1* that lead to activation of JA-mediated resistance in pea against *Erysiphe pisi*. Several mechanisms of bio-priming mediated stress management classified and they are as follows:

- Mycoparasitism
- Antibiosis
- Induced phenolic production
- Activation of antioxidant production
- Vigorous growth of the plant
- Increased uptake and nutrient status
- Induced expression of plant defense-related genes
- Systemic defense activation

8.5 Conclusion and Future Prospects

Microbial seed priming or seed bio-priming is one of the most economical methods which require small volume of inoculums. Microbial seed priming enhances plant growth promotion by producing different phytohormones as well as enhances nutritional value of the plant products. Seed bio-priming also induces defense responses in plants against biotic and abiotic stresses. But it is important to note that different crop seeds show variation in shape, size, and nature of seed coat. Due to this, it is important to standardize the microbial concentration and time of priming for each crop. Few available records explain that some crops enhance plant growth even at very low microbial concentration. This will not minimize the application of higher microbial concentration but also reduce the cost of seed priming. Several aspects such as proper formulation development and mechanistic insight of the defense responses as well as plant growth promotion require elaboration in near future. Seed bio-priming will be the best tool to overcome the environmental stresses in a sustainable way under the continuously increasing demand of chemical-free produces in the present scenario.

Acknowledgments Authors are thankful to the Department of Biotechnology, New Delhi, India, for awarding project grant (BT/PR5990/AGR/5/587/2012). BKS and HBS are also thankful to the Indian Council of Agricultural Research, New Delhi, for providing financial assistance through the grant ICAR-NBAIM/AMAAS/2014-15/73. The authors are also grateful to Technology Development Component of the ICAR seed project.

References

Arkhipova TN, Veselov SU, Melentiev AI, Martynenko EV, Kudoyarova GR (2005) Ability of bacterium *Bacillus subtilis* to produce cytokinins and to influence the growth and endogenous hormone content of lettuce plants. Plant Soil 272:201–209

Atzorn R, Crozier A, Wheeler CT, Sandberg G (1988) Production of gibberellins and indole-3-acetic acid by *Rhizobium phaseoli* in relation to nodulation of *Phaseolus vulgaris* roots. Planta 175:532–538

Azcon R, Barea JM (1975) Synthesis of auxins, gibberellins and cytokinins by *Azotobacter vinelandii* and *Azotobacter beijerinckii* related to effects produced on tomato plants. Plant Soil 43:609–619

Bal HB, Nayak L, Das S et al (2013) Isolation of ACC deaminase PGPR from rice rhizosphere and evaluating their plant growth promoting activity under salt stress. Plant Soil 366:93–105

Bennett AJ, Whipps JM (2008) Dual application of beneficial microorganisms to seed during drum priming. Appl Soil Ecol 38:83–89

Bisen K, Keswani C, Mishra S, Saxena A, Rakshit A, Singh HB (2015) Unrealized potential of seed biopriming for versatile agriculture. In: Rakshit A, Singh HB, Sen A (eds) Nutrient use efficiency: from basics to advances. Springer, New Delhi, pp 193–206

Bjorkman T, Blanchard LM, Harman GE (1998) Growth enhancement of *shrunken-2 (sh2)* sweet corn by *Trichoderma harzianum* 1295-22: effect of environmental stress. J Am Soc Hortic Sci 123:35–40

Bomke C, Tudzynski B (2009) Diversity, regulation, and evolution of the gibberellin biosynthetic pathway in fungi compared to plants and bacteria. Phytochemistry 70:1876–1893

Callan NW, Mathre D, Miller JB (1990) Bio-priming seed treatment for biological control of *Pythium ultimum* preemergence damping-off in *sh-2* sweet corn. Plant Dis 74:368–372

Callan NW, Mathre DE, Miller JB (1991) Field performance of sweet corn seed bio-primed and coated with *Pseudomonas fluorescens* AB254. Hortic Sci 26:1163–1165

Chakraborty U, Chakraborty BN, Chakraborty AP et al (2013) Water stress amelioration and plant growth promotion in wheat plants by osmotic stress tolerant bacteria. World J Microbiol Biotechnol 29:789–803

Chevalier F, Rossignol M (2011) Proteomic analysis of *Arabidopsis thaliana* ecotypes with contrasted root architecture in response to phosphate deficiency. J Plant Physiol 168:1885–1890

Contreras-Cornejo HA, Macías-Rodríguez L, Cortés-Penagos C, López-Bucio J (2009) *Trichoderma virens*, a plant beneficial fungus, enhances biomass production and promotes lateral root growth through an auxin-dependent mechanism in *Arabidopsis*. Plant Physiol 149:1579–1592

Garcia de Salamone IE, Hynes RK, Nelson LM (2001) Cytokinin production by plant growth promoting rhizobacteria and selected mutants. Can J Microbiol 47:404–411

Glick BR (2012) Plant growth-promoting bacteria: mechanisms and applications. Scientifica (Cairo) 2012:1–15

Hammond-Kosack KE, Jones JD (1996) Resistance gene-dependent plant defense responses. Plant Cell 8:1773

Harman GE, Taylor AG (1988) Improved seedling performance by integration of biological control agents at favorable pH levels with solid matrix priming. Phytopathology 78:520–525

Harman GE, Petzoldt R, Comis A, Chen J (2004) Interactions between *Trichoderma harzianum* strain T22 and maize inbred line Mo17 and effects of these interactions on diseases caused by *Pythium ultimum* and *Colletotrichum graminicola*. Phytopathology 94:147–153

Howell SH, Lall S, Che P (2003) Cytokinins and shoot development. Trends Plant Sci 8:453–459

Jaderlund L, Arthurson V, Granhall U et al (2008) Specific interactions between arbuscular mycorrhizal fungi and plant growth promoting bacteria: as revealed by different combinations. FEMS Microbiol Lett 287:174–180

Jain A, Singh A, Singh S, Singh HB (2013) Microbial consortium-induced changes in oxidative stress markers in pea plants challenged with *Sclerotinia sclerotiorum*. J Plant Growth Regul 32:388–398

Jain A, Singh A, Chaudhary A, Singh S et al (2014) Modulation of nutritional and antioxidant potential of seeds and pericarp of pea pods treated with microbial consortium. Food Res Int 64:275–282

Jain A, Singh A, Singh S, Singh HB (2015) Biological management of *Sclerotinia sclerotiorum* in pea using plant growth promoting microbial consortium. J Basic Microbiol 55:961–972

Jha A, Saxena J, Sharma V (2013) An investigation on phosphate solubilization potential of agricultural soil bacteria as affected by different phosphorus sources, temperature, salt and pH. Commun Soil Sci Plan 44:2443–2458

Joo GJ, Kim YM, Kim JT, Rhee IK, Kim JH, Lee IJ (2005) Gibberellins-producing rhizobacteria increase endogenous gibberellins content and promote growth of red peppers. J Microbiol 43(6):510–515

Kang SM, Joo GJ, Hamayun M, Na CI, Shin DH, Kim HY, Hong JK, Lee IJ (2009) Gibberellin production and phosphate solubilization by newly isolated strain of *Acinetobacter calcoaceticus* and its effect on plant growth. Biotechnol Lett 31:277–281

Keswani C, Mishra S, Sarma BK, Singh SP, Singh HB (2014) Unraveling the efficient applications of secondary metabolites of various *Trichoderma* spp. Appl Microbiol Biotechnol 98:533–544

Keswani C, Bisen K, Singh V, Sarma BK, Singh HB (2016) Formulation technology of biocontrol agents: present status and future prospects. In: Arora NK, Mehnaz S, Balestrini R (eds) Bioformulations: for sustainable agriculture. Springer, New Delhi, pp 35–52

Kohler J, Hernandez JA, Caravaca F, Roldan A (2009) Induction of antioxidant enzymes is involved in the greater effectiveness of a PGPR versus AM fungi with respect to increasing the tolerance of lettuce to severe salt stress. Environ Exp Bot 65:245–252

Krouk G, Crawford NM, Coruzzi GM, Tsay YF (2010) Nitrate signaling: adaptation to fluctuating environments. Curr Opin Plant Biol 13:265–272

Kumar A, Maurya BR, Raghuwanshi R (2014) Isolation and characterization of PGPR and their effect on growth, yield and nutrient content in wheat (*Triticum aestivum* L.). Biocatal Agric Biotechnol 3:121–228

Kumar G, Maharshi A, Patel J, Mukherjee A, Singh HB, Sarma BK (2016) Trichoderma: a potential fungal antagonist to control plant diseases. SATSA Mukhapatra 21:206–218

Lavakush YJ, Verma JP, Jaiswal DK et al (2014) Evaluation of PGPR and different concentration of phosphorus level on plant growth, yield and nutrient content of rice (*Oryza sativa*). Ecol Eng 62:123–128

Leibfried A, To JP, Busch W, Stehling S, Kehle A, Demar M, Kieber JJ, Lohmann JU (2005) WUSCHEL controls meristem function by direct regulation of cytokinin-inducible response regulators. Nature 438:1172

Maathuis FJM (2009) Physiological functions of mineral macronutrients. Curr Opin Plant Biol 12:250–258

Mahmood A, Turgay OC, Farooq M, Hayat R (2016) Seed biopriming with plant growth promoting rhizobacteria: a review. FEMS Microbiol Ecol 92 (fiw112)

Martínez-Morales LJ, Soto-Urzúa L, Baca BE, Sánchez-Ahédo JA (2003) Indole-3-butyric acid (IBA) production in culture medium by wild strain *Azospirillum brasilense*. FEMS Microbiol Lett 228:167–173

Masciarelli O, Llanes A, Luna V (2014) A new PGPR co-inoculated with *Bradyrhizobium japonicum* enhances soybean nodulation. Microbiol Res 169:609–615

Mauch-Mani B, Baccelli I, Luna E, Flors V (2017) Defense priming: an adaptive part of induced resistance. Annu Rev Plant Biol 68:485–512

Mayak S, Tirosh T, Glick BR (1999) Effect of wild-type and mutant plant growth-promoting rhizobacteria on the rooting of mung bean cuttings. J Plant Growth Regul 18:49–53

McQuilken MP, Halmer P, Rhodes DJ (1998) Application of microorganisms to seeds. In: Formulation of microbial biopesticides. Springer, Dordrecht, pp 255–285

Meena S, Rakshit A, Meena VS (2016) Effect of seed bio-priming and N doses under varied soil type on nitrogen use efficiency (NUE) of wheat (*Triticum aestivum* L.) under greenhouse conditions. Biocatal Agric Biotechnol 12:172–178

Meena S, Rakshit A, Singh HB, Meena VS (2017) Effect of nitrogen levels and seed bio-priming on root infection, growth and yield attributes of wheat in varied soil type. Biocatal Agric Biotechnol 12:172–178

Nadeem SM, Zaheer ZA, Naveed M et al (2013) Mitigation of salinity-induced negative impact on the growth and yield of wheat by plant growth-promoting rhizobacteria in naturally saline conditions. Ann Microbiol 63:225–232

Nieto KF, Frankenberger WT (1989) Biosynthesis of cytokinins by *Azotobacter chroococcum*. Soil Biol Biochem 21:967–972

Ortíz-Castro R, Contreras-Cornejo HA, Macías-Rodríguez L, López-Bucio J (2009) The role of microbial signals in plant growth and development. Plant Signal Behav 4:701–712

Pandey PK, Yadav SK, Singh A, Sarma BK, Mishra A, Singh HB (2012) Cross-species alleviation of biotic and abiotic stresses by the endophyte *Pseudomonas aeruginosa* PW09. J Phytopathol 160:532–539

Patel JS, Singh A, Singh HB, Sarma BK (2015) Plant genotype, microbial recruitment and nutritional security. Front Plant Sci 6:608. https://doi.org/10.3389/fpls.2015.00608

Patel JS, Sarma BK, Singh HB, Upadhyay RS, Kharwar RN, Ahmed M (2016) *Pseudomonas fluorescens* and *Trichoderma asperellum* enhance expression of Gα subunits of the pea heterotrimeric G-protein during *Erysiphe pisi* infection. Front Plant Sci 6:1206. https://doi.org/10.3389/fpls.2015.01206

Patten CL, Glick BR (2002) Role of *Pseudomonas putida* indoleacetic acid in development of the host plant root system. Appl Environ Microbiol 68:3795–3801

Paulucci NS, Gallarato LA, Reguera YB et al (2015) *Arachis hypogaea* PGPR isolated from Argentine soil modifies its lipids components in response to temperature and salinity. Microbiol Res 173:1–9

Raja N (2013) Biopesticides and biofertilizers: ecofriendly sources for sustainable agriculture. J Biofertil Biopestic. https://doi.org/10.4172/2155-6202.1000e112

Rakshit A, Pal S, Meena S, Manjhee B, Rai S, Rai A, Bhowmick MK, Singh HB (2014) Seed bio-priming: a potential tool in integrated resource management. SATSA Mukhaptra Ann Tech 18:94–103

Rakshit A, Singh HB, Sen A (2015a) Nutrient use efficiency: from basics to advances. Springer, New Delhi

Rakshit A, Sunita K, Pal S, Singh A, Singh HB (2015b) Bio-priming mediated nutrient use efficiency of crop species. In: Rakshit A, Singh HB and Sen A (eds) Nutrient use efficiency: from basics to advances. Springer, New Delhi, pp 181–191

Ray S, Yadav SK, Sarma BK, Singh HB, Singh S (2017) How the microbial consortium enhances the stress tolerance in plants during various environmental conditions. In: Singh SS (ed) Plants and microbes in ever changing environment. Nova Science Publisher, New York, pp 71–82

Reddy PP (2013) Bio-priming of seeds. In: Reddy PP (ed) Recent advances in crop protection. Springer, New Delhi, pp 83–90

Revillas JJ, Rodelas B, Pozo C, Martinez-Toledo MV, González-Lopez J (2000) Production of B-group vitamins by two *Azotobacter* strains with phenolic compounds as sole carbon source under diazotrophic and adiazotrophic conditions. J Appl Microbiol 89:486–493

Rojas-Tapias D, Moreno-Galvan A, Pardo-Diaz S et al (2012) Effect of inoculation with plant growth promoting bacteria (PGPB) on amelioration of saline stress in maize (*Zea mays*). Appl Soil Ecol 61:264–272

Sarkar A, Patel JS, Yadav S, Sarma BK, Srivastava JS, Singh HB (2014) Studies on rhizosphere-bacteria mediated biotic and abiotic stress tolerance in chickpea (*Cicer arietinum* L.). Vegetos 27(1):158–169

Sarma BK, Yadav SK, Singh DP, Singh HB (2012) Rhizobacteria mediated induced systemic tolerance in plants: prospects for abiotic stress management. In: Maheshwari DK (ed) Bacteria in agrobiology: stress management. Springer, Berlin, pp 225–238

Sarma BK, Yadav SK, Singh S, Singh HB (2015) Microbial consortium mediated plant defense against phytopathogens: readdressing for enhancing efficacy. Soil Biol Biochem 87:25–33

Sharif RS (2012) Study of nitrogen rates effects and seed biopriming with PGPR on quantitative and qualitative yield of safflower (*Carthamus tinctorius* L.). Tech J Eng Appl Sci 2:162–166

Singh HB (2016) Seed biopriming: a comprehensive approach towards agricultural sustainability. Indian Phytopathol 69:203–209

Singh A, Jain A, Sarma BK, Upadhyay RS, Singh HB (2013a) Rhizosphere microbes facilitate redox homeostasis in *Cicer arietinum* against biotic stress. Ann Appl Biol 163:33–46

Singh A, Sarma BK, Upadhyay RS, Singh HB (2013b) Compatible rhizosphere microbes mediated alleviation of biotic stress in chickpea through enhanced antioxidant and phenylpropanoid activities. Microbiol Res 168:33–40

Singh A, Jain A, Sarma BK, Upadhyay RS, Singh HB (2014) Rhizosphere competent microbial consortium mediates rapid changes in phenolic profiles in chickpea during *Sclerotium rolfsii* infection. Microbiol Res 169:353–360

Singh V, Upadhyay RS, Sarma BK, Singh HB (2016a) Seed bio-priming with *Trichoderma asperellum* effectively modulate plant growth promotion in pea. Int J Agric Environ Biotech 9:361–365

Singh V, Upadhyay RS, Sarma BK, Singh HB (2016b) Trichoderma asperellum spore dose depended modulation of plant growth in vegetable crops. Microbiol Res 193:74–86

Singh BN, Dwivedi P, Sarma BK, Singh GS, Singh HB (2018) Trichoderma asperellum T42 reprograms tobacco for enhanced nitrogen utilization efficiency and plant growth when fed with N nutrients. Frontiers in Plant Science 9:163

Sinha RK, Valani D, Chauhan K, Agarwal S (2014) Embarking on a second green revolution for sustainable agriculture by vermiculture biotechnology using earthworms: reviving the dreams of Sir Charles Darwin. Int J Agric Health Saf 1:50–64

Spaepen S, Vanderleyden J (2011) Auxin and plant-microbe interactions. Cold Spring Harb Perspect Biol 3:001438

Taller BJ, Wong TY (1989) Cytokinins in *Azotobacter vinelandii* culture medium. Appl Environ Microbiol 55:266–267

Timmusk S, Nicander B, Granhall U, Tillberg E (1999) Cytokinin production by *Paenibacillus polymyxa*. Soil Biol Biochem 31:1847–1852

Tsavkelova EA, Klimova SY, Cherdyntseva TA, Netrusov AI (2006) Microbial producers of plant growth stimulators and their practical use: a review. Appl Biochem Microbiol 42:117–126

United Nations Department of Economic and Social Affairs (UN DESA) (2013) World population projected to reach 9.6 billion by 2050. UN DESA, New York. http://www.un.org/en/development/desa/news/population/un-report-world-population-projected to-reach-9-6- billion-by-2050.html

Valencia-Cantero E, Hernández-Calderón E, Velázquez-Becerra C, López-Meza JE, Alfaro-Cuevas R, López-Bucio J (2007) Role of dissimilatory fermentative iron-reducing bacteria in Fe uptake by common bean (*Phaseolus vulgaris* L.) plants grown in alkaline soil. Plant Soil 291:263–273

Wright B, Rowse H, Whipps JM (2003) Microbial population dynamics on seeds during drum and steeping priming. Plant Soil 255:631–640

Yadav SK, Dave A, Sarkar A, Singh HB, Sarma BK (2013) Co-inoculated Biopriming with *Trichoderma, Pseudomonas* and *Rhizobium* improves crop growth in *Cicer arietinum* and *Phaseolus vulgaris*. Intern J Agri Environ Biotech 6(2):255–259

Yadav SK, Singh S, Singh HB, Sarma BK (2017) Compatible rhizosphere-competent microbial consortium adds value to the nutritional quality in edible parts of chickpea. J Agric Food Chem 65(30):6122–6130. https://doi.org/10.1021/acs.jafc.7b01326

Younesi O, Moradi A (2014) Effects of plant growth-promoting rhizobacterium (PGPR) and arbuscular mycorrhizal fungus (AMF) on antioxidant enzyme activities in salt-stressed bean (*Phaseolus vulgaris* L.). Agric (Polnohospodˇ arstvo) 60:10–21

Potential of Biopriming in Enhancing Crop Productivity and Stress Tolerance

9

Ahmad Mahmood and Ryota Kataoka

Abstract

Appropriate application of beneficial microbes to the seed or soil is essential for their efficacy and efficiency. Several methods like soil application, seed inoculation, root dipping, foliar application, and seed coating or covering have been employed, yet with the increase in concerns regarding survival of microbes on seed surface, and increasing focus on endophytes has led to search for potential methods ensuring the survival and colonization of the seed by the required microorganisms. To overcome the problems of lack of uniformity in seed emergence, poor seedling vigor, and establishment, researchers have come up with the technique of seed priming, which employing activation of physiological processes prior to sowing has certain advantages. Incorporation of hydration or immersion of seeds in microbial suspension, and/or seed covering in case of fungi, followed by incubation for a predetermined duration, and subsequent drying termed as biopriming have shown better survival, and colonization of desired microbes on/in the seeds providing plant growth promoting, and stress tolerance activities when used in many crops. A diversity of biopriming methods have been used with varying potential of plant growth-promoting activities as well as promising results in biocontrol of pathogens.

Keywords

Seed biopriming · Plant growth-promoting microorganisms · Crop growth · Stress tolerance

A. Mahmood (✉) · R. Kataoka
Faculty of Life and Environmental Sciences, University of Yamanashi, Yamanashi, Japan
e-mail: rkataoka@yamanashi.ac.jp

© Springer Nature Singapore Pte Ltd. 2018

A. Rakshit, H. B. Singh (eds.), *Advances in Seed Priming*,
https://doi.org/10.1007/978-981-13-0032-5_9

Abbreviation

CFUs Colony forming units

9.1 Introduction

Increasing food requirement for continuously on the rise population has forced the scientific and farming community to look for enhancing crop yields within the available resources (Seufert et al. 2012) besides avoiding damage to the environment (Goswami et al. 2016). The productivity of crops is hindered by several factors, where poor stand establishment (Grassbaugh and Bennett 1998) and incidence of seed-borne or early pathogens on seedling stage lead to reduced crop yields (Orzolek 1991). Enhancing the production on the cost of environment is also discouraged, so there is need to search for sustainable production technology to meet the requirements. Pre-sowing techniques come as an opportunity to enhance the stand establishment inclusive of enabling the plant to cope the stress without compromising on yield, coupled with no or minute deleterious effect on the surroundings.

Activation of physiological processes without the emergence of the radicle of seed through seed priming has gained importance in crop production. Initially, the technique aimed to bring up uniform seed germination (Parera and Cantliffe 1994) and better stand establishment, but with the time and research, it has been used for enhancing the crop growth jointly with stress resistance (Hossain et al. 2015), where its economical and effective application has after 'been debated' Ashraf and Foolad (2005). The priming treatment has been used extensively in crops like barley, lentil, rice, and sunflower, for speeding up their germination and emergence (Saglam et al. 2010) and growth, viz., stand establishment, also under stress conditions like drought (Ahmed et al. 2016; Farooq et al. 2013; Kaya et al. 2006; Rouhollah 2013; Sun et al. 2010; Tabatabaei 2013; Yan 2015), salinity (Abraha and Yohannes 2013; Ellouzi et al. 2017; Jisha and Puthur 2016), heavy metal stress (Galhaut et al. 2014; Moulick et al. 2016), low temperature or chilling (Ahmad et al. 2015; Yadav et al. 2011), and high temperature (Parera et al. 1993). Seed priming includes different techniques, namely, hydropriming, osmopriming, thermopriming, matrix priming, hormonal priming, osmohardening, halopriming, and biopriming (Farooq et al. 2017; Jisha et al. 2013).

Seed priming enhances the plant growth and stress tolerance through different mechanisms (Shehab et al. 2010). The mechanisms in enhancing plant growth (uniform and quick germination, better plant establishment) and stress tolerance (activation of enzymes (Farhad et al. 2011; Posmyk et al. 2009), enhancing the water uptake (Farooq et al. 2009), and accumulation of sugar and proline within the seedlings (Yan 2015)) are notable. The enhanced and uniform germination is a response to reduced imbibition duration (Brocklehurst and Dearman 2008), activation of the enzymes (Lee and Kim 2000), and enhanced metabolism (Farooq et al. 2006). Similarly, for the stress tolerance, activation of the cellular defense mechanisms and enzymes (Hussain et al. 2016) prior to the stress in the field leads to better stress tolerance in the plants, be it biotic or abiotic. Activation of antioxidant mechanism in tobacco through seed priming has been reported (Xu et al. 2011).

The environmental and sustainability factors are major motives behind the preference of biological or organic fertilizers over the chemical ones. The biological fertilizers also termed as biofertilizers include certain microorganisms, plant growth-promoting microorganisms (PGPM), including plant growth-promoting bacteria (PGPB), plant growth-promoting fungi (PGPF), and plant growth-promoting cyanobacteria (PGPC). Potential of beneficial microorganisms in enhancing crop productivity and combating stress conditions has been widely discussed and reviewed (Ahemad and Kibret 2014; Glick 2012; Goswami et al. 2016; Hayat et al. 2010; Santoyo et al. 2016; Souza et al. 2015). PGPB include the bacteria showing positive effects on the plants, either free living in the rhizosphere or endophytic, living within the plants without showing symptoms of any damage, and enhance the plant growth and stress tolerance through different mechanisms, viz., symbiotic and nonsymbiotic nitrogen fixation; facilitation of nutrient uptake including phosphorus, potassium, and iron; release of certain metabolites involved in plant growth promotion and stress tolerance; and remediation of organic and inorganic pollutants (Ahemad and Kibret 2014; Meena et al. 2015; Santoyo et al. 2016). Similarly, PGPF have showed different mechanisms of PGP, including supply of nutrients to the plants, mobilizing micronutrients, increasing the surface area of the roots, and release of certain metabolites (Kumar 2016), and also reduce the infestation of plant diseases (Hossain et al. 2017), where arbuscular mycorrhizal fungi (AMF) and *Trichoderma* hold important position. PGPC are responsible for enhancing the crop growth through nitrogen fixation and release of metabolites, improving soil fertility by soil aggregation, and increasing water-holding capacity (Shariatmadari et al. 2013). Efficient application of these PGPM is necessary for taking better advantage of all the PGP mechanisms showed by them (Compant et al. 2005; Fan et al. 2012; Ongena and Jacques 2008; Weller 2007).

Application of biofertilizers mainly bacteria has long been carried out through seed inoculation by the farming community, whereas instances of root dipping, soil incorporation, and foliar spray can also be observed, usually at research level. All the application methods have different merits and demerits (Mahmood et al. 2016) where there is need for a more effective application method, viz., survival of the bacteria collectively with efficiency of the latter. Among all the methods used, the bacteria are applied to the seeds either directly or in a formulation and are usually used immediately after application; however their survival remains inefficient (Wright et al. 2003a) so application of beneficial bacteria through priming is an attractive alternative. It has also been observed that a number of microorganisms increased during the process of priming, and the survival rate was not affected even by the subsequent drying (Wright et al. 2003b).

Biopriming or biological seed priming involves biological material for seed priming (Farooq et al. 2017). It has been explained as application of bacteria to the seeds during the hydration process as carried out in other seed priming techniques (Prasad et al. 2016), where application of fungus through seed biopriming has also been carried out. Among other biological materials used for seed priming, aqueous extracts of the plants have been used (Aiyaz et al. 2015; Ghezal et al. 2016). This technique has been defined and explained differently by different researchers (Ashraf and Foolad 2005; McDonald 1999) and simply involves application of beneficial microbes to the seed in integration with the seed hydration (Rawat et al. 2011; Singh et al. 2016b), where application to the seedlings or plantlets has also been reported (Harish et al. 2008), also termed as biohardening (Panigrahi et al. 2016).

Application of endophytic bacteria or fungi through biopriming has potential benefits, as it allows sufficient time and environment for successful colonization of the seeds. The incubation duration allows the beneficial microbes to colonize the seed surface, also enter the seeds where needed. Biopriming and other seed-coating techniques have long been advocated for uniform application of biocontrol agents on the seeds (Bennett et al. 1992; Harman and Taylor 1988; Khan 1992; Kubik 1995; Lewis et al. 1987; Pill 1995; Warren and Bennett 1997), but the drying phase lacked information (Bennett 1998), which has been differently introduced by other researchers, usually drying in the shade for 24 h or overnight (Abuamsha et al. 2010; Carrozzi et al. 2012; Nandakumar et al. 2001). A 100% bacterial population increase was observed on the seeds of chickpea after 20-h incubation on sterilized vermiculite in talc-based formulation (Vidhyasekaran and Muthamilan 1995) which indicates that seed biopriming provides a microclimate favorable for the proliferation of microbes on the seeds.

Keeping in view the potential of biopriming in crop production, this chapter focuses on development of the technique, different methods used, and potential of this technique in enhancing plant growth, and productivity, and its role in ameliorating biotic and abiotic stresses.

9.2 Methods Used in Biopriming

Beneficial microorganisms can be applied to the soil, seed, and seedlings in different ways, mostly associated with their required function. In case of biological control, bacteria are generally coated on the seeds or the seedlings which create a competition environment and stop the pathogenic microbes' adhesiveness with the seeds or the seedlings. In general, when the aim of bacterial application is just to introduce the beneficial bacteria in the soil environment, then conventional inoculation method is accompanied with other methods discussed by Malik and Williams (2008). Uniformity in germination and better stand establishment options when considered, biopriming is favored method. Biopriming has been practiced and explained by different researchers (Bennett et al. 2009; Callan et al. 1991; Chakraborty et al. 2011; Gururani et al. 2012; Mirshekari et al. 2012; Moeinzadeh et al. 2010; Sharifi 2011; Sharifi and Khavazi 2011; Sharifi 2012) in several ways but needs to be explored and discussed further.

The term biopriming was first introduced by Callan et al. (1990), where they coated the sweet corn seeds with bacteria and immersed the seeds in warm water for imbibition of water up to 35–40%. Almost all of the later methods employed the soaking or immersion of the seeds in microbial suspension or similar treatments, except for the biopriming of seedlings or plantlets (Harish et al. 2008; Panigrahi et al. 2016). Different incubation or hydration procedures have been reported, where the use of powdered lignite or coal has been reported by Harman and Taylor (1988) and plastic bag by Callan et al. (1990), both including moist conditions. The priming treatment should be carried out after surface disinfecting the seeds (Singh et al. 2016b), as there is a chance of pathogens presence on the seeds, which can reduce the proliferation of desired microbe on the seed by inducing competition (Wright et al. 2003b).

Among the methods reported, trash was removed from the wheat seed and was surface sterilized for 1 min in 70% ethanol and then treated in 5% sodium hypochlorite solution for 40 min, followed by rinsing six times with autoclaved distilled water. Seeds were further dried at room temperature and were subjected to priming in bacterial suspension (Saber et al. 2012). Abuamsha et al. (2010) primed the rapeseed seeds in a bacterial suspension having $\log_{10}$ 11 CFUs (colony forming units) mL^{-1}, which were prepared from 24 h TSB cultures by centrifugation at 25,000 × g for 10 min and subsequent dilution with the culture supernatant. Bacterial suspension was applied at the ratio 1 mL to 1 g seed, and the suspension kept on shaking at 150 rpm at 22 °C for 4 h followed by drying the seeds at 28 °C overnight. In another method, surface-sterilized seeds with 1% NaOCl for 1 min followed by rinsing with sterilized water were immersed in *Azospirillum brasilense* Sp245 inoculum containing 10^7 bacteria per seed suspended in 66 m mol L^{-1} phosphate buffer, pH 7 for 2 h at 20 °C. The seeds were then dried at 20 °C under airflow (Carrozzi et al. 2012). Similarly, surface disinfection of the seeds was carried out by 10% H_2O_2 according to method given by Miche and Balandreau (2001), and seeds were further soaked in bacterial suspensions containing 10^7 bacteria mL^{-1} for 2 h at 28 °C (Kasim et al. 2013). In another method explained by Gholami et al. (2009), they surface sterilized the maize seeds with 0.02% sodium hypochlorite for 2 min, rinsed thoroughly in sterile distilled water followed by coating of seeds with 20% Arabic gum, and then suspended the seeds in bacterial perlite mixture until the seeds were uniformly coated (Firuzsalari et al. 2012; Sharifi and Khavazi 2011; Sharifi et al. 2011).

For instance, Reddy (2013) suggested the soaking of the seeds in the water for 12 h followed by mixing of the bacteria at the rate of 10 g kg^{-1} seed. Then a heap of the treated seeds should be made which should be covered with moist jute bag to ensure enough humidity. Seeds should be further incubated in highly humid conditions for 48 h at temperature range of 25–32 °C. Bio-agent thus added gets very good conditions for the growth and forms a protective layer around the seed which restricts the attachment of pathogenic bacteria to the seeds. This method is mainly characterized to be used in biocontrol. In another report, trash was removed from the wheat seed and the former was surface sterilized for 1 min in 70% ethanol and then treated with 5% sodium hypochlorite solution for 40 min, followed by rinsing six times with autoclaved distilled water. Seeds were further dried at room temperature and were subjected to priming in bacterial suspension (Saber et al. 2012).

The other two methods widely used in seed biopriming include talc-based formulation and drum priming. In the talc-based priming, the bacterial suspension is mixed with talc powder, augmented also with calcium carbonate and carboxymethyl cellulose followed by subsequent drying of the seeds in shade for 24 h (Nandakumar et al. 2001). In drum priming, used commercially for pesticide application to the seeds and initially introduced by (Rowse 1996), the seeds are soaked in the aqueous bacterial suspension in the continuously shaking drum for predetermined time and suspension volume, followed by drying (Wright et al. 2003a).

9.3 Role of Biopriming in Increasing Crop Productivity

Biopriming plays its role in improving seed viability, germination, plant vigor, growth, and yield (Prasad et al. 2016) and can significantly enhance the seed germination and uniformity in emergence of seedlings (Bhatt et al. 2015) along with stand establishment (Nakkeeran et al. 2005). Like the other seed priming treatments, biopriming helps in starting the physiological processes pre-sowing and helps in multiplication of PGPR in the area surrounding the seed (Taylor and Harman 1990). Several researchers have employed the biopriming in enhancing the crop growth and productivity. Table 9.1 presents the studies documenting positive effects of biopriming with different bacterial and fungal strains on various crop plants. Different applications on barley, carrot, maize, safflower, and sunflower have been reviewed previously by Mahmood et al. (2016).

Application of *Pseudomonas aureofaciens* through drum priming system enhanced the stand establishment in tomato (Warren and Bennett 1997). Application of various other PGPM in different crops through biopriming as compiled in Table 9.1 presents that wide application of *Trichoderma* species to plants for plant growth promotion and biocontrol has long been practiced (Altomare et al. 1999). Application of *Trichoderma* has gained more importance in plant stress tolerance, especially biotic stress; however its application for PGP has also been documented. Applied to pea seeds as biopriming agent, *Trichoderma* significantly enhanced the plant growth parameters, with biopriming concluded as an efficient method (Singh et al. 2016a). Among other biopriming applications of *Trichoderma*, it increased the growth of wheat, furthermore nitrogen uptake and recovery, and agronomic and physiological use efficiency and also performed well even under 75% of the recommended dose of the fertilizer (Meena et al. 2016). Similarly, in an experiment featuring six crops, namely, brinjal, chili, guar, okra, ridge gourd, and tomato, Singh et al. (2016b) observed that the seed germination depended on specific dose of *Trichoderma* spores, where the latter enhanced plant growth and induced systemic resistance-related enzyme activity. Other than that, response of snapdragon to the biopriming with *Trichoderma* and *Bacillus subtilis* was documented, where the authors observed increased germination and vegetative and reproductive growth (Bhargava et al. 2015). In another experiment, *Trichoderma* and *Pseudomonas* application through biopriming with varying duration was investigated, and *Pseudomonas* enhanced the seedling growth and vigor better than *Trichoderma*, but both proved to be way better when compared with non-primed seeds (Ananthi et al. 2014). Application of *Trichoderma* through seed biopriming also enhanced the enzyme activity through release of certain metabolites in maize plant (López-Coria et al. 2016). Finally, the application of two PGPR strains through priming enhanced

Table 9.1 Role of seed biopriming with different plant growth-promoting microorganisms in promoting plant growth

Strain	Crop	Crop response	Experiment	Priming duration	References
Trichoderma asperellum	Maize	Increased root length, shoot length, plasma membrane, and H^+-ATPase activity	Laboratory	1.5 h	López-Coria et al. (2016)
Trichoderma harzianum	Wheat	Increased plant height, chlorophyll content, root length, and effective tillers	Pot	30 min	Meena et al. (2016)
Trichoderma asperellum	Pea	Increased shoot length, root length, number of leaves, shoot fresh weight, root fresh weight, shoot dry weight, and root dry weight	Pot	–	Singh et al. (2016a)
Trichoderma asperellum	Brinjal, chili, guar, okra, ridge gourd, tomato	Increased seed germination, radicle length, number of leaves, shoot length, root length, chlorophyll, phenylpropanoid, and lignifications activity	Laboratory and pot	24 h	Singh et al. (2016b)
Trichoderma harzianum Bacillus subtilis	Snapdragon	Increased germination, number of leaves, shoot length, root length, plant height, spreading of the plant, number of spikes, total number of florets per spike, and flowering duration	Laboratory	24 h	Bhargava et al. (2015)
Trichoderma viride or Pseudomonas fluorescens	Chili	Increased germination, root length, shoot length, biomass, and seedling vigor index	Laboratory	3 h, and 12 h	Ananthi et al. (2014)
Azotobacter chroococcum Azospirillum lipoferum	Barley	Increased 1000 grain weight, grain yield, biological yield, and harvest index	Field	–	Mirshekari et al. (2012)

– Indicates no data given, h stands for hours, and min stands for minutes

the growth and yield of barley at different fertilizer levels (Mirshekari et al. 2012). Promotion of plant growth and productivity by the PGPR and PGPF applied through priming has been linked to PGP activities, and mechanisms of growth promotion by beneficial bacteria and fungi have been extensively reviewed (Goswami et al. 2016). Major microbial contributions toward enhanced plant productivity include nitrogen fixation; enhanced solubility, and availability of phosphorus, potassium, and iron; production of hormones, vitamins, enzymes, and organic acids from bacteria; facilitation of nutrient uptake through hyphal networks; and release of several metabolites from fungi, inclusive of quicker seed emergence, more vigorous seedlings, and better stand establishment due to seed priming treatments.

9.4 Role of Biopriming in Stress Tolerance

The crop plants are prone to seed- and soilborne diseases, resulting in heavy yield losses, and are usually controlled chemically; however, biopriming, being an environment-friendly, economical, and effective technique (Ashraf and Foolad 2005), is being focused for control of biotic and abiotic stresses. Bennett (1998) advocated the use of microorganisms as biopriming agents for enhanced seedling growth under stress conditions. An increase of tenfold on the seeds was observed in the population of applied antagonist by Callan et al. (1990), which ultimately helps to control the plant pathogens in the rhizosphere. Biopriming with PGPB can provide systemic resistance against multitude of pathogens (Compant et al. 2005), in combination with uniform germination in stress conditions (Singh et al. 2003), which suggests its use against different type of stresses.

9.4.1 Role of Biopriming Against Biotic Stress

Seed- and soilborne diseases pose threat to crop productivity and are often dealt with chemical pesticides, which are not sufficient against pathogenic attack on late stages of the plant from aerial parts. Under these circumstances, biopriming serves as an attractive approach, in which microbes keep on multiplying and form a biofilm around the root surface supplementary to putting competition by occupying the space, which helps reduced or no infestation by the pathogens, even in the later stages of the plant (Prasad et al. 2016). The biopriming with plant beneficial bacteria also helps in initializing the systemic resistance in the plants.

From the initial reports, it was observed that the seeds of carrot already infested with *Alternaria radicina* and *Alternaria dauci* when bioprimed with antagonistic fungi *Clonostachys rosea* reduced the incidence of these pathogens (Jensen et al. 2001). Similar instances including different bacterial and fungal biopriming agents against several pathogens including *Colletotrichum* (Begum et al. 2010), *Fusarium* (Mnasri et al. 2017; Nayaka et al. 2010; Srivastava et al. 2010), *Rhizoctonia*, *Sclerotium* (El-Mougy and Abdel-Kader 2008), and *Verticillium* (Rybakova et al. 2016) have been compiled in Table 9.2.

Table 9.2 Role of seed biopriming with different plant growth-promoting microorganisms in combating biotic stress

Strain	Crop	Pathogen/disease controlled	Crop response	Experiment	Priming duration	References
Rhizospheric bacteria, Endophytic bacteria	Wheat	*Fusarium culmorum*	Increased germination and seedling vigor, decreased disease incidence	Pot	24 h	Mnasri et al. (2017)
Serratia spp. *Paenibacillus* spp.	Oilseed rape and cauliflower	*Verticillium dahlia*, *Verticillium longisporum*	Increased root weight, weight of the green parts, and germination	Laboratory and pot	4 h	Rybakova et al. (2016)
Trichoderma harzianum, *Trichoderma virens*, *Pseudomonas aeruginosa*	Soybean	*Colletotrichum truncatum*	Increased seed germination and seedling establishment	Field	12 h	Begum et al. (2010)
Trichoderma harzianum	Maize	*Fusarium verticillioides*	Increased seed germination, vigor index, field emergence, yield, and thousand seed weight	Pot	24 h	Nayaka et al. (2010)
Pseudomonas fluorescens, *Trichoderma harzianum*	Tomato	*Fusarium oxysporum*	Increased seed germination, decreased germination time, and disease incidence	Pot and field	24 h	Srivastava et al. (2010)
Trichoderma viride, Trichoderma harzianum, Bacillus subtilis, Pseudomonas fluorescens, Trichoderma hamatum, Bacillus cereus	Faba bean	*Rhizoctonia solani, Fusarium solani, Sclerotium rolfsii*	Seed storage experiment[a]	Greenhouse and field	16 h	El-Mougy and Abdel-Kader (2008)

[a]The experiment investigated the efficacy of priming agents against disease causing pathogens in different storage durations

– Indicates no data given, h stands for hours, and min stands for minutes

Similar to the crop productivity, biopriming with *Trichoderma* has been used extensively in controlling plant pathogens. The seeds of soybean were bioprimed with *Trichoderma harzianum*, *Trichoderma virens*, and *Pseudomonas aeruginosa* separately and in consortium against damping-off causing *Colletotrichum truncatum*, and it was observed that population of bacteria on the seeds increased more than four times after 12 h, where germinating hyphae were observed in case of fungi, and reduced the disease incidence up to 97.2% together giving healthy seedlings (Begum et al. 2010). In case of biopriming through immersion of the maize seeds in conidial suspension of *Trichoderma harzianum* against *Fusarium*, it was observed that *Trichoderma* pure culture was more effective in controlling disease incidence, whereas talc formulation stood second when compared with no priming and fungicide applied treatments (Nayaka et al. 2010). Similarly, another experiment employing *Trichoderma*, *Pseudomonas*, and *Glomus* spp., tomato wilt was significantly decreased up to 74%, where seed germination time decreased and seed germination was enhanced (Srivastava et al. 2010). In a recent report, biopriming of wheat seeds with free-living and endophytic bacteria resulted in antagonism through contact and production of volatile compounds, thus reducing the disease incidence and also enhancing the stand establishment (Mnasri et al. 2017). Application of *Serratia* and *Paenibacillus* spp. through biopriming helped control the pathogen *Verticillium* in rapeseed and cauliflower where the authors recommended use of non-sterile soil for related experiments (Rybakova et al. 2016). In a different experiment, investigating the efficacy of biopriming agents against root rot pathogens of faba bean, the seeds were stored for long durations, and it was concluded that disease incidence was completely controlled up to 2 and 3 months of storage; however less protection was observed when the seeds were stored for 4 and 6 months (El-Mougy and Abdel-Kader 2008).

The biopriming agents successfully colonize the seed surface and inner parts in case of endophytes, which compete with pathogens for space and nutrients, thus leading to decreased incidence of the disease-causing organisms. For instance, the seed-inhabiting microbes also release certain antibiotics, organic acids, and metabolites to limit the approach of foreign organisms. Additionally, the preference showed by the plant through certain exudates also helps the applied microorganisms retain the spermosphere combined with xylem and phloem of the plant roots and stem. Briefly, the microbial antagonistic activities include ACC deaminase assisted regulation of plant ethylene levels in response to pathogenic infections (Van Loon 2007), release of siderophores and antibiotic metabolites (Beneduzi et al. 2012), production of enzymes responsible for pathogenic fungus cell lysis (Maksimov et al. 2011; Neeraja et al. 2010), and antagonism in case of space and nutrients (Kamilova et al. 2005).

9.4.2 Role of Biopriming Against Abiotic Stress

Biopriming significantly enhances the stress tolerance in the plants through various mechanisms as discussed above. Biopriming has gained more importance as a biocontrol agent due to its effectivity against pathogenic microbes; however use of biopriming agents against abiotic stress conditions can also be found. The use of biopriming against drought, osmotic, and salinity stress, as compiled in Table 9.3,

Table 9.3 Role of seed biopriming with different plant growth-promoting microorganisms in combating abiotic stress

Strain	Crop	Stress ameliorated	Crop response	Experiment	Priming duration	References
Trichoderma lixii	Maize	Salinity	Increased length, fresh and dry weights of root/shoots, relative water content, soluble protein, proline, chlorophyll, and carotenoid content	Pot	1 h	Pehlivan et al. (2017)
			Decreased lipid peroxidation, hydrogen peroxide, and lipid peroxidation			
Thalassobacillus denorans Oceanobacillus kapialis	Rice	Salinity	Increased germination, root and shoot lengths, fresh and dry weight of seedlings, chlorophyll and carotenoid contents, total nitrogen and protein contents, Ca^{+2} and K^+ ion concentration	Pot	30 min	Shah et al. (2017)
			Decreased Na^+ ion concentration			
Enterobacter spp.	Tomato	Osmotic/drought	Increased germination percentage and germination rate	Laboratory	24 h	Bhatt et al. (2015)
Citricoccus zhacaiensis	Onion	Osmotic	Increased germination, seedling vigor, and germination rate	Laboratory	24 h	Selvakumar et al. (2015)
Trichoderma harzianum	Wheat	Drought	Decreased proline, malondialdehyde, and hydrogen peroxide concentration, increased total phenolics content, and L-phenylalanine ammonia-lyase activity	Pot	24 h	Shukla et al. (2015)
Pseudomonas aeruginosa	Sunflower	Salinity *Also biocontrol against *Macrophomina phaseolina*	Increased germination, root length, shoot length, fresh weight, dry weight	Laboratory	10 min	Tewari and Arora (2014)
Trichoderma harzianum	Wheat	Salinity	Increased germination, root and shoot lengths, chlorophyll and membrane stability index, proline content, and phenolics concentration, Decreased salinity and malondialdehyde	Pot	–	Rawat et al. (2011)

– indicates no data given, h stands for hours, and min stands for minutes

has been observed in onion (Selvakumar et al. 2015), sunflower (Tewari and Arora 2014), tomato (Bhatt et al. 2015), and wheat (Rawat et al. 2011; Shukla et al. 2015).

Among abiotic stress amelioration by biopriming, *Trichoderma* has been used in controlling salinity and drought stresses, where *Trichoderma lixii* when used in maize (Pehlivan et al. 2017) and *Trichoderma harzianum* in wheat (Rawat et al. 2011) showed better physiological and morphological parameters when compared with untreated control. Use of *Trichoderma harzianum* as biopriming agent against drought stress was studied in wheat crop, where it was reported that no application of water from 4 to 13 days resulted in increased concentration of stress-related enzymes and metabolites such as phenolics and decreased concentration of hydrogen peroxide, malondialdehyde, and proline (Shukla et al. 2015). The bacterial applications have been employed in combating the salinity, osmotic, and drought stress. *Thalassobacillus denorans* and *Oceanobacillus kapialis* isolates from salt mine showing halophilic behavior enhanced the growth of fine rice variety under varying salinity concentrations and showed improvement in morphological and physiological parameters after 15 and 28 days, respectively, when applied through biopriming (Shah et al. 2017). In a laboratory experiment, tomato seeds were treated with different *Enterobacter* strains and under five different levels of osmotic stress, viz., 0, −0.2, −0.4, −0.6, −0.8, and −1.0 MPa, and it was observed that seedling vigor was enhanced in combination with germination (Bhatt et al. 2015). In another similar instance, with an exception of stress levels, i.e., 0, −0.2, −0.4, −0.6, and −0.8 MPa, *Citricoccus zhacaiensis* showing production of hormones, ACC deaminase, and ammonia and solubilization of zinc and phosphate under stress conditions, when applied in onion, enhanced the seedling growth under stress conditions (Selvakumar et al. 2015). Additionally, application of *Pseudomonas aeruginosa* having the capability of producing exopolysaccharides at a concentration as high as 2000 mM NaCl, through seed biopriming in sunflower, significantly enhanced plant growth, along with control of pathogen *Macrophomina phaseolina* under stress conditions (Tewari and Arora 2014).

Abiotic stress tolerance can be due to several mechanisms where microorganisms contribute toward the drought tolerance by production of abscisic acid and gibberellic acid (Cohen et al. 2009). Other physiological mechanisms helping the plants tolerating drought stress include modifications in root architecture, photosynthetic and photo-protective pigments, increase in proline content, and decreased stomatal conductance (Beckett et al. 2012; Piccoli and Bottini 2013). Similarly, the 1-aminocyclopropane-1-carboxylate (ACC) deaminase production by the PGPB leads to regulation of ethylene levels within the plants, helping them cope better with salinity stress. Additionally, the stimulation of the tree for exudation by the PGPM to release organic acids for chelating sodium ions in the soil (Li et al. 2007), and siderophore production can also ameliorate the effects of certain stress conditions. The PGPR also change the selectivity of certain ions like sodium, potassium, and calcium in the plants to tolerate the salinity (Hamdia et al. 2004).

9.5 Limitations and Future Prospects

Storage and short shelf life of the bioprimed seeds are a limitation of this technology. Instances of reduced population of bacteria or fungi on the seed indicate that there is need for appropriate storage technology (Callan et al. 1997; Warren and Bennett 1999), which may include environment regulation along with use of additives for long-term survival of microbes. Investigation of seed storage duration on the population of biopriming agents on the seeds concluded lesser protection after seed storage for 4–6 months (El-Mougy and Abdel-Kader 2008), so enhancing the shelf life of bioprimed seed lots needs to be explored. Secondly, the contamination from the endophytic pathogens remains a problem, although the population of beneficial endophytes applied can help limit the formers' activity, yet the seed disinfection prior to biopriming should include this aspect and should be further investigated. Similarly, the introduction of on-farm seed biopriming would help the farmers apply desired microbes when necessary, but for this, priming technique should be as easy as hydropriming which will encourage farmers to use and also will reduce the chances of contamination, desiccation, and seed damage.

9.6 Conclusion

Application of beneficial microbes with the aim of enhancing crop productivity has long been practiced, and with the time and research, new methods of application have been introduced, including biopriming. Biopriming, an amalgamation of seed priming with application of plant beneficial fungi and bacteria, can significantly improve seed germination and emergence, seedling establishment, crop growth, and yield parameters under normal and stress conditions. Its use has been investigated in diverse field crops and has shown potential benefits, with diversity of methods reported.

References

Abraha B, Yohannes G (2013) The role of seed priming in improving seedling growth of maize (*Zeamays* L.) under salt stress at field conditions. Agric Sci 4:666–672

Abuamsha R, Salman M, Ehlers RU (2010) Improvement of seed bio-priming of oilseed rape (*Brassica napus* ssp. oleifera) with *Serratia plymuthica* and *Pseudomonas chlororaphis*. Biocontrol Sci Tech 21:199–213. https://doi.org/10.1080/09583157.2010.537311

Ahemad M, Kibret M (2014) Mechanisms and applications of plant growth promoting rhizobacteria: current perspective. J King Saud Univ – Sci 26:1–20. https://doi.org/10.1016/j.jksus.2013.05.001

Ahmad I, Basra SMA, Hussain S, Hussain SA, Rehman A, Ali A (2015) Priming with ascorbic acid, salicylic acid and hydrogen peroxide improves seedling growth of spring maize at suboptimal temperature. J Environ Agric Sci 3:14–22

Ahmed M, Qadeer U, Ahmed ZI, Hassan F-u (2016) Improvement of wheat (Triticum aestivum) drought tolerance by seed priming with silicon. Arch Agron Soil Sci 62:299–315. https://doi.org/10.1080/03650340.2015.1048235

Aiyaz M, Divakara ST, Chandranayaka S, Niranjana SR (2015) Efficacy of seed hydropriming with phytoextracts on plant growth promotion and antifungal activity in maize. Int J Pest Manag 61:153–160. https://doi.org/10.1080/09670874.2015.1025116

Altomare C, Norvell W, Björkman T, Harman G (1999) Solubilization of phosphates and micronutrients by the plant-growth-promoting and biocontrol fungus *Trichoderma* harzianum Rifai 1295-22. Appl Environ Microbiol 65:2926–2933

Ananthi M, Selvaraju P, Sundaralingam K (2014) Effect of bio-priming using bio-control agents on seed germination and seedling vigour in chilli (Capsicum annuum L.) 'PKM 1'. J Hortic Sci Biotechnol 89:564–568. https://doi.org/10.1080/14620316.2014.11513121

Ashraf M, Foolad MR (2005) Pre-sowing seed treatment—a shotgun approach to improve germination, plant growth, and crop yield under saline and non-saline conditions. Adv Agron 88:223–271. https://doi.org/10.1016/S0065-2113(05)88006-X

Beckett M, Loreto F, Velikova V, Brunetti C, Di Ferdinando M, Tattini M, Calfapietra C, Farrant JM (2012) Photosynthetic limitations and volatile and non-volatile isoprenoids in the poikilochlorophyllous resurrection plant Xerophyta humilis during dehydration and rehydration. Plant Cell Environ 35:2061–2074. https://doi.org/10.1111/j.1365-3040.2012.02536.x

Begum MM, Sariah M, Puteh AB, Zainal Abidin MA, Rahman MA, Siddiqui Y (2010) Field performance of bio-primed seeds to suppress Colletotrichum truncatum causing damping-off and seedling stand of soybean. Biol Control 53:18–23. https://doi.org/10.1016/j.biocontrol.2009.12.001

Beneduzi A, Ambrosini A, Passaglia LMP (2012) Plant growth-promoting rhizobacteria (PGPR): their potential as antagonists and biocontrol agents. Genet Mol Biol 35:1044–1051

Bennett MA (1998) The use of biologicals to enhance vegetable seed quality. Seed Technology 20:198–208

Bennett M, Fritz VA, Callan NW (1992) Impact of seed treatments on crop stand establishment. Hort Technol 2:345–349

Bennett AJ, Mead A, Whipps JM (2009) Performance of carrot and onion seed primed with beneficial microorganisms in glasshouse and field trials. Biol Control 51:417–426

Bhargava B, Gupta YC, Dhiman SR, Sharma P (2015) Effect of seed priming on germination, growth and flowering of snapdragon (Antirrhinum majus L.). Natl Acad Sci Lett 38:81–85. https://doi.org/10.1007/s40009-014-0298-4

Bhatt RM, Selvakumar G, Upreti KK, Boregowda PC (2015) Effect of biopriming with enterobacter strains on seed germination and seedling growth of tomato (Solanum lycopersicum L.) under osmotic stress. Proc Natl Acad Sci, India Section B: Biol Sci 85:63–69. https://doi.org/10.1007/s40011-014-0333-8

Brocklehurst PA, Dearman J (2008) Interaction between seed priming treatments and nine seed lots of carrot, celery and onion II. Seedling emergence and plant growth. Ann Appl Biol 102:583–593

Callan NW, Mathre DE, Miller JB (1990) Bio-priming seed treatment for control of Pythium ultimum pre emergence damping-off in sh-2 sweet corn. Plant Dis 74:368–372

Callan NW, Mathre DE, Miller JB (1991) Field performance of sweet corn seed bio-primed and coated with Pseudomonas fluorescens AB254. Hortscience 26:1163–1165

Callan NW, Mathre DE, Miller IB, Vavrina CS (1997) Biological seed treat-ments, factors affecting their efficacy. Hortic Sci 32:197–183

Carrozzi LE, Creus CM, Barassi CA et al (2012) Reparation of aged lettuce (*Lactuca sativa*) seeds by osmotic priming and Azospirillum brasilense inoculation. Botany 90:1093–1102

Chakraborty AP, Dey P, Chakraborty B et al (2011) Plant growth promotion and amelioration of salinity stress in crop plants by a salt-tolerant bacterium. Recent Res Sci Technol 3:61–70

Cohen AC, Travaglia CN, Bottini R, Piccoli PN (2009) Participation of abscisic acid and gibberellins produced by endophytic Azospirillum in the alleviation of drought effects in maize. Botany 87:455–462

Compant S, Duffy B, Nowak J, Clément C, Barka EA (2005) Use of plant growth-promoting bacteria for biocontrol of plant diseases: principles, mechanisms of action, and future prospects. Appl Environ Microbiol 71:4951–4959. https://doi.org/10.1128/AEM.71.9.4951-4959.2005

Ellouzi H, Sghayar S, Abdelly C (2017) H2O2 seed priming improves tolerance to salinity; drought and their combined effect more than mannitol in Cakile maritima when compared to Eutrema salsugineum. J Plant Physiol 210:38–50. https://doi.org/10.1016/j.jplph.2016.11.014
El-Mougy NS, Abdel-Kader MM (2008) Long-term activity of bio-priming seed treatment for biological control of faba bean root rot pathogens. Australas Plant Pathol 37:464–471. https://doi.org/10.1071/ap08043
El-Mougy N, Abdel-Kader M (2008) Long-term activity of bio-priming seed treatment for biological control of faba bean root rot pathogens. Australas Plant Pathol 37:464–471
Fan B, Borriss R, Bleiss W, Wu X (2012) Gram-positive rhizobacterium *Bacillus* amyloliquefaciens FZB42 colonizes three types of plants in different patterns. J Microbiol 50:38–44. https://doi.org/10.1007/s12275-012-1439-4
Farhad W, Cheema MA, Saleem MF, Saqib M (2011) Evaluation of drought tolerance in maize hybrids. Int J Agric Biol Eng 13(4):523–528
Farooq M, Basra SMA, Hafeez K (2006) Seed invigoration by osmohardening in coarse and fine rice. Seed Sci Technol 34:181–187. https://doi.org/10.15258/sst.2006.34.1.19
Farooq M, Wahid A, Kobayashi N, Fujita D, Basra SMA (2009) Plant drought stress: effects, mechanisms and management. Agron Sustain Dev 29:185–212
Farooq M, Irfan M, Aziz T, Ahmad I, Cheema SA (2013) Seed priming with ascorbic acid improves drought resistance of wheat. J Agron Crop Sci 199:12–22
Farooq M et al (2017) Effects, tolerance mechanisms and management of salt stress in grain legumes. Plant Physiol Biochem. https://doi.org/10.1016/j.plaphy.2017.06.020
Firuzsalari SM, Mirshekari B, Khochebagh SB (2012) Effect of seed inoculation with bio-fertilizer on germination and early growth of corn. Int Res J App Basic Sci 3:1097–1102
Galhaut L, de Lespinay A, Walker DJ, Bernal MP, Correal E, Lutts S (2014) Seed priming of Trifolium repens L. improved germination and early seedling growth on heavy metal-contaminated soil. Water Air Soil Pollut 225:1905
Ghezal N et al (2016) Improvement of Pisum sativum salt stress tolerance by bio-priming their seeds using Typha angustifolia leaves aqueous extract. S Afr J Bot 105:240–250. https://doi.org/10.1016/j.sajb.2016.04.006
Gholami A, Shahsavani S, Nezarat S (2009) The effect of plant growth promoting rhizobacteria (PGPR) on germination, seedling growth and yield of maize. Proc World Acad Sci Eng Technol 49:19–24
Glick BR (2012) Plant growth-promoting bacteria: mechanisms and applications. Australas Sci 2012:15. https://doi.org/10.6064/2012/963401
Goswami D, Thakker JN, Dhandhukia PC, Tejada Moral M (2016) Portraying mechanics of plant growth promoting rhizobacteria (PGPR): a review. Cogent Food Agric 2:1127500. https://doi.org/10.1080/23311932.2015.1127500
Grassbaugh EM, Bennett MA (1998) Factors affecting vegetable stand establishment. Sci Agric 55:116–120
Gururani MA, Upadhyaya CP, Baskar V et al (2012) Plant growthpromoting rhizobacteria enhance abiotic stress tolerance in Solanum tuberosum through inducing changes in the expression of ROS-scavenging enzymes and improved photosynthetic performance. J Plant Growth Regul 32:245–258
Hamdia MAE-S, Shaddad MAK, Doaa MM (2004) Mechanisms of salt tolerance and interactive effects of Azospirillum brasilense inoculation on maize cultivars grown under salt stress conditions. Plant Growth Regul 44:165–174. https://doi.org/10.1023/B:GROW.0000049414.03099.9b
Harish S, Kavino M, Kumar N, Saravanakumar D, Soorianathasundaram K, Samiyappan R (2008) Biohardening with plant growth promoting rhizosphere and endophytic bacteria induces systemic resistance against banana bunchy top virus. Appl Soil Ecol 39:187–200. https://doi.org/10.1016/j.apsoil.2007.12.006
Harman GE, Taylor AG (1988) Improved seedling performance by integration of biological control agents at favorable pH levels with solid matrix priming. Phytopathology 78:520–525
Hayat R, Ali S, Amara U, Khalid R, Ahmed I (2010) Soil beneficial bacteria and their role in plant growth promotion: a review. Ann Microbiol 60:579–598. https://doi.org/10.1007/s13213-010-0117-1

Hossain MA et al (2015) Hydrogen peroxide priming modulates abiotic oxidative stress tolerance: insights from ROS detoxification and scavenging. Front Plant Sci 6:420. https://doi.org/10.3389/fpls.2015.00420
Hossain MM, Sultana F, Hyakumachi M (2017) Role of ethylene signalling in growth and systemic resistance induction by the plant growth-promoting fungus Penicillium viridicatum in Arabidopsis. J Phytopathol: 1–10. https://doi.org/10.1111/jph.12577
Hussain S, Khan F, Hussain HA, Nie L (2016) Physiological and biochemical mechanisms of seed priming-induced chilling tolerance in rice cultivars. Front Plant Sci 7:116. https://doi.org/10.3389/fpls.2016.00116
Jensen B, Povlsen FV, Knudsen I, Jensen SDF (2001) Combining microbial seed treatment with priming of carrot seeds for control of seed borne Alternaria spp. In: Biocontrol agents: mode of action and interaction with other means of control: proceedings of the phytopathogens WG meeting, 2001, pp 197–201
Jisha KC, Puthur JT (2016) Seed priming with BABA (β-amino butyric acid): a cost-effective method of abiotic stress tolerance in Vigna radiata (L.). Wilczek Protoplasma 253:277–289. https://doi.org/10.1007/s00709-015-0804-7
Jisha KC, Vijayakumari K, Puthur JT (2013) Seed priming for abiotic stress tolerance: an overview. Acta Physiol Plant 35:1381–1396. https://doi.org/10.1007/s11738-012-1186-5
Kamilova F, Validov S, Azarova T, Mulders I, Lugtenberg B (2005) Enrichment for enhanced competitive plant root tip colonizers selects for a new class of biocontrol bacteria. Environ Microbiol 7:1809–1817
Kasim WA, Osman ME, Omar MN, Abd El-Daim IA, Bejai S, Meijer J (2013) Control of drought stress in wheat using plant-growth-promoting bacteria. J Plant Growth Regul 32:122–130. https://doi.org/10.1007/s00344-012-9283-7
Kaya MD, Okcu G, Atak M, Cıkıhı Y, Kolsarıcı O (2006) Seed treatments to overcome salt and drought stress during germination in sunflower (*Helianthus annuus* L.). Eur J Agron 24(4):291–295
Khan AA (1992) Preplant physiological seed conditioning. Hort Reviews 14:131–181
Kubik KK (1995) Seed priming and survival of desiccated Enterobacter cloacae inoculated during seed priming. PhD thesis, University of Nebraska, Lincoln
Kumar VV (2016) Plant growth-promoting microorganisms: interaction with plants and soil. In: Hakeem KR, Akhtar MS, Abdullah SNA (eds) Plant, soil and microbes: volume 1: implications in crop science. Springer, Cham, pp 1–16. https://doi.org/10.1007/978-3-319-27455-3_1
Lee SS, Kim JH (2000) Total sugars, a-amylase activity, and emergence after priming of normal and aged rice seeds. Korean J Crop Sci 45:108–111
Lewis JA, Papavizas GC, Connick WJ Jr (1987) Preparation of pellets containing fungi and nutrient for control of soilborne plant pathogens. Google Patents
Li WC, Ye ZH, Wong MH (2007) Effects of bacteria on enhanced metal uptake of the Cd/Zn-hyperaccumulating plant, Sedum alfredii. J Exp Bot 58:4173–4182
López-Coria M, Hernández-Mendoza J, Sánchez-Nieto S (2016) Trichoderma asperellum induces maize seedling growth by activating the plasma membrane H+-ATPase. Mol Plant-Microbe Interact 29:797–806
Mahmood A, Turgay OC, Farooq M, Hayat R (2016) Seed biopriming with plant growth promoting rhizobacteria: a review. FEMS Microbiol Ecol 92:1–13. https://doi.org/10.1093/femsec/fiw112
Maksimov IV, Abizgil'dina RR, Pusenkova LI (2011) Plant growth promoting rhizobacteria as alternative to chemical crop protectors from pathogens (Review). Appl Biochem Microbiol 47:333–345
Mallik MAB, Williams RD (2008) Plant growth promoting rhizobacteria and mycorrhizal fungi in sustainable agriculture and forestry. In: Zeng RS, Mallik AU, Luo SM (eds) Allelopathy in sustainable agriculture and forestry. Springer, New York, pp 321–345. https://doi.org/10.1007/978-0-387-77337-7_17
McDonald MB (1999) Seed deterioration: physiology, repair and assessment. Seed Sci Technol 27:177–237

Meena VS, Maurya BR, Verma JP, Aeron A, Kumar A, Kim K, Bajpai VK (2015) Potassium solubilizing rhizobacteria (KSR): isolation, identification, and K-release dynamics from waste mica. Ecol Eng 81:340–347. https://doi.org/10.1016/j.ecoleng.2015.04.065

Meena SK, Rakshit A, Meena VS (2016) Effect of seed bio-priming and N doses under varied soil type on nitrogen use efficiency (NUE) of wheat (Triticum aestivum L.) under greenhouse conditions. Biocatal Agric Biotechnol 6:68–75. https://doi.org/10.1016/j.bcab.2016.02.010

Miche L, Balandreau J (2001) Effects of rice seed surface sterilization with hypochlorite on inoculated Burkholderia vietnamiensis. Appl Environ Microbiol 67:3046–3052

Mirshekari B, Hokmalipour S, Sharifi RS, Farahvash F, Gadim A (2012) Effect of seed biopriming with plant growth promoting rhizobacteria (PGPR) on yield and dry matter accumulation of spring barley (Hordeum vulgare L.) at various levels of nitrogen and phosphorus fertilizers. J Food Agric Environ 10:314–320

Mnasri N, Chennaoui C, Gargouri S, Mhamdi R, Hessini K, Elkahoui S, Djébali N (2017) Efficacy of some rhizospheric and endophytic bacteria in vitro and as seed coating for the control of Fusarium culmorum infecting durum wheat in Tunisia. Eur J Plant Pathol 147:501–515. https://doi.org/10.1007/s10658-016-1018-3

Moeinzadeh A, Sharif-Zadeh F, Ahmadzadeh M et al (2010) Biopriming of sunflower (*Helianthus annuus* L.) seed with Pseudomonas fluorescens for improvement of seed invigoration and seedling growth. Aust J Crop Sci 4:564

Moulick D, Ghosh D, Chandra Santra S (2016) Evaluation of effectiveness of seed priming with selenium in rice during germination under arsenic stress. Plant Physiol Biochem 109:571–578. https://doi.org/10.1016/j.plaphy.2016.11.004

Nakkeeran S, Fernando WD, Siddiqui ZA (2005) Plant growth promoting rhizobacteria formulations and its scope in commercialization for the management of pests and diseases. In: PGPR: biocontrol and biofertilization. Springer, Dordrecht, pp 257–296

Nandakumar R, Babu S, Viswanathan R, Sheela J, Raguchander T, Samiyappan R (2001) A new bio-formulation containing plant growth promoting rhizobacterial mixture for the management of sheath blight and enhanced grain yield in rice. BioControl 46:493–510. https://doi.org/10.1023/a:1014131131808

Nayaka S, Niranjana S, Uday Shankar A, Niranjan Raj S, Reddy M, Prakash H, Mortensen C (2010) Seed biopriming with novel strain of Trichoderma harzianum for the control of toxigenic Fusarium verticillioides and fumonisins in maize. Arch Phytopathol Plant Protect 43:264–282

Neeraja C, Anil K, Purushotham P, Suma K, Sarma P, Moerschbacher BM, Podile AR (2010) Biotechnological approaches to develop bacterial chitinases as a bioshield against fungal diseases of plants. Crit Rev Biotechnol 30:231–241

Ongena M, Jacques P (2008) *Bacillus* lipopeptides: versatile weapons for plant disease biocontrol. Trends Microbiol 16:115–125. https://doi.org/10.1016/j.tim.2007.12.009

Orzolek MD (1991) Establishment of vegetables in the field. HortTechnology 1:78–81

Panigrahi S, Aruna Lakshmi K, Mir N, Taiyebi M (2016) Biopriming of micropropagated plantlets using various bacterial strains in different combinations-induces the systemic resistance. In: Avadhanam S, Jyothsna G, Kashyap A (eds) Next generation DNA led technologies. Springer, Singapore, pp 113–122. https://doi.org/10.1007/978-981-287-670-6_13

Parera CA, Cantliffe DJ (1994) Presowing seed priming. Hortic Rev 16:109–141

Parera CA, Qiao P, Cantliffe DJ (1993) Enhanced celery germination at stress temperature via solid matrix priming. Hortscience 28:20–22

Pehlivan N, Yesilyurt AM, Durmus N, Karaoglu SA (2017) Trichoderma lixii ID11D seed biopriming mitigates dose dependent salt toxicity in maize. Acta Physiol Plant 39:79. https://doi.org/10.1007/s11738-017-2375-z

Piccoli P, Bottini R (2013) Abiotic stress tolerance induced by endophytic PGPR. In: Aroca R (ed) Symbiotic endophytes. Springer, Berlin/Heidelberg, pp 151–163. https://doi.org/10.1007/978-3-642-39317-4_8

Pill WG (1995) Low water potential and presowing germination treatments to improve seed quality. In: Basra AS (ed) Seed quality, basic mechanisms and agricultural implications. Haworth Press, New York

Posmyk MM, Bałabusta M, Janas KM (2009) Melatonin applied by osmopriming, as phytobiostimulator improving cucumber (*Cucumissativus* L.) seedlings growth at abiotic stresses conditions. In: Li S, Wang Y, Cao F, Huang P, Zhang Y (eds) Progress in environmental science and technology vol II A. Science Press, Beijing, pp 362–369

Prasad SR, Kamble UR, Sripathy KV, Udaya Bhaskar K, Singh DP (2016) Seed bio-priming for biotic and abiotic stress management. In: Singh DP, Singh HB, Prabha R (eds) Microbial inoculants in sustainable agricultural productivity: vol. 1: research perspectives. Springer, New Delhi, pp 211–228. https://doi.org/10.1007/978-81-322-2647-5_12

Rawat L, Singh Y, Shukla N, Kumar J (2011) Alleviation of the adverse effects of salinity stress in wheat (*Triticum aestivum* L.) by seed biopriming with salinity tolerant isolates of *Trichoderma harzianum*. Plant Soil:347–387. https://doi.org/10.1007/s11104-011-0858-z

Reddy PP (2013) Recent advances in crop protection. Springer, New Delhi, p 83

Rouhollah A (2013) Drought stress tolerance of barley (*Hordeum vulgare L.*) affected bypriming with PEG. Int J Farming and Allied Sci 2(20):803–808

Rowse H (1996) Drum priming: a non-osmotic method of priming seeds. Seed Sci Technol 24:281–294

Rybakova D, Schmuck M, Wetzlinger U, Varo-Suarez A, Murgu O, Müller H, Berg G (2016) Kill or cure? The interaction between endophytic *Paenibacillus* and *Serratia* strains and the host plant is shaped by plant growth conditions. Plant Soil 405:65–79. https://doi.org/10.1007/s11104-015-2572-8

Saber Z, Pirdashti H, Esmaeili M et al (2012) Response of wheat growth parameters to co-inoculation of plant growth promoting rhizobacteria (PGPR) and different levels of inorganic nitrogen and phosphorus. World Appl Sci J 16:213–219

Saglam S, Day S, Kaya G, Gürbüz A (2010) Hydropriming increases germination oflentil (*Lens culinaris* Medik.) under water stress. Notulae Scientia Biologicae 2:103–106

Santoyo G, Moreno-Hagelsieb G, del Carmen Orozco-Mosqueda M, Glick BR (2016) Plant growth-promoting bacterial endophytes. Microbiol Res 183:92–99. https://doi.org/10.1016/j.micres.2015.11.008

Selvakumar G, Bhatt RM, Upreti KK, Bindu GH, Shweta K (2015) Citricoccus zhacaiensis B-4 (MTCC 12119) a novel osmotolerant plant growth promoting actinobacterium enhances onion (Allium cepa L.) seed germination under osmotic stress conditions. World J Microbiol Biotechnol 31:833–839. https://doi.org/10.1007/s11274-015-1837-y

Seufert V, Ramankutty N, Foley JA (2012) Comparing the yields of organic and conventional agriculture. Nature 485:229–232. https://doi.org/10.1038/nature11069

Shah G et al (2017) Halophilic bacteria mediated phytoremediation of salt-affected soils cultivated with rice. J Geochem Explor 174:59–65. https://doi.org/10.1016/j.gexplo.2016.03.011

Shariatmadari Z, Riahi H, Seyed Hashtroudi M, Ghassempour A, Aghashariatmadary Z (2013) Plant growth promoting cyanobacteria and their distribution in terrestrial habitats of Iran. Soil Sci Plant Nutr 59:535–547. https://doi.org/10.1080/00380768.2013.782253

Sharifi RS (2011) Study of grain yield and some of physiological growth indices in maize (Zea mays L.) hybrids under seed biopriming with plant growth promoting rhizobacteria (PGPR). J Food Agric Environ 189:3–4

Sharifi RS (2012) Study of nitrogen rates effects and seed biopriming with PGPR on quantitative and qualitative yield of Safflower (*Carthamus tinctorius* L.). Tech J Eng Appl Sci 2:162–166

Sharifi RS, Khavazi K (2011) Effects of seed priming with plant growth promotion rhizobacteria (PGRP) on yield and yield attribute of maize (*Zeamays* L.) hybrids. J Food Agric Environ 9:496–500

Sharifi RS, Khavazi K, Gholipouri A (2011) Effect of seed priming with plant growth promoting Rhizobacteria (PGPR) on dry matter accumulation and yield of maize (Zea mays L.) hybrids. Int Res J Biochem Bioinf1:76–83

Shehab GG, Ahmed OK, El-Beltagi HS (2010) Effects of various chemical agents for alleviation of drought stress in rice plants (*Oryza sativa* L.). Notulae Botanicae Horti Agrobotanici Cluj-Napoca 38(1):139–148

Shukla N, Awasthi RP, Rawat L, Kumar J (2015) Seed biopriming with drought tolerant isolates of *Trichoderma* harzianum promote growth and drought tolerance in Triticum aestivum. Ann Appl Biol 166:171–182. https://doi.org/10.1111/aab.12160

Singh US, Zaidi NW, Joshi D, Varshney S, Khan T (2003) Current status of Trichoderma as biocontrol agent. In: Ramanujam B, Rabindra RJ (eds) Current status of biological control of plant diseases using antagonistic organisms in India. Project Directorate of Biological Control, Bangalore, pp 13–48

Singh V, Upadhyay R, Sarma B, Singh H (2016a) Seed bio-priming with Trichoderma asperellum effectively modulate plant growth promotion in pea. Int J Agric Environ Biotechnol 9:361

Singh V, Upadhyay RS, Sarma BK, Singh HB (2016b) Trichoderma asperellum spore dose depended modulation of plant growth in vegetable crops. Microbiol Res 193:74–86. https://doi.org/10.1016/j.micres.2016.09.002

Souza R, Ambrosini A, Passaglia LMP (2015) Plant growth-promoting bacteria as inoculants in agricultural soils. Genet Mol Biol 38:401–419

Srivastava R, Khalid A, Singh US, Sharma AK (2010) Evaluation of arbuscular mycorrhizal fungus, fluorescent Pseudomonas and Trichoderma harzianum formulation against Fusarium oxysporum f. sp. lycopersici for the management of tomato wilt. Biol Control 53:24–31. https://doi.org/10.1016/j.biocontrol.2009.11.012

Sun YY, Sun YJ, Wang MT, Li XY, Guo X, Hu R, Jun MA (2010) Effects of seed priming on germination and seedling growth under water stress in rice. Acta Agron Sin 36(11):1931–1940

Tabatabaei SA (2013) Effect of osmo-priming on germination and enzyme activity in barley (*Hordeum vulgare* L.) seeds under drought stress conditions. J Stress Physiol Biochem 9(4):25–31

Taylor AG, Harman GE (1990) Concept and technologies of selected seed treatments. Annu Rev Phytopathol 28:321–339

Tewari S, Arora NK (2014) Multifunctional exopolysaccharides from pseudomonas aeruginosa PF23 involved in plant growth stimulation, biocontrol and stress amelioration in sunflower under saline conditions. Curr Microbiol 69:484–494. https://doi.org/10.1007/s00284-014-0612-x

Van Loon LC (2007) Plant responses to plant growth-promoting rhizobacteria. Eur J Plant Pathol 119:243–254

Vidhyasekaran P, Muthamilan M (1995) Development of formulations of *Pseudomonas fluorescens* for control of chickpea wilt. Plant Dis 79:782–786

Warren JE, Bennett MA (1997) Seed hydration using the drum priming system. Hortscience 32:1220–1221

Warren JE, Bennett MA (1999) Bio-osmopriming tomato (*Lycopersicon esculentum* Mill.) seeds for improved stand establishment. Seed Sci Technol 27:489–499

Weller DM (2007) *Pseudomonas* biocontrol agents of soilborne pathogens: looking back over 30 years. Phytopathology 97:250–256. https://doi.org/10.1094/PHYTO-97-2-0250

Wright B, Rowse H, Whipps J (2003a) Application of beneficial microorganisms to seeds during drum priming. Biocontrol Sci Tech 13:599–614

Wright B, Rowse H, Whipps JM (2003b) Microbial population dynamics on seeds during drum and steeping priming. Plant Soil 255:631–640. https://doi.org/10.1023/a:1026055112679

Xu S, Hu J, Li Y, Ma W, Zheng Y, Zhu S (2011) Chilling tolerance in Nicotiana tabacum induced by seed priming with putrescine. Plant Growth Regul 63:279–290. https://doi.org/10.1007/s10725-010-9528-z

Yadav PV, Kumari M, Ahmed Z (2011) Seed priming mediated germination improvement and tolerance to subsequent exposure to cold and salt stress in capsicum. Res J Seed Sci 4:125–136

Yan M (2015) Seed priming stimulate germination and early seedling growth of Chinese cabbage under drought stress. S Afr J Bot 99:88–92. https://doi.org/10.1016/j.sajb.2015.03.195

Stimulating Plant Tolerance Against Abiotic Stress Through Seed Priming

10

Mona Gergis Dawood

Abstract

The seed priming process is a physiological method that involves seed hydration (usually within 10–20% of full imbibition) and effective enough for enhancement of seed germination, early seedling growth, and yield under stressed and non-stressed conditions but insufficient to allow radicle protrusion. Primed seeds germinate faster and more uniformly than the non-primed ones. Seed priming is influenced by many factors such as aeration, light, temperature, time, and seed quality and induced a set of biochemical changes in the seed which are required for initiating the germination process. These changes include activation of enzymes, breaking dormancy, metabolism of germination inhibitors, and imbibition. The positive effects of priming on the germination performance of many species are attributed to the induction of biochemical mechanisms of cell repair: the resumption of metabolic activity can restore cellular integrity, through the synthesis of nucleic acids (DNA and RNA) and proteins and the improvement of the antioxidant defense system. Several methods of seed priming were successfully used in agriculture for seed conditioning to accelerate the germination rate and improve the seedling uniformity such as seed priming with water (hydropriming), plant growth regulators, beta-aminobutyric acid, 5-aminolevulinic acid, osmoprotectant, melatonin, chitosan, plant extract, polyethylene glycol, and inorganic salts. It is worthy to mention that all these methods showed pronounced effect on germination, seedling growth, and yield of different crops under normal or stress conditions.

Keywords

Seed soaking · Plant resistance · Abiotic stress · Germination quality

M. G. Dawood (✉)
Botany Department, National Research Centre, Cairo, Egypt

© Springer Nature Singapore Pte Ltd. 2018

A. Rakshit, H. B. Singh (eds.), *Advances in Seed Priming*,
https://doi.org/10.1007/978-981-13-0032-5_10

10.1 Introduction

In nature, plants often face the challenge of severe environmental conditions, which include various biotic and abiotic stresses that exert adverse effects on plant growth and development causing considerable losses in the crop productivity. Plants are subjected to a variety of abiotic stress such as salinity, drought, high temperature, low temperature, etc. that reduced germination rate and seedling growth with significant variations from crop to crop (Hamidi and Safarnejad 2010). Abiotic stress causes many physiological and biochemical changes in the seedlings, which include the generation of reactive oxygen species (ROS), leading to membrane damage and cell leakage and destruction of photosynthetic components (Jisha and Puthur 2014). ROS are known to exacerbate imbalance between light absorption and its utilization by inhibiting Calvin-Benson cycle activity (Logan et al. 2006). ROS also reduce content and activity of ribulose bisphosphate carboxylase oxygenase (RUBISCO) enzyme which leads to higher electron flux to O_2 coupled with reduced CO_2 accumulation (Zhou et al. 2006; Ahmad et al. 2015), and they can be extremely reactive with several cellular constituents such as proteins, lipids, and nucleic acids (Hasanuzzaman et al. 2013), which in turn results in negative effects on metabolism and cellular structures (França et al. 2007; Mouradi et al. 2016). Free radical oxidations and protein enzymatic dehydrogenation and aldehyde oxidation might contribute to seed quality reduction (Ghassemi-Golezani et al. 2010).

In many plants, germination, seedling growth, and subsequent crop yield can be inhibited by salinity stress (Soccio et al. 2010; Carvalho et al. 2011). *Salinity* has an adverse effect on seed germination and seedling growth of several crops either by creating an osmotic potential in the rhizosphere of the plant that inhibits the absorption of water or creates toxic effect to the roots and whole crop because of Na^+ and Cl^- (Khajeh-Hosseini et al. 2003; Munns and Tester 2008). Moreover, salinity has a negative effect on plasma membrane by affecting on its permeability which in turn modulates the pattern of ion leakage (Sairam et al. 2002). Salt tolerance of plants can be increased by treatment of seeds with NaCl solution prior to sowing (Sivritepe et al. 2003). Farhoudi et al. (2007) reported that priming of canola seed decreased the absorption of harmful ion and cell damage of canola seedling under salinity stress.

Drought is one of the most important environmental factors limiting plant growth and productivity. Soltani et al. (2008) mentioned that as the drought severity increased, the germination rate linearly decreased in unprimed cotton seeds, but primed seeds had lower reduction slope. Moreover, Khatami et al. (2015) stated that some priming treatments were sufficient to invigorate the corn seed germination under drought stress, since hormone priming under moderate drought and osmopriming under severe drought were the best methods for seed improvement.

Low-temperature conditions decreased plant growth rate because of inhibition of photosynthesis and increasing photooxidative injury of the photosystems. Photooxidative damage caused lipid peroxidation and degradation of chlorophyll and carotene. Plants exert many physiological and biochemical changes under low-temperature conditions that make them survive under these conditions (Xin and Browse 2000). In general, physiological and biochemical dysfunctions induced by low temperature in plants are not equal and most of them can be convert if plant return to optimum condition before appearance of damage, since disturbances of physiological and biochemical functions are reversible.

Heat stress is often defined as the rise in temperature beyond a threshold level for a period of time sufficient to cause irreversible damage to plant growth and development. In general, a transient elevation in temperature, usually 10–15 °C above ambient, is considered heat shock or heat stress. However, heat stress is a complex function of intensity (temperature in degrees), duration, and rate of increase in temperature. The extent to which it occurs in specific climatic zones depends on the probability and period of high temperatures occurring during the day and/or the night (Wahid et al. 2007a).

The time from sowing to plant establishment is a crucial period in crop growth (Bray 1995), with a direct impact on final yield and quality (Gupta et al. 2008) especially under abiotic stress. So, several ways for enhancing the plant tolerance toward abiotic stress have been experimented like breeding of plants and developing transgenics (Jisha et al. 2013). Moreover, some studies on different plants showed that seed priming treatments have effectively increased germination and seedling growth either at normal conditions or stressed conditions (Bradford 1986). Since, efficient seed germination, rapid and uniform seedlings emergence lead to successful culture establishment (Chen and Arora 2011).

10.2 Seed Priming

Seed priming is a simple, safe, economic, and effective approach for enhancement of seed germination, early seedling growth, and yield under stressed and non-stressed conditions (Sedghi et al. 2010). Seed priming is a form of seed preparation in which seeds are presoaked before planting with a certain solution that allows partial hydration but not germination and redried to original moisture content (Ahmad et al. 2012). The seed priming process is a physiological method that involves seed hydration (usually within 10–20% of full imbibition) (Pill 1995), sufficient to permit pre-germinative metabolic events to proceed, but insufficient to allow radicle protrusion (Bradford 1986). Primed seeds germinate faster and more uniformly than the non-primed ones.

During priming, the germination process is not completed, but metabolic activities for radical protrusion may be initiated (Heydecker et al. 1973). At the cellular level, few processes have been described to act during priming, some of these being activation of cell cycle (De Castro et al. 2000) and mobilization of storage proteins (Gallardo et al. 2001). The priming process induces the rate of seed germination and is associated with the initiation of germination-related processes (Soeda et al. 2005) and repair processes (Sivritepe and Dourado 1995) and increases various free radical-scavenging enzymes, such as superoxide dismutase, catalase, and peroxidase (Gallardo et al. 2001).

Several seed priming methods were successfully used in agriculture for seed conditioning to accelerate the germination rate and improve the seedling uniformity (Nouman et al. 2012; Aghbolaghi and Sedghi 2014; Bagheri 2014; Lara et al. 2014). Moreover, seed priming helps many crops to neutralize the adverse effects of abiotic stress (Ashraf and Foolad 2005; Zhang et al. 2012; Hameed et al. 2013; Jisha et al. 2013; Jisha and Puthur 2016).

10.2.1 There are Three Principal Methods of Seed Priming

1. *Hydropriming (water-only treatments)*: soaking of seeds with water overnight and then drying before sowing markedly improve plant stand, establishment, vigor, and the final yield (Harris et al. 1999). Seeds are submerged in water with or without aeration. Therefore, water is freely available to seeds, its uptake only being governed by the affinity of the seed tissue for water (Taylor et al. 1998). As a main drawback, seed germination can proceed until radicle protrusion. Thus, the process needs to be stopped at a precise moment, before phase III begins. Another disadvantage is that seeds are not equally hydrated, which results in a nonuniform activation of the physiological processes necessary to synchronize and improve germination (McDonald 2000).
2. *Solid matrix priming or matri-priming*: hydrated solid matrices soaked with osmotic solutions (McDonald 2000), such as hydrated sand (Hu et al. 2005), peat, and vermiculite (Taylor et al. 1998), or cotton soaked with osmotic solutions; involves the use of a wet organic or inorganic material (Parera and Cantliffe 1994), which simulates the natural imbibition processes taking place in the soil (McDonald 2000). The substrate must possess the given characteristics: low matric potential, high seed safety, high specific surface (i.e., high surface to volume ratio), negligible water solubility, high adhesiveness to seed surface, and high capacity to retain water (Khan 1991).
3. *Osmotic priming (osmopriming)*: osmopriming is the process that involves the use of osmotic solutions with a low water potential to control seed water uptake. The most common substances used for osmopriming are polyethylene glycol (PEG), inorganic salts, mannitol, glycerol, and plant hormones (Taylor et al. 1998; Foti et al. 2002; Tiryaki and Buyukcingil 2009; Afzal et al. 2013; Farooq et al. 2015).

The three methods may be grouped into two categories: non-controlled water uptake (hydropriming) and controlled water uptake (osmopriming and solid matrix priming) (Taylor et al. 1998).
Bio-priming: Additional method includes seed coating with bacteria (bio-priming or bio-osmopriming), e.g., *Trichoderma* spp. (Pill et al. 2009; Begum et al. 2010) and *Pseudomonas aureofaciens* (Warren and Bennett 1999). Bio-priming has been able to control damping-off of seedlings in sweet corn (*Zea mays* L.) (Callan et al. 1990),

cucumber (*Cucumis melo* L.) (Pill et al. 2009), pea (*Pisum sativum* L.), and soybean (*Glycine max* (L.) Merr.) (Taylor et al. 1994).

10.2.2 Factors Affecting Seed Priming Process

Seed priming is influenced by many factors such as aeration, light, temperature, time, and seed quality.

1. *Aeration* is considered an important step to assist seed respiration (Bujalski and Nienow 1991) and seed viability and contributes to synchronize the emergence (Heydecker et al. 1975) and ensures a safer seed habitat. However, the effect of aeration varies according to species: in onion, aeration of the PEG solution increased the germination percentage, compared to non-aerated treatment (Heydecker and Coolbear 1977; Bujalski et al. 1989). By contrast, no difference was observed in the germination of lettuce between aerated and non-aerated K_3PO_4 priming (Cantliffe 1981).
2. *Light* effect is widely varied according to species. Khan et al. (1978) mentioned that illumination during priming of celery seeds may reduce dormancy. On the other hand, the best results with lettuce were obtained with priming in the dark (Cantliffe et al. 1981).
3. *Temperature* is another important variable, as it affects the speed of chemical reactions. Treatment temperature varied between 15 (several cases) and 30 °C (rice in hydropriming, Basra et al. 2005). Temperatures of about 15 °C during priming were shown to improve the overall seed performance in most species (Bradford 1986), whereas lower temperatures slowed the germination processes, requiring longer times to achieve the same results (McDonald 2000). The range of temperatures normally used in priming varies between 15 and 20 °C.
4. *Treatment duration* is widely varied according to species and experiment, from a minimum of 8 h (sunflower primed in a salt solution, Wahid et al. 2008) to a maximum of 14 days (four ornamental species, Finch-Savage et al. 1991). Treatment duration mainly depends on the type of osmotic solution, osmotic potential, temperature, and crop species. It also depends on the specific time and likelihood of radicle protrusion: long priming can more easily lead to this occurrence, creating irreversible damage during drying-back (Parera and Cantliffe 1994).
5. *Seed quality* is a key aspect influencing the effects of priming. A vigorous seed, free from pathogens is an essential requisite for a good priming result (Cantliffe et al. 1987), in contrast to the belief that this technique may improve the performance of seeds of intrinsic modest quality. Other seed characteristics may influence priming process. For instance, osmopriming with PEG solution is not suitable for seed treatment of sorghum with high content of tannin, because tannins can be removed with the solution treatment and determine a reduction of germination (Patanè et al. 2008). In fact, tannins reduce seed susceptibility to insects, birds, and mold diseases and protect from weathering (Beta et al. 1999).

In this specific case, it is advisable to adopt treatment solutions different from PEG or others technique, such as bio-priming (Patanè et al. 2008). Moreover, Bradford (1986) mentioned that composition of solution and osmotic potential had pronounced effect on seed priming.

10.3 Triphasic Model of Seed Imbibition

In orthodox seeds the dry seed, ready for germination, exhibits a triphasic pattern of water uptake (Bewley and Black 1978). *Phase I* is the rapid water uptake that is largely a consequence of the matric forces exerted by the seed. During this phase, DNA and mitochondria are repaired and proteins are synthesized using existing messenger ribonucleic acid (mRNA) (McDonald 2000). *Phase II* is a lag phase, in which seed water potential is in balance with that of the environment. In this phase the major metabolic changes preparing the embryo for germination occur, including the synthesis of mitochondria and proteins by new mRNA. Thus, phase II is also called activation phase. *Phase III* is a second rapid uptake of water occurs and the radicle emerges so called visible germination (Bewley and Black 1978; Bradford 1995). Phases I and II represent the most delicate phases for the process of germination and are crucial for a successful seed priming (Bewley 1997). The triphasic model has deep implications for seed viability. The seed tolerates a return to the initial moisture necessary for storage, a process known as drying-back or redrying, when it is in phase I or II, whereas phase III is too advanced to allow a drying-back without seed damage (Taylor et al. 1998). According to the triphasic model, the start of germination is associated with a rapid synthesis of RNA and proteins, to carry out the repairing processes before the beginning of DNA replication (Osborne 1983). Seed priming typically involves an extension of phase II, which in turn permits the completion of more repair processes (Bray 1995), and allows the drying-back, which is necessary when the final sowing is postponed (industrial seed production). The postponement of phase III involved in priming plus redrying results in a better seed performance under favorable conditions.

10.4 Biochemical Changes Induced by Priming

Seed priming induced a set of biochemical changes in the seed which are required for initiating the germination process. These changes include activation of enzymes, breaking dormancy, metabolism of germination inhibitors, and imbibition to start the germination process (Ajouri et al. 2004; Farooq et al. 2010). Some or all of these changes that precede germination are triggered by seed priming and persist following the redrying of seeds (Asgedom and Becker 2001). Thus, upon sowing, primed seed can rapidly imbibe and revive the seed metabolism, resulting in higher germination percentage (Rowse 1995). Moreover, seed priming may repair some damage in the membrane caused by deterioration and result in better germination pattern and higher vigor level compared with non-primed seeds (Ruan and Xue 2002).

Priming showed stimulatory effects in the early stages of germination by mediation of cell division in germinating seeds (Hassanpouraghdam et al. 2009). The increase in speed of germination and germination percentage may be due to the modification of physiological and biochemical nature of seed embryo and its associated structures, i.e., pre-enlargement of the embryo (Austin et al. 1969), and biochemical changes like enzyme activation; gibberellin-like substances may be released during phase II of germination which triggers the synthesis of hydrolytic enzymes that causes the early availability of high-energy compounds (Basra et al. 2005) and vital biomolecules to the germinating seedling (Renugadevi and Vijayageetha 2006).

The positive effects of priming on the germination performance of many species are attributed to the induction of biochemical mechanisms of cell repair: the resumption of metabolic activity can restore cellular integrity, through the synthesis of nucleic acids (DNA and RNA) and proteins and the improvement of the antioxidant defense system (Bewley and Black 1994; Di Girolamo and Barbanti 2012).

10.4.1 Effects on DNA

A strong increase in DNA synthesis occurs at the end of germination (phase III) in both primed and unprimed wheat grains, as shown by Dell'Aquila and Taranto (1986). Thereafter, an increase in DNA was observed 14 days after the seed priming, when the seed had entered the irreversible germination phase (Bray 1995). The positive effects of priming on DNA are offered by a study on *Brassica oleracea* L., where an aerated hydration determined an advance in DNA synthesis (Thornton et al. 1993). In spite of its crucial role, the amount of DNA which is needed in the repair processes is only 20–30% of the total DNA synthesized during priming. The rest is mainly represented by mitochondrial DNA; in fact, the number of mitochondria was shown to rapidly increase during priming in leek seeds (Ashraf and Bray 1993). The enhancement of DNA replication during priming depends on species, cultivar, seed lot quality (Lantieri et al. 1994), and treatment conditions (Ozbingol et al. 1999).

10.4.2 Effects on RNA

Priming allows the recovery of rRNA integrity (Coolbear et al. 1990), in turn ensuring a correct coding of amino acids for the synthesis of proteins during seed germination. Bray et al. (1989) proved that this accumulation involved rRNA (ribosomal RNA, 85% of total RNA) in a turnover between degradation of damaged rRNA and synthesis of new rRNA, while the level of mRNA (messenger RNA, 0.5% of total RNA) remained constant. rRNA is as much necessary to repair cell damages as DNA.

10.4.3 Effects on Protein Synthesis

Protein synthesis is an essential requisite for germination and starts after few minutes from hydration (Cheung et al. 1979). Bray (1995) showed that the amount of synthesized protein observed 2 days after germination in primed leek seed was the same as that observed 4 days after germination in unprimed seed. Chen et al. (2012) observed an increase of dehydrin in spinach (*Spinacia oleracea* L.) during osmopriming. This increase was also observed during germination of primed seeds in chilling stress and desiccation stress, suggesting that osmopriming may play a positive role in the tolerance to these stresses.

10.4.4 Effects on Enzymes

Osmopriming induce the synthesis and activation of enzymes catalyzing the breakdown and mobilization of reserve substances (Varier et al. 2010). Sung and Chang (1993) mentioned that enzymes responsible for the mobilization of reserve carbohydrates (α- and β-amylases) and lipids (isocitrate lyase) are activated due to priming process. This effect is associated with the water deficit induced by osmopriming, which is supposed to determine a mobilization of reserve proteins (Varier et al. 2010).

In fact, priming appears to strengthen defense system (antioxidant enzymes): seed priming was associated with an increase in CAT expression in *Arabidopsis* (Gallardo et al. 2001) and sunflower (*Helianthus annuus* L.) (Kibinza et al. 2011) and maize seeds (Chiu et al. 2002). In sunflower seeds, Bailly et al. (1998, 2000) showed that osmopriming with PEG led to an increase of SOD and CAT, in response to the rise of metabolic activity during priming, which is responsible for a secondary production of AOS from mitochondrial respiration and/or lipid peroxidation. It appears, therefore, that the defense system of the antioxidant enzymes is enhanced in response to a higher amount of potential threats.

Several methods of seed priming were successfully used in agriculture for seed conditioning to accelerate the germination rate and improve the seedling uniformity as seed priming with water, plant growth regulators, beta-aminobutyric acid, 5-aminolevulinic acid, osmoprotectant, melatonin, chitosan, plant extract, polyethylene glycol, and inorganic salts. It is worthy to mention that all these methods showed pronounced effect on germination, seedling growth, and yield of different crops under normal conditions or stressful conditions.

10.5 Seed Priming with Water (Hydropriming)

Soaking seeds with water overnight and then drying before sowing markedly improved seedling emergence, plant growth, establishment, vigor, and the final yield as reported by Harris et al. (1999) and Ahmad and Shad (2010). On the other hand, Harris et al. (2002) and Basu et al. (2005) reported that effects of hydropriming persisted only till early vegetative growth of maize. The primed seed germinated soon after planting compared with untreated dry seed, since the early emergence and its effect on early maturity of primed seed may be due to the completion of pre-germinative metabolic activities that making the seed ready for radical protrusion (Rajpar and Wright 2000). Moreover, Rajpar et al. (2006) reported that primed seeds overnight took significantly fewer days to emerge and reach maturity when compared to the untreated dry seed, whereas 36 h primed seeds showed poor germination and 48 h primed seeds inhibited germination. This inhibition may be attributed to the long period of priming that led to excess water in the seeds and greater reduction in the O_2 availability to the embryo. This was reported by Perry and Harrison (1974) who stated that excess of water during priming had inhibitory effect on seed germination of *Beta vulgaris*. The reduction in O_2 availability could lead to an inhibition of ethylene synthesis (Beyerjr et al. 1985) and consequently seed germination. Dawood (2005) demonstrated that the changes in the chemical composition and anti-nutritional factors under the effect of soaking of canola seeds with water for 24 h depend on the variety and caused a marked decrease in tannins and glucosinolates, total polyphenols, and phytic acid of canola meal. In this regard, Kibinza et al. (2011) showed that priming of sunflower seeds improves germination percentages due to the significant drop in H_2O_2 accumulation and restoration of catalase activity that protects stressed seeds against damages caused by ROS activities. Moreover, Kester et al. (1997) reported that seed priming is able to increase protein content in plant tissue via conferring protection to the cellular proteins damaged, improving the performance of the protein synthesis system, and also increase in protein L-isoaspartyl methyltransferase enzyme that repairs plant tissue protein.

10.6 Seed Priming with Plant Growth Regulators

Priming seeds with plant growth regulators has been reported to be beneficial to growth stages and yield of some crops and known as a strategy to alleviate the harmful effect of environmental stresses (Bahrani and Pourreza 2012; Jisha et al. 2013). For instance, soaked rye seeds with gibberellic acid increased germination percentage under water stress conditions (Ansari et al. 2013). The germination of pretreated pepper seeds with salicylic acid improved significantly under high salinity levels (Khan et al. 2009). In addition, Nascimento (2004) documented that seed priming with ethylene minimizes the effect of high temperatures on lettuce seed germination. Seed priming with brassinosteroids (BR) increased peroxidase (POD), superoxide dismutase (SOD), and catalase (CAT) activities in *Medicago sativa* seedlings under salinity stress (Zhang et al. 2007). Priming *Vicia faba* seed with salicylic acid improved salinity tolerance due to enhanced activities of CAT, ascorbate peroxidase (APX), POD, and glutathione reductase (GR) (Azooz 2009). Priming seeds of *Agropyron elongatum* with gibberellin (GA) and abscisic acid (ABA) induced CAT and SOD activities under drought conditions when compared to unprimed seeds (Eisvand et al. 2010). Maize seeds primed with hormones exhibited acquired abiotic stress tolerance through a more responsive antioxidative system (Afzal et al. 2008;

Farooq et al. 2008). Moreover, phytohormonal priming can enhance seed germination through enhancing some enzymes such as amylase activities and protease that hydrolyzed starch and protein molecules into simple forms available for the embryo to germinate (Miransari and Smith 2014). Hormonal priming was found also to reduce the reactive oxygen species (ROS) produced under $PbCl_2$ stress, and it can alleviate the stress by controlling the oxidative damage on seed germination and embryo growth under stress conditions (Sajedi et al. 2011). Although hydropriming and hormonal priming on soybean can improve seedling establishment and field performance, hormonal priming using auxin, cytokinin, and GA_3 were the most appropriate priming treatments for soybean seeds grown under Pb stress conditions (Abu-Muriefah 2017). ROS produced under stress conditions often interact with the phytohormone priming, and this interaction between ROS and phytohormone could be antagonistic or synergistic (Golldack et al. 2013).

10.6.1 Auxin

Generally, auxins are the best hormones to use because they are nontoxic to plants over a wide range of concentration and effective in promoting root system of large number of plant species. Auxins might regulate cell elongation, cell division, tissue swelling, formation of adventitious roots, callus initiation, and induction of embryogenesis at very low concentrations (Vanderhoef and Dute 1981). The principal auxin in plants is indole-3-acetic acid (IAA) that produced mainly in the shoot apex bud and young leaves of plants. Other meristematic tissues, flowers, fruits, and young seeds have also been shown to be sites of this hormone production. IAA has wide range of effects on many processes such as cell division, vascular tissue differentiation, root initiation, flowering, fruit setting, ripening, senescence, and gravitropism (MacDonald 1997). Furthermore, IAA stimulated cell elongation and apical dominance (Wang et al. 2001), increased photosynthetic activities (Naeem et al. 2004), and activated the translocation of carbohydrates during their synthesis (Awan et al. 1999).

It is well known that IBA is an auxinic hormone and plays a regulatory role in many vital processes within plants such as growth, development, and vascular and pollen formation. Its function on growth, including embryo development, is controlled by its transport which is regulated by transcriptional factors (Hayashi 2012). Another important function of the IBA is its role in cell elongation in the growing embryo (Hauvermale et al. 2012). Auxin by itself is not a necessary hormone for seed germination (Hentrich et al. 2013); however, according to the analyses regarding the expression of auxin-related genes, auxin is present in the seed radicle tip during and after seed germination.

10.6.2 Cytokinins

It was found that cytokinins are able to stimulate seed germination by alleviating the stresses such as drought, salinity, and heavy metal as well as oxidative stress (Peleg and Blumwald 2011; Miransari and Smith 2014). Cytokinins have the ability to regulate a range of cell activities including cell division and seed germination. Heyl et al. (2012) mentioned that cytokinins were active at all stages of germination and enhance the activity of meristematic cells in epicotyls and hypocotyls. These hormones have potent effects on plant physiology and are intimately involved in the regulation of cell division, apical dominance, chloroplast development, anthocyanin production, and maintenance of the source-sink relationship (Hutchkinson and Kileber 2002). In addition, cytokinins are regarded as the most important senescence-retarding hormones, and their application has been demonstrated to prevent the degradation of chlorophyll and photosynthetic proteins as well as reverse leaf and fruit abscission (Pospíšilová et al. 2000). Kinetin is a synthetic cytokinin known to significantly improve plant growth and development even grown under environmental stress. It stimulates leaf expansion and development of reproductive organs and delays senescence (Shah 2007).

10.6.3 Gibberellins

Gibberellins are diterpenoid, regulating seed germination and plant growth through its antagonistic effects with ABA. Abu-Muriefah (2017) stated that GA_3 priming was found to enhance seed germination, maybe through its effect on stored food within seeds, and makes it available for embryos during germination processes. The endosperm within seeds becomes available to the embryo via the activities of some hydrolase enzymes. It is well known that GA_3 stimulates the synthesis and production of the hydrolases, especially α-amylase, resulting in the germination of seeds. In this regard, Yamaguchi (2008) found that gibberellins were able to induce a range of enzymes necessary for seed germination including amylase, protease, and glucanase. Moreover, seed germination is often controlled through suppression effects of excess ABA on the expansion of embryo organs caused by inhibition of GA_3 effects on the growth of radicle and hypocotyl (Voegele et al. 2011). In addition, gibberellins are essential in seed germination for the production of mannanase that is necessary for seed germination (Wang et al. 2005).

10.6.4 Abscisic Acid (ABA)

ABA inhibited the activity of many enzymes involved in germination of pigeon pea, barley grains, and soybean as reported by Sneideris et al. (2015), Staroske et al. (2016), and Abu-Muriefah (2017), respectively. Although ABA had negative effects on seed germination process, it positively affects seed dormancy and plant activities under biotic and abiotic stresses (Popko et al. 2010). Chiu et al. (2016) stated that high ABA concentrations can inhibit seed germination in many species probably due to the inhibitory effect of ABA on the activity and/or the synthesis of some enzymes involved in the degradation of endosperm cells such as α-amylase within seeds, which considered an important process for seed germination. Moreover, the inhibition of seed germination at high levels of ABA was through inhibition of the radicle expansion and suppression of some transcriptional factors, which can negatively affect the process of seed germination (Graeber et al. 2010).

10.6.5 Ethylene

Ethylene regulates plant responses under different stress conditions and controls many processes in plants, including seed germination (Keunen et al. 2016) and embryo radicle growth (Baskin et al. 2003). It was found that ethylene concentration increased during seed germination process of many species including wheat, corn, soybean, and rice (Zapata et al. 2004). 1-Aminocyclopropane-1-carboxylic acid (ACC), the precursor of ethylene, can enhance seed radicle emergence through the production of ethylene which is produced in the radicle (De Poel and der Straeten 2014). In this regard, seed priming with ethylene precursor ACC (1-aminocyclopro pane-1-carboxylic acid) increased the rate of germination in lettuce seeds (Nascimento 2004), but it didn't affect significantly the rate of germination in ryegrass (Tiryaki et al. 2004). However, under abiotic stress conditions and GA_3-deficient mutants, ethylene can play the same role of GA_3; therefore ethylene priming makes seeds able to germinate completely at such conditions (Matilla and Matilla-Vazquez 2008).

10.6.6 Salicylic Acid

Salicylic acid (SA) acts as an endogenic phytohormone from phenolic compounds (among the group of ortho-hydroxyl benzoic acid), having the ability of antioxidant defense system, and regulates various physiological and biochemical processes in plant such as stomata conductivity (Hayat et al. 2010), activity of photosynthesis pigments (Hayat et al. 2005), maintenance of tissue water contents and reduced membrane permeability (Farooq et al. 2008), adjustment of the activity of antioxidant enzymes (Carvalho et al. 2011), and tolerance to environmental stresses (Kabiri et al. 2012). In addition, Sakhabutdinova et al. (2003) reported that salicylic acid treatments maintain IAA and cytokinin levels in the plant tissues, which enhanced the cell division.

Regarding priming process, it was reported that salicylic acid pretreatment produced a higher total biomass and seed vigor index (Kabiri et al. 2012; Sharifzadeh et al. 2013) and increased seedling field emergence because the salicylic acid treatment prevented the decrease in indoleacetic acid and cytokinin content completely which reduces inhibition of plant growth (Afzal et al. 2006; Ansari and Sharifzadeh 2012).

Seed priming with salicylic acid induced salinity tolerance (Kumar et al. 2010; Ahmad et al. 2012, 2015) via increasing SOD activity which quenches oxygen radicals (Gautam and Singh 2009; Orabi et al. 2015) and reducing membrane permeability and leakage of ions like NO^{3-} in wheat and canola (Wahid et al. 2007b; Sakr and Arafa 2009) and enhancing chlorophyll contents and sugars which is necessary for osmotic regulations as well as retards ethylene biosynthesis (Hamid et al. 2008). Moreover, Dellali et al. (2012) recognized that salicylic acid priming at the seedling growth stage alleviated salt-induced oxidative stress by reducing malondialdehyde (MDA) and H_2O_2 content.

Furthermore, seed priming with salicylic acid improved chilling tolerance and increased germination (Sedghi et al. 2010), via activation of antioxidants, maintenance of tissue water contents, and reduced membrane permeability (Farooq et al. 2008), as well as protected structure of plant cell (Horvath et al. 2007). Farooq et al. (2008) showed that maize seed priming by salicylic acid solution improved germination characteristics and seedling root and shoot growth under both normal and low temperature conditions through activation of antioxidant enzymes system, including catalase (CAT), superoxide dismutase (SOD), and ascorbate peroxides (APX). Meanwhile, Pouramir-Dashtmian et al. (2014) revealed that seed priming with salicylic acid solution improved rice seedling growth under low temperature condition via increasing proline, soluble sugar, and protein content.

Regarding optimum concentration, it was found that seed priming with either 20 mg L^{-1} acetylsalicylic acid or salicylic acid seems to be the suitable concentration that showed maximum seed invigoration and better performance in maize through inducing SOD activity and improving nutrient contents in root and shoot as reported by Ahmad et al. (2015). In addition, Hayat et al. (2005) reported that soaking of wheat grains with low concentrations of salicylic acid significantly promoted growth of wheat seedlings. Pretreatment of barley seeds with 1150–1252 μmolar salicylic acid significantly increased the number of seed in spike and thousand seed weight and thus enhanced seedling emergence in field condition (Khaliliaqdam and Mir-Mahmoodi 2013.

10.7 Seed Priming with Beta-Aminobutyric Acid (BABA)

Recently, nonprotein amino acids like β-aminobutyric acid (BABA) were employed in seed priming of various crops against biotic and abiotic stress (Worrall et al. 2012). BABA is known as a potent inducer of resistance in plants against nematodes (Oka et al. 1999), microbial pathogens (Cohen 2002), insects (Hodge et al. 2005), and abiotic stress (Jakab et al. 2005; Zimmerli et al. 2008). According to Zhong et al. (2014), BABA can bring plants into a sensitization state in which defenses are not expressed, but are able to react more rapidly and/or more strongly to various stress. BABA-induced priming functions by the interaction with several hormones like salicylic acid (SA), abscisic acid (ABA), and ethylene (Jakab et al. 2005) or by causing a cascade of signaling processes mediated through H_2O_2 (Cohen et al. 2010). Jisha and Puthur (2016) mentioned that priming rice grain with BABA reduced MDA content of seedlings through reducing the lipid peroxidation of biomembranes and increasing antioxidant enzyme activities and thus improved the PEG-6000 and NaCl stress tolerance of rice seedlings.

10.8 Seed Priming with 5-Aminolevulinic Acid (ALA)

5-Aminolevulinic acid (ALA) is an aliphatic precursor in the biosynthesis of all porphyrin compounds such as chlorophyll and heme (Wang et al. 2004). ALA is a biodegradable herbicide and insecticide, but it has promotional effects on the growth and photosynthesis of crops and vegetables and is harmless to humans, animals, and crops (Hotta et al. 1997a, b). Rae-hyun and Song (2007) demonstrated that the application of *Rhodopseudomonas* sp. which produced indole-3-acetic acid and 5-aminolevulinic acid could increase the germination percentage of tomato seed by 30.2%. ALA is able not only to act as an antioxidant enzyme promoter but also to stimulate nitrate reductase activity (Beyzaei et al. 2014) and the respiration rate and ATP synthesis (Fu et al. 2014). ALA not only promotes the growth and yield of crop plants but also acts as an alleviator of oxidative damage and enhances the activity of antioxidant enzymes such as superoxide dismutase (SOD), peroxidase (POD), and ascorbate peroxidase (APX) to encounter radicals under stress conditions, i.e., drought stress and salinity stress (Li et al. 2011; Liu et al. 2011; Zhang et al. 2013a). Korkmaz et al. (2010) reported that seed soaking with 25 ppm ALA enhanced the germination rate and seedling uniformity, thereby providing chilling stress tolerance of pepper seedling. Kanto et al. (2015) suggested that rice seed priming with ALA at 0.5 μg mL^{-1} could ameliorate the germination capacity even after exposure to the accelerated ageing treatment by enhancing the activity of antioxidant enzymes to scavenge ROS during rice seedling growth under accelerated ageing.

10.9 Seed Priming with Osmoprotectants

Many plants tolerate stress by production of different types of organic solutes called osmoprotectants (compatible solutes or osmolytes) which lower the osmotic potential and attract water molecules into the cell and ultimately maintain the cell turgor. These compatible solutes including soluble sugars, sugar alcohols, proline, glycine-betaine, etc. are low molecular weight, highly soluble in water, and nontoxic to plant even at higher cytosolic concentration (Hoque et al. 2007).

Generally, these compatible solutes protect plants from stress injury through different means, including protection of cytoplasm and chloroplasts from Na^+ damage and scavenging of reactive oxygen species (Smirnoff and Cumbes 1989), stabilization of proteins and protecting membrane structure (Bohnert and Jensen 1996), maintaining the osmotic balance, and general maintenance of physiological stability of plants under stressful conditions (Farooq et al. 2009).

10.9.1 Glycinebetaine (GB)

Glycinebetaine (GB) is a quaternary ammonium compound, an amino acid derivative, and regarded as one of the most effective compatible solutes that protect plants from injury of abiotic stresses. GB application improves growth, survival, and tolerance of a wide variety of plants under various stress conditions (Ashraf and Foolad 2007) by regulating a number of physiological and biochemical processes (Qureshi et al. 2013), maintaining turgor pressure (Agboma et al. 1997), enhancing net CO_2 assimilation rate (Lopez et al. 2002), protecting the functional proteins and enzymes (e.g., Rubisco) and lipids of the photosynthetic apparatus, and maintaining electron flow through thylakoid membranes (Allakhverdiev et al. 2003) and regulation of photosynthetic machinery and ion homeostasis (Raza et al. 2014). Further, GB induces defense response in crops against reactive oxygen species (ROS) produced due to biotic and abiotic stresses and plays a vital role in the process of osmotic adjustment in many crops under environmental stresses (Gadallah 1999). Moreover, it may act as an antitranspirant which allowed the plant to access more water for a long period and facilitates photosynthesis as reported by Agboma et al. (1997).

The main role of GB in plants exposed to saline soil is considered to be maintaining osmotic regulation in cells (Gadallah 1999). It was suggested that GB supports various transporters for normal functioning under salt stress. From this, it can be proposed that GB has a protective effect in discriminating Na^+ against K^+ under saline states (Mansour 1998). The second possible role of GB is that it might enhance vacuolar efficiency in the roots of salt-stressed plants for accumulating more Na^+ as reported by Rahman et al. (2002). Gadallah (1999) reported that GB improves germination and seedling growth of many crops under stressful condition. Plants treated with GB also maintain higher antioxidative enzyme activities that minimize oxidative stress (Ma et al. 2006). The application of GB has an important role in water availability for the imbibition of germinating seeds during limited water. Under salinity conditions, GB increased seed germination by enhancing the osmotic pressure (Sakamoto and Murata 2000) and alleviated lipid peroxidation and facilitated the protection of membrane functions (Hu et al. 2012), protected the photosynthetic machinery by stabilizing the activity of repair proteins (Murata et al. 2007), and provided a direct protective effect on antioxidant enzyme activities under salinity stress (Hoque et al. 2007).

Cuin and Shabala (2005) suggested that compatible solutes, and GB in particular, regulate the net fluxes of Na^+ and K^+ across the plasma membrane at the cellular level in response to NaCl stress. Priming safflower seeds with GB (60 mM) was the most effective treatment that enhanced the plant's tolerance to salt stress during the germination stage. Dawood and Sadak (2014) mentioned that soaking canola seeds with different concentrations with GB caused significant increases in IAA, proline, and total soluble sugars and significant decreases in MDA, H_2O_2, and antioxidant enzymes (POX, PPO, SOD, CAT, APX, and NR) in canola plants irrigated with different levels of water relative to corresponding controls. All GB treatments caused significant increases in seed yield, oil, carbohydrate, protein, total phenolic content, tannins, and antioxidant activity of the yielded seeds and nonsignificant increases in flavonoids in the yielded canola seeds either in plants irrigated with 75% FC or 50% FC relative to corresponding controls. The increases in seed yield/plant due to 20 mM GB were 30.80% and 60.28% at 75% FC and 50% FC, respectively, relative to corresponding controls. The fatty acid profile of canola oils showed different responses to GB treatments either in unstressed plants or drought-stressed plants. Oleic and linoleic acids were increased accompanied by decreases in linolenic and erucic acids under the interaction effect of GB treatments and drought stress (75% FC and 50% FC), and these results led to decreases in total saturated fatty acid and increases in unsaturated fatty acid relative to corresponding controls. Generally, 20 mM GB was the most pronounced and effective treatment in alleviating the deleterious effect of moderate or severe drought stress on canola plants. Meanwhile, 60 mM GB had the most pronounced effect on tolerance to salinity stress in safflower seedling. The GB-increased tolerance to salt in safflower was mainly related to increased CAT and SOD activities and the prevention of cell membrane damage as a result of reduced lipid peroxidation and improved ion homeostasis under salinity stress condition (Alasvandyari et al. 2017).

10.9.2 Proline

It is evident from different reports that application of proline induces abiotic stress tolerance in plants (Ali et al. 2007; Taie et al. 2013; Dawood et al. 2014). Its further role in salinity appears to involve the induction of salt-responsive genes, with the resultant formation of new proteins which may improve the adaptation to salinity stress (Khedr et al. 2003). Moreover, proline may be having a role in stabilization of cellular proteins and membranes in presence of high concentrations of osmotic stress. Proline accumulation in plants could be only useful as a possible drought injury sensor instead of its role in stress tolerance mechanism (Jahari et al. 2010). In addition, Vendruscolo et al. (2007) reported that proline is involved in tolerance mechanisms against oxidative stress, and this is the main strategy of plants to avoid detrimental effects of water stress. Proline plays an important role as a sink for energy to regulate redox potentials (Simiroff and Cumbes 1989), alleviates salt stress induced by enhancement in oxygenase and carboxylase activities of Rubisco (Sivakumar et al. 2000), and protects plants from free radical that induced damage by quenching of singlet oxygen (Matysik et al. 2002). Several functions are proposed for the accumulation of proline in tissues exposed to salinity stress: osmotic adjustment (Voetberg and Sharp 1991), C and N reserves for growth after stress relief (Hellmann et al. 2000), detoxification of excess ammonia (Skopelitis et al. 2006), stabilization of proteins and membranes (Mansour 1998), protection of macromolecules from denaturation (Hamilton and Heckathorn 2001), osmoprotection (Kishor et al. 1995), free radical scavenging (Chen and Dickman 2005), antioxidation (Hoque et al. 2007), and regulation of cytosolic acidity (Sivakumar et al. 2000).

Taie et al. (2013) mentioned that pre-sowing seed treatment with proline seems to enhance faba bean salt tolerance via improving growth parameters, photosynthetic pigments, soluble carbohydrate, and total carbohydrate; meanwhile phenolic, proline, Na^+, and Cl^- contents were decreased relative to their corresponding salinity controls. Proline treatments induced over expression for new protein bands with high density.

10.10 Seed Priming with Melatonin

Melatonin (N-acetyl-5-methoxytryptamine) is an indolic compound (biogenic indoleamine) structurally related with other important substances, such as tryptophan, serotonin, indole-3-acetic acid (IAA), etc. Several authors hypothesized that melatonin may possess some auxin-like effects and may act as a regulatory molecule in plants (Van Tassel et al. 2001). Many evidences have shown melatonin could alleviate biotic and abiotic stresses such as temperature, salinity, light, drought, and pathogen (Li et al. 2012; Zhang et al. 2013a, b, 2014; Meng et al. 2014). Moreover, melatonin also could regulate the expression of a large number of genes involved in plant stress defense (Weeda et al. 2014; Zhang et al. 2014). Melatonin application could be a good bio-stimulator improving not only seed germination and seedling/plant growth but also crop production especially under stress conditions (Janas and Posmyk 2013). Melatonin is soluble in both water and lipid so it may act as a universal hydrophilic and hydrophobic antioxidant (Janas and Posmyk 2013). Tan et al. (2007) mentioned that elevated levels of melatonin probably protect plants against water and soil pollutants through acting as a direct free radical scavenger and as an indirect antioxidant. One melatonin molecule may scavenge up to ten free radicals (Tan et al. 2007), which contrasts with the classic antioxidants that typically detoxify one radical per molecule. Its antioxidant activity may manifest itself in several ways: (1) direct free radical scavenging, (2) elevating the antioxidant enzyme activity, (3) protecting antioxidant enzymes from oxidative damage, (4) increasing the efficiency of mitochondrial transport chain, and (5) reducing the generation of free radicals (Tan et al. 2010). Paredes et al. (2009) reported that melatonin functions in plants can be recognized into three categories: growth promoters as auxins, antioxidants for free radicals which serve as a first-line defense against oxidative stress, and other functions (signal molecules for circadian maintenance, regulation of flower development, or maintenance of developmental stages in fruit tissues). Hernandez-Ruiz et al. (2004) mentioned that a higher concentration of melatonin (200 μM) had no significant effect or even inhibitory effect on seed germination. However, lower concentrations of melatonin (50 or 100 μM) promoted seed germination as mentioned by Wei et al. (2015).

Pretreatment of melatonin attenuates cold-induced apoptosis in carrot suspension cells (Lei et al. 2004) and decreased lipid peroxidation caused by toxic copper ion in red cabbage seedlings (Posmyk et al. 2008). Melatonin prevented chlorophyll degradation during the senescence of barley leaves (Arnao and Hernandez-Ruiz 2009) and protected membrane structures against peroxidation during chilling stress and recovery in cucumber seeds (Posmyk et al. 2009). Posmyk et al. (2008) reported that the pre-sowing seed treatment with melatonin-protected red cabbage seedlings against toxic Cu ion concentrations as well as melatonin application to cucumber seeds had a beneficial effect on seed germination, the growth of seedlings, and crop production of plants especially those subjected to cold stress (Posmyk et al. 2009) and water stress (Zhang et al. 2013a, b). Jiang et al. (2016) mentioned that seed priming with 0.8 mM melatonin alleviates the salinity damage to maize by improving SOD, CAT, and PAL activities, relative water content, and proline and total phenolic contents and decreasing membrane relative electrolyte leakage and lipid peroxidation product. Szafrańska et al. (2012) and Jiang et al. (2016) mentioned that either seed priming or exogenous application with melatonin significantly increased PAL activities and total phenolic content. Janas and Posmyk (2013) showed that hydroprimed or osmoprimed seeds of corn (*Zea mays* L.), mung bean (*Vigna radiata* L.), and cucumber (*Cucumis sativus* L.) with melatonin had higher crop yield than the control ones under field conditions, since the production of corn, cucumber, and mung bean primed with melatonin was about 10–25% greater in comparison to those primed without melatonin and is dependent on plant species. Zhang et al. (2013b) reported that 100 μM melatonin alleviated polyethylene glycol-induced inhibition of cucumber seed germination, showing the greatest germination rate and photosynthetic rate, and at the same time significantly reduced chlorophyll degradation. Furthermore, the ultrastructure of chloroplasts in water-stressed cucumber leaves was improved after melatonin treatment, thus reversing the effect of water stress. Szafrańska et al. (2014) mentioned that hydropriming *Vigna radiata* seeds with melatonin (50 μM L^{-1}) increased level of melatonin in roots derived from hydroprimed seeds with melatonin by sevenfold higher than roots derived from non-primed seeds. Wei et al. (2015) concluded that coating soybean seeds with melatonin promoted soybean plant growth, increased yield, and improved salinity stress tolerance. Recently, Dawood and EL-Awadi (2015) concluded that melatonin treatments (100 and 500 mM) improved growth parameters, relative water content, photosynthetic pigments, and total carbohydrate, total phenolic, indoleacetic acid, K^+, and Ca^{+2} contents and reduced the levels of compatible solutes and Na^+ and Cl^- contents in leaf tissues of faba bean plants irrigated with diluted seawater (3.85 and 7.69 dS/m). Melatonin at 500 mM had a more pronounced effect in alleviating the adverse effects of the two salinity levels on the performance of faba bean plants than 100 mM melatonin. The beneficial effects of melatonin treatments in alleviating the harmful effect of salinity stress on the growth parameters were more pronounced in the plants grown under the higher salinity level (S2 = 7.69 dS/m) than those grown under lower salinity level (S1 = 3.85 dS/m) relative to corresponding controls. The increases in total photosynthetic pigments were 22.31%, 12.87%, and 15.85% in the plants treated with 500 mM melatonin and irrigated with tap water (S0) and diluted seawater at lower (S1) and higher (S2) concentrations, respectively, as compared with corresponding controls.

Jiang et al. (2016) suggest that seed priming with 0.8 mM melatonin significantly improved germination energy, germination percentage, seedling vigor index, shoot and root lengths, seedling fresh and dry weights, K+ content, relative water content, proline and total phenolic contents, superoxide dismutase, and catalase and phenylalanine ammonia lyase activities and significantly decreased mean emergence time, Na+ content, electrolyte leakage, and malondialdehyde content compared with untreated seeds under salinity stress; thereby seed priming with melatonin alleviates the salinity damage to maize.

10.11 Seed Priming with Plant Extract

The allelopathic action of various natural compounds on the growth and development of many plants may be inhibitory or stimulatory depending on their concentration in the surrounding medium and on their physiological activity within plants (El-Daly and Soliman 1997). Allelochemicals like phenolic compounds, flavonoids, terpenoids, alkaloids, steroids, and other compounds, sometimes have a greater allelopathic effect than individual compound alone. Einhellig (1995) mentioned that these metabolites may be selective in their action or plants may be selective in their responses. Moreover, allelochemicals which inhibit the growth of some species at certain concentrations may stimulate the growth of the same or different species at different concentrations (Narwal 1994). These allelochemicals promote or inhibit the crop growth based on species-specific concentrations (Ambika et al. 2003). Phenolic compounds can act on enzymes, phytohormone activity, and mineral content (Einhellig 2004). Saponins are readily soluble in water which may enhance the nutrient absorption (Satish et al. 2007). Alkaloids, phenolic compounds, and saponins protect the plants against pathogens and also produce antioxidant activity.

Leaves of different plants contain alkaloids and phenolic compounds which protect the plants against pathogens and also produce antioxidant activity (Satish et al. 2007). Rathinavel and Dharmalingam (1999) mentioned that the presence of bioactive substances in the leaf extracts enhanced lipid utilization and enzyme activity that increased dry weight and development of seedling to reach autotropic stage, enabling them to produce relatively more quantity of dry matter which is discerning the cause for the hike in vigor index by hardening treatment. It is generally assumed that physiologically active substances might have activated the embryo and other associated structures which resulted in the absorption of more water due to cell wall elasticity and development of stronger and efficient root system, and that would have ultimately resulted in higher vigor index (Rangaswamy et al. 1993). Many researchers also reported the benefits of seed hardening with different medicinal plants leaf extract to overcome the adverse condition (Renugadevi et al. 2008; Kamaraj and Padmavathi 2012).

It is worthy to mention the most important plants containing allelochemicals as follows: *Chlorophytum* leaves are high in saponin, carbohydrate, proteins, and various alkaloids (Chakraborthy et al. 2014). Leaves of *Vinca rosea* are also high in alkaloids and phenolic compounds (Tiong et al. 2013). Neem leaves contain flavonoids, steroids, carbohydrates, glycosides, antiquinone, terpenoides, and alkaloids (Raphael 2012). The seeds of fenugreek contain lysine and L-tryptophan-rich proteins, mucilaginous fiber, and other rare chemical constituents such as saponins, coumarin, fenugreekine, nicotinic acid, sapogenins, phytic acid, scopoletin, and trigonelline (Bukhari et al. 2008). Furthermore, intercropping fenugreek with faba bean can reduce *Orobanche crenata* infection (Fernández-Aparicio et al. 2006).

Guava leaf contains volatile oil such as quercetin, avicularin, guaijaverin, etc. (Morant et al. 2008). Previous studies on the chemical composition of guava leaves have identified chemical products belonging to the groups with allelopathic properties (Monteiro and Vieira 2002) such as terpenoids, flavonoids, coumarins, and cyanogenic acids, among others (Gutiérrez et al. 2008). *Some studies have already identified guava allelopathic effects on other species, the effect of guava fruit extracts on cucumber germination* (*Cucumis sativus*) (Bovey and Diaz-Colon 1968) as well as the effect of guava root exudates on lettuce (*L. sativa*) germination and root growth and the root growth of bristly foxtail (*Setaria verticillata*) (Brown et al. 1983). Chapla and Campos (2010) reported that allelopathic effect of the guava leaf aqueous extract on the germination and growth of lettuce occurred only at 20% concentration. Regarding *Lantana camara, their* leaves, roots, and fruits contain allelochemicals mainly phenolics, flavonoids, tannins, and carbohydrates as mentioned by Gopie-shkhanna and Kannabiran (2007). Moreover, Basu and Hazra (2006) indicated that lantana plant has strong antioxidant activities. Yi et al. (2005) reported the presence of several phenolic compounds in lantana leaf extract identified by HPLC as salicylic, gentisic, β-resorcylic, vanillic, caffeic, ferulic, and p-hydroxybenzoic acids, coumarin, and 6-methylcoumarin. Ahmed et al. (2007) found that different concentrations of aqueous leaf extracts of lantana caused significant inhibitory effect on germination, root and shoot elongation, and development of lateral roots of receptor crops. The inhibitory effect was much pronounced in root and lateral root development rather than shoot and germination. Earlier, Dawood and Taie (2009) soaked lupine seeds with freshly prepared aqueous leaf extract (5, 10, 15% w/v) of eucalyptus and lantana separately for 24 h. The phenolic compounds and flavonoids (allelochemicals) are detected in higher amounts in aqueous leaf extract of eucalyptus than that of lantana. Aqueous leaf extract of either lantana or eucalyptus caused a gradual decrease in germination % of both sweet and bitter lupine varieties. Their effect was proportional to the extract concentration. Germination % increased by extending the germination period. Phenolic and alkaloid contents of 9-day-old seedlings increased by all treatments. All applied treatments had positive effect on photosynthetic pigments and yield as well as protein, carbohydrate, ash, and phenolic contents of the yielded seeds, whereas alkaloid % decreased. In respect to oil content, it was decreased by lantana treatments. Moreover, Dawood et al. (2012) mentioned that incorporated 10% fenugreek seeds or 20% guava leaves or 20% lantana leaves into the soil caused significant increases in the photosynthetic pigments (chlorophyll a, chlorophyll b and carotenoid) and the total carbohydrate content of sunflower leaf tissues accompanied by a significant decrease in the total phenolic content. The head weight, seed weight/head, and 100-seed weight were increased in the following order: fenugreek treatment > lantana treatment > guava treatment. It was noted that fenugreek treatment is the most effective treatment in increasing oil % followed by guava treatment, while the lowest increase resulted from lantana treatment. Fenugreek treatment caused the highest decrease in (C16:0 + C18:0) accompanied by the highest increase in (C18:1 + C18:2). Regarding the total essential amino acids, it was noticed that fenugreek treatment showed a noticeable increase, whereas lantana and guava treatments showed a decrease in total essential amino acids. Phuwiwat et al. (2012) concluded that leaves of *Melia azedarach* contained water soluble allelochemicals that caused inhibition of both water uptake and α-amylase activity of *E. crus-galli* during germination process. Prabha et al. (2016) concluded that seed priming with 2% leaf extract of *Chlorophytum*, *Agel*, *Azadirachta*, and *Vinca* were found suitable to enhance seedling vigor and reduce mortality rate in the presence of *Fusarium oxysporum* in tomato. Seed priming with these plant leaf extracts is cost-effective and easy to apply at farmer's field and is suitable under "organic farming" framework.

10.12 Seed Priming with Chitosan

Chitosan is an abundant and relatively cheaper cationic polysaccharide obtained as waste material during seafood processing (Guan et al. 2009). Chitosan application to peanut seeds increased levels of germination percentage, energy, lipase activity, gibberellic acid (GA3), and indoleacetic acid (IAA) (Zhou et al. 2002). Chitosan has been proven to act as a positive factor in enhancing shoot and root length, fresh and dry weights of shoots, and roots and leaf area in bean plants watered with the chitosan solution (Sheikh and AL-Malki 2011). Moreover, it is evident that chitosan has some potential under stressful environment. Chitosan-coated seeds have been reported to demonstrate accelerated germination and tolerance to stress in hybrid rice (Ruan and Xue 2002) and enhanced vigor of maize seedlings (Shao et al. 2005). Furthermore, chitosan priming improved maize germination and seedling growth in relation to physiological changes under low-temperature stress (Guan et al. 2009) and improved the germination of plants under drought stress (Suchada et al. 2007). Moreover, chitosan priming under stress resulted in highly improved germination index and reduced germination time to promote early seedling establishment and synchronized growth in rice (Suchada et al. 2007) and maize (Guan et al. 2009). Sadeghi et al. (2011) mentioned that the improvement in germination and vigor of soybean plant was probably due to the reserve mobilization of food material, activation and resynthesis of some enzymes, and DNA and RNA synthesis which started during osmotic priming. There is possibility that similar germination-responsive genes may be activated because of chitosan priming under osmotic stress. Hameed et al. (2014) observed that chitosan priming treatments not only improved seed germination but also enhanced wheat seedling growth under osmotic stress induced by PEG. Its beneficial effects on germination and seedling vigor provided evidence that chitosan is a promising seed priming agent for the improvement of osmotic stress tolerance in wheat.

10.13 Seed Priming with Polyethylene Glycol (PEG)

Polyethylene glycol (PEG) as an inert material can prevent embryo toxicity problems during priming (Cantliffe 1983). The large size of PEG molecule (6000–8000 mw) also prevents its penetration into seed tissues, avoiding lowering the osmotic potential (Brocklehurst and Dearman 1984). Osmopriming with polyethylene glycol is most commonly used to induce either osmotic stress varying between −0.5 MPa (sweet corn, Ghiyasi et al. 2008) and −2 MPa (sunflower, Bailly et al. 2000; sugar beet, Capron et al. 2000) in plants or water-deficit condition because it is not naturally produced in the plant tissue and cannot penetrate into cell from the media. Soaking seeds before sowing with PEG solution helps to initiate the membrane repairing systems and metabolic preparation for germination via controlling the water absorption rate of seeds (Jisha et al. 2013). Massarat et al. (2014) concluded that priming with PEG 6000 had beneficial effects on germination and seedling establishment of corn seeds under drought and saline conditions, whereas the major disadvantage resulting from the use of PEG is the reduction of oxygen in the solution, because of its viscosity (Mexal et al. 1975), so aerating the solution during PEG osmopriming can overcome this problem (Bujalski and Nienow 1991).

10.14 Seed Priming with Inorganic Salts

10.14.1 NaCl

It has been shown that NaCl seed priming could be used as an adaptation method to improve salt tolerance of seeds. Sivritepe et al. (2003) and Yildirim et al. (2011) concluded that NaCl seed priming improves seed germination, seedling emergence, and growth under saline conditions. The benefits of NaCl seed priming did not persist beyond the seedling stage in cucumber (Passam and Kakouriotis 1994), while Cano et al. (1991) concluded that NaCl seed priming had positive effects on mature plants and on yield of tomato. According to Cano et al. (1991), the higher salt tolerance of plants from primed seeds seems to be the result of a higher capacity for osmotic adjustment since plants from primed seeds have more Na^+ and Cl^- ions in their roots and more sugars and organic acids in leaves than plants from non-primed seeds. Moreover, priming canola seeds with NaCl solution increased seedling cell membrane stability and decreased seedling damage under salinity condition due to increased seedling K^+ and proline content (Farhoudi et al. 2007). Massarat et al. (2014) observed that priming corn seeds with NaCl had beneficial effects on germination and seedling establishment under drought and saline conditions. Omami (2005) suggested that priming of amaranth seeds with NaCl improve cell membrane stability and decrease MDA production and increased salt tolerance by promoting K^+ and Ca^{2+} accumulation.

10.14.2 KNO3

The beneficial effects of seed priming with KNO_3 solution on seed germination under salinity conditions have been observed in sunflower (Kaya and Day 2008;

Farhoudi 2014) and muskmelon (Nascimento 2003). Singh and Rao (1993) reported that KNO_3 priming effectively improved germination, seedling growth, and seedling vigor index of the seeds of sunflower varieties with low germination. Improving seedling sunflower growth by KNO_3 priming under salinity condition can suggest nontoxify of KNO_3 due to ion accumulation in the embryo (Kaya et al. 2006). Mauromicale and Cavallaro (1996) mentioned that seed priming of herbage grasses with KNO_3 solution decreased the mean germination time as compared to PEG solution because of KNO_3 did not have toxicity and did not prevent water uptake. Farhoudi (2014) showed that priming of sunflower seeds with 0.6 and 0.9 MPa KNO_3 solution was more effective than non-priming seeds under salinity condition and demonstrated its potential effect in improving tolerance to salinity by increasing seedling fresh weight and POX enzyme activity accompanied by decrease in seedling MDA concentration compared to non-priming seeds.

10.14.3 Sodium Silicate

Evidences have proved that sodium silicate treatment improved the cell membrane stability by reducing the lipid peroxidation in different plants under abiotic stresses (Liang et al. 2007; Pei et al. 2010; Wang et al. 2011). Sodium silicate resulted in improved germination, growth, and antioxidant enzyme activities and reduced lipid peroxidation during drought stress in wheat (Pei et al. 2010; Ali et al. 2012). Moreover, sodium silicate treatment lowered down the oxidative stress by enhancement of antioxidant production (glutathione reductase, catalase, peroxidase, and superoxide dismutase) during drought stress in wheat, barley, and soybean plants (Liang et al. 2003; Gong et al. 2005; Miao et al. 2010; Wang et al. 2011). Moreover, sodium silicate application has been reported to enhance the germination and nutrient use leading to better seedling development in soybean (Miao et al. 2010). Hameed et al. (2013) mentioned that seed priming with sodium silicate not only improved the seed germination and seedling vigor but also enhanced the wheat seedling growth under water-deficit stress induced by PEG.

It is worthy to mention that the accumulation of salts in the seed could be toxic (Bradford 1995), reduce the osmotic potential, and induce a high water absorption during treatment (Parera and Cantliffe 1994), resulting in a more likely radicle protrusion. In addition, effects of inorganic salts on germination are different from those carried out by PEG, depending on seed species. For instance, osmopriming with inorganic salts was toxic to sorghum seeds (Haigh and Barlow 1987), whereas it was effective as PEG in asparagus (Pill 1995) and performed better as PEG in tomato (Mauromicale and Cavallaro 1997). The difference in the response of different species to salts or PEG may be due to a selective semipermeable layer that surrounds the embryo: when this layer is present, it allows the absorption of water, but prevents salt diffusion; when it is absent, ions can be absorbed and cause embryo damages (Welbaum et al. 1998). For example, tomato (*Solanum lycopersicum* L.), melon (*Cucumis melo* L.), lettuce (*Lactuca sativa* L.), and *Capsicum annuum* seeds possess this layer and may be safely subjected to osmopriming with inorganic salts

(Welbaum and Bradford 1990). Conversely, this treatment is harmful to broccoli and cabbage seeds (*Brassica oleracea* L.), which lack this layer (Taylor et al. 1997).

References

Abu-Muriefah SS (2017) Phytohormonal priming improves germination and antioxidant enzymes of soybean (*Glycine max*) seeds under lead (Pb) stress. Biosci Res 14(1):42–56

Afzal I, Basra SMA, Farooq M, Nawaz A (2006) Alleviation of salinity stress in spring wheat by hormonal priming with ABA, salicylic acid and ascorbic acid. Int J Agric Biol 80(1):23–28

Afzal I, Basra SMA, Shahid M, Farooq M, Saleem M (2008) Priming enhances germination of spring maize (*Zea mays* L.) under cool conditions. Seed Sci Technol 36:497–503

Afzal I, Basra SMA, Cheema MA, Farooq M, Jafar MZ, Shahid M, Yasmeen A (2013) Seed priming: a shotgun approach for alleviation of salt stress in wheat. Int J Agric Biol 15:1199–1203

Agboma PC, Jones MGK, Peltonen-Sainio P, Rita H, Pehu E (1997) Exogenous glycinebetaine enhances grain yield of maize, sorghum and wheat grown under two supplementary watering regimes. J Agron Crop Sci 178:29–37

Aghbolaghi MA, Sedghi M (2014) The effect of osmo and hormone priming on germination and seed reserve utilization of millet seeds under drought stress. J Stress Physiol Biochem 10(1):214–221

Ahmad K, Shad KK (2010) Effect of leaf area on dry matter production in aerated mungbean seed. Int J Plant Physiol Biochem 2(4):52–61

Ahmad I, Khaliq T, Ahmad A, Basra SMA, Hasnain Z, Ali A (2012) Effect of seed priming with ascorbic acid, salicylic acid and hydrogen peroxide on emergence, vigor and antioxidant activities of maize. Afr J Biotechnol 11:1127–1137

Ahmad I, Basra SMA, Hussain S, Hussain SA, Hafez-ur-Rehman AR, Ali A (2015) Priming with ascorbic acid, salicylic acid and hydrogen peroxide improves seedling growth of spring maize at suboptimal temperature. J Environ Agric Sci Res 3:14–22

Ahmed R, Uddin BM, Khan AS, Mukul AS, Hossain KM (2007) Allelopathic effects of *Lantana camara* on germination and growth behavior of some agricultural crops in Bangladesh. J Forest Res 18(4):301–304

Ajouri A, Haben A, Becker M (2004) Seed priming enhances germination and seedling growth of barley under conditions of P and Zn deficiency. J Plant Nutr Soil Sci 167:630–636

Alasvandyari F, Mahdavi B, Hosseini SM (2017) Glycinebetaine affects the antioxidant system and ion accumulation and reduces salinity-induced damage in safflower seedlings. Arch Biol Sci 69(1):139–147

Ali Q, Ashraf M, Athar HR (2007) Exogenously applied proline at different growth stages enhances growth of two maize cultivars grown under water deficit conditions. Pak J Bot 39(4):1133–1144

Ali A, Basra SMA, Iqbal J, Hussain S, Subhani MN, Sarwar M, Haji A (2012) Silicon mediated biochemical changes in wheat under salinized and non-salinzed solution cultures. Afr J Biotechnol 11:606–615

Allakhverdiev SI, Hayashi H, Nishiyama Y, Ivanovo AG, Aliev JA, Klimov VV, Murata NN, Carpemtier R (2003) Glycinebetaine protects the D1/D2/Cytb559 complex of photosystem II against photo-induced and heat-induced inactivation. J Plant Physiol 160:41–49

Ambika SR, Poonima S, Palaniraj R, Sati SC, Narwal SS (2003) Allelopathic plants. 10. *Lantana camara* L. Allelopathy J 12(2):147 (C.F. http://www.allelopathy-journal.com)

Ansari O, Sharifzadeh F (2012) Does gibberelic acid (GA), salicylic acid (SA) and ascorbic acid (ASc) improve Mountain Rye (*Secale montanum*) seeds germination and seedlings growth under cold stress? Int Res J Appl Basic Sci 3(8):1651–1657

Ansari O, Azadi MS, Sharif-Zadeh F, Younesi E (2013) Effect of hormone priming on germination characteristics and enzyme activity of mountain rye (*Secale montanum*) seeds under drought stress conditions. J Stress Physiol Biochem 9:61–71

Arnao MB, Hernandez-Ruiz J (2009) Protective effect of melatonin against chlorophyll degradation during the senescence of barley leaves. J Pineal Res 46:58–63

Asgedom H, Becker M (2001) Effects of seed priming with nutrient solutions on germination, seedling growth and weed competitiveness of cereals in Eritrea. In: Proc. Deutscher Tropentag, University of Bonn and ATSAF, Magrraf Publishers Press, Weickersheim, p 282

Ashraf M, Bray CM (1993) DNA synthesis in osmoprimed leek (*Allium porrum* L.) seed. Seed Sci Res 3:15–23

Ashraf M, Foolad MR (2005) Pre-sowing seed treatment: a shotgun approach to improve germination, plant growth and crop yield under saline and non-saline conditions. Adv Agron 88:223–271

Ashraf M, Foolad MR (2007) Roles of glycinebetaine and proline in improving plant abiotic stress resistance. Environ Exp Bot 59:206–216

Austin R, Longden PC, Hutchinso J (1969) Some effects of hardening on carrot seed. Ann Bot 33:883–895

Awan IU, Baloch MS, Sadozai NS, Sulemani MZ (1999) Stimulatory effect of GA_3 and IAA on ripening process, kernel development and quality of rice. Pak J Biol Sci 2:410–412

Azooz MM (2009) Salt stress mitigation by seed priming with salicylic acid in two faba bean genotypes differing in salt tolerance. Int J Agric Biol 11:343–350

Bagheri MZ (2014) The effect of maize priming on germination characteristics, catalase and peroxidase enzyme activity and total protein content under salt stress. Int J Biosci 4(2):104–112

Bahrani A, Pourreza J (2012) Gibberellin acid and salicylic acid effects on seed germination and seedlings growth of wheat under salt stress condition. World Appl J 18(5):633–641

Bailly C, Benamar A, Corbineau F, Côme D (1998) Free radical scavenging as affected by accelerated ageing and subsequent priming in sunflower seeds. Physiol Plant 104:646–652

Bailly C, Benamar A, Corbineau F, Côme D (2000) Antioxidant systems in sunflower (*Helianthus annuus* L.) seeds as affected by priming. Seed Sci Res 10:35–42

Baskin CC, Baskin JM, Chester EW, Smith M (2003) Ethylene as a possible cue for seed germination of *Schoenoplectus hallii* (Cyperaceae), a rare summer annual of occasionally flooded sites. Am J Bot 90(4):620–627

Basra SMA, Farooq M, Tabassum R, Ahmed N (2005) Physiological and biochemical aspects of pre-sowing seed treatments in fine rice (*Oryza sativa* L.). Seed Sci Technol 33:623–628

Basu S, Hazra B (2006) Evaluation of nitric oxide scavenging activity, in vitro and ex vivo, of the selected medicinal plants traditionally used in inflammatory diseases. Phytother Res 20:896–900

Basu S, Sharma SP, Daldani M (2005) Effect of hydropriming on field mergence, crop performance and seed yield of maize parental lines during winter and spring- summer season. Seeds Res 33(1):24–27

Begum MM, Sariah M, Puteh AB, Abidin MAZ, Rahman MA, Siddiqui Y (2010) Field performance of bio-primed seeds to suppress *Colletotrichum truncatum* causing damping-off and seedling stand of soybean. Biol Control 53:18–23

Beta T, Roney LW, Marovatsanga LT, Taylor RN (1999) Phenolic compounds and kernel characteristics of Zimbabwean sorghums. J Sci Food Agric 79:1003–1110

Bewley JD (1997) Seed germination and dormancy. Plant Cell 9:1055–1066

Bewley JD, Black M (1978) Physiology and biochemistry of seeds in relation to germination. In: Development, germination and growth, vol 1. Springer, New York, pp 305–375

Bewley JD, Black M (1994) Seeds: physiology of development and germination, 2nd edn. Plenum Press, New York

Beyerjr EM, Morgan PW, Yang SF (1985) Ethylene. In: Wilkins M (ed) Advanced plant physiology. Pitman, London, pp 111–126

Beyzaei Z, Sherbakov RA, Averina NG (2014) Response of nitrate reductase to exogenous application of 5-aminolevulinic acid in barley plants. J Plant Growth Regul 33:745–750

Bohnert HJ, Jensen RJ (1996) Strategies for engineering water-stress tolerance in plants. Trends Biotechnol 14:89–97

Bovey RW, Diaz-Colon JD (1968) Occurrence of plant growth inhibitors in tropical and subtropical vegetation. Physiol Plant 22:253–259

Bradford KJ (1986) Manipulation of seed water relations via osmotic priming to improve germination under stress conditions. Hortic Sci 21:1105–1112

Bradford KJ (1995) Water relations in seed germination. In: Kigel J, Galili G (eds) Seed development and germination. Marcel Dekker, New York, pp 351–396

Bray CM (1995) Biochemical processes during the osmopriming of seeds. In: Kigel J, Galili G (eds) Seed development and germination. Marcel Dekker, New York, pp 767–789

Bray CM, Davison PA, Ashraf M, Taylor RM (1989) Biochemical changes during osmopriming of leek seeds. Ann Bot 63:185–193

Brocklehurst PA, Dearman J (1984) A comparison of different chemicals for osmotic treatments of vegetable seed. Ann Appl Biol 105:391–398

Brown RL, Tang CS, Nishimoto RK (1983) Growth inhibition from guava roots exudates. Hortic Sci 18:316–318

Bujalski W, Nienow AW (1991) Large-scale osmotic priming of onion seeds: a comparison of different strategies for oxygenation. Sci Hortic 46:13–24

Bujalski W, Nienow AW, Gray D (1989) Establishing the large scale osmotic priming of onion seeds by using enriched air. Ann Appl Biol 115:171–176

Bukhari SB, Bhanger MI, Memon S (2008) Antioxidative activity of extracts from fenugreek seeds (*Trigonella foenum-graecum*). Pak J Environ Chem 9:78–83

Callan NW, Mathre DE, Miller JB (1990) Bio-priming seed treatment for biological control of *Pythium ultimum* preemergence damping-off in sh2 sweet corn. Plant Dis 74:368–372

Cano EA, Bolarin MC, Perez-Alfocea F, Caro M (1991) Effect of NaCl priming on increased salt tolerance in tomato. J Hortic Sci 66:621–628

Cantliffe DJ (1981) Priming of lettuce for early and uniform emergence under conditions of environmental stress. Acta Hortic 122:29–38

Cantliffe DJ (1983) Sowing primed seed. Am Veg Grow 31:42–43

Cantliffe DJ, Shuler KD, Guedes AC (1981) Overcoming seed thermos-dormancy in a heat sensitive romaine lettuce by seed priming. HortScience 16:196–198

Cantliffe DJ, Elbala M, Guedes A, Odell GB, Perkins-Veazie P, Schultheis JR, Seale DN, Shuler KD, Tanne I, Watkins JT (1987) Improving stand establishment of direct-seeded vegetables in Florida. Proc Fla State Hortic Soc 100:213–216

Capron I, Corbineaua F, Dacher F, Job C, Come D, Job D (2000) Sugar beet seed priming: effects of priming conditions on germination, solubilization of 11-S globulin and accumulation of LEA proteins. Seed Sci Res 10:243–254

Carvalho RF, Piotto FA, Schmidt D, Peters LP, Monterio CC, Azevedo RA (2011) Seed priming with hormones dose not alleviate induced oxidative stress in maize seedling subjected to salt stress. Sci Agric (Piracicaba, Braze) 68:598–602

Chakraborthy GS, Aeri V, Verma P, Singh S (2014) Phytochemical and antimicrobial studies of *Chlorophytum borivilianum*. Pharmacophore 5:258–261

Chapla TE, Campos JB (2010) Allelopathic evidence in exotic guava (*Psidium guajava* L.). Braz Arch Biol Technol 53:1359–1362

Chen K, Arora R (2011) Dynamics of the antioxidant system during seed osmopriming, post-priming germination, and seedling establishment in Spinach (*Spinacia oleracea*). Plant Sci 180:212–220

Chen C, Dickman MB (2005) Proline suppresses apoptosis in the fungal pathogen *Colletotrichum trifolii*. Proc Natl Acad Sci U S A 102:3459–3464

Chen K, Fessehaie A, Arora R (2012) Dehydrin metabolism is altered during seed osmopriming and subsequent germination under chillin and desiccation in *Spinacia oleracea* L. cv. Bloomsdale: possible role in stress tolerance. Plant Sci 183:27–36

Cheung CP, Wu J, Suhadolnik RJ (1979) Dependence of protein synthesis on RNA synthesis during the early hours of germination of wheat embryos. Nature 277:66–67

Chiu KY, Chen CL, Sung JM (2002) Effects of priming temperature onstorability of primed sh-2 sweet corn seed. Crop Sci 42:1996–2003

Chiu RS, Saleh Y, Gazzarrini S (2016) Inhibition of FUSCA3 degradation at high temperature is dependent on ABA signaling and is regulated by the ABA/GA ratio. Plant Signal Behav 11(11):e1247137.11

Cohen YR (2002) ß-aminobutyric acid-induced resistance against plant pathogens. Plant Dis 86(5):448–457

Cohen Y, Rubin AE, Kilfin G (2010) Mechanisms of induced resistance in lettuce against *Bremia lactucae* by DL-β-aminobutyric acid (BABA). Eur J Plant Pathol 126:553–573

Coolbear P, Slater RJ, Bryant JA (1990) Changes in nucleic acid levels associated with improved germination performance of tomato seeds after low-temperature pre-sowing treatments. Ann Bot 65:187–195

Cuin TA, Shabala S (2005) Exogenously supplied compatible solutes rapidly ameliorate NaCl-induced potassium efflux from barley roots. Plant Cell Physiol 46:1924–1933

Dawood MG (2005) Study of some changes in chemical composition during germination of some canola varieties. Submitted in partial fulfillment of the requirements for the degree Ph.D. in biochemistry, Faculty of Science, Cairo University, Egypt

Dawood MG, EL-Awadi ME. (2015) Alleviation of salinity stress on *vicia faba* L. plants via seed priming with melatonin. Acta Biol Colomb 20:223–235

Dawood MG, Sadak MS (2014) Physiological role of glycinebetaine in alleviating the deleterious effects of drought stress on canola plants (*Brassica napus* L.). Middle East J Agric Res 3(4):943–954

Dawood MG, Taie HAA (2009) Aqueous leaf extract of lantana and eucalyptus on germination, yield and chemical constituents of lupine seeds. J Biol Chem Environ Sci 4(4):167–185

Dawood MG, El-Awadi ME, El-Rokiek KG (2012) Physiological impact of fenugreek, guava and lantana on the growth and some chemical parameters of sunflower plants and associated weeds. J Am Sci 8(6):166–174

Dawood MG, Taie HAA, Nassar RMA, Abdelhamid MT, Schmidhalter U (2014) The changes induced in the physiological, biochemical and anatomical characteristics of *Vicia faba* by the exogenous application of proline under seawater stress. South Afr J Bot 2014(93):54–63

De Castro RD, van Lammeren AA, Groot SP, Bino RJ, Hilhorst HW (2000) Cell division and subsequent radicle protrusion in tomato seeds are inhibited by osmotic stress but DNA synthesis and formation of microtubular cytoskeleton are not. Plant Physiol 122:327–336

Monteiro C de A, Vieira EL (2002) Substâncias alelopáticas. In: Castro, P. R. de C. e, Sena, J. O. A. de and Kluge, R. A. Introdução à fisiologia do desenvolvimento vegetal

De Poel VB, der Straeten VD (2014) 1-aminocyclopropane-1-carboxylic acid (ACC) in plants: more than just the precursor of ethylene. Front Plant Sci 5:640

Dell'Aquila A, Taranto G (1986) Cell division and DNA-synthesis during osmo-priming treatment and following germination in aged wheat embryos. Seed Sci Technol 14:333–341

Dellali H, Maalej EM, Boughanmi NG, Haouala R (2012) Salicylic acid priming *Hedysarum carnosum* and *Hedysarum coronarium* reinforces NaCl tolerance at germination and the seedling growth stage. Aust J Crop Sci 6(3):407–414

Di-Girolamo G, Barbanti L (2012) Treatment conditions and biochemical processes influencing seed priming effectiveness. Ital J Agron 7:178–188

Einhellig FA (1995) Mechanism of action of allelochemical in allelopathy. In: Allelopathy organisms, processes and application. American Chemical Society, Washington, USA, p 96–116

Einhellig FA (2004) Mode of allelochemical action of phenolic compounds. In: Macias FA, JCG G, JMG M, Cutler HG (eds) Allelopathy: chemistry and mode of action of allelochemicals. CRC Press, Boca Raton, pp 217–238

Eisvand HR, Tavakkol-Afshari R, Sharifzadeh F, Maddah-Arefi H, Hesamzadeh-Hejaz SM (2010) Effects of hormonal priming and drought stress on activity and isozyme profiles of antioxidant enzymes in deteriorated seed of tall wheatgrass (*Agropyron elongatum* host). Seed Sci Technol 38:280–297

El-Daly FA, Soliman MH (1997) Effect of different concentration of lupine seed extract on the growth criteria and pigmentation of soybean plant at different growth stages. Egypt J Physiol Sci 21(2):187–196

Farhoudi R (2014) Evaluation effect of KNO_3 seed priming on seedling growth and cell membrane damage of sunflower (*Heliantus annus*) under salt stress. Rep Opin 6(8):6–11
Farhoudi R, Sharifzadeh F, Poustini K, Makkizadeh MT, Kochakpour M (2007) The effects of NaCl priming on salt tolerance in canola (*Brassica napus*) seedlings grown under saline conditions. Seed Sci Technol 35:754–759
Farooq M, Aziz T, Basra SMA, Cheema A, Rehman MH (2008) Chilling tolerance in hybrid maize induced by seed priming with salicylic acid. J Agron Crop Sci 194:161–168
Farooq M, Wahid A, Kobayashi N, Fujita D, Basra SMA (2009) Plant drought stress: effects, mechanisms and management. Agron Sustain Dev 29:185–212
Farooq M, Basra SMA, Wahid A, Ahamad N (2010) Changes in nutrient homeostasis and reserve metabolism during rice seed priming: consequences for seedling emergence and growth. Agric Sci China 9:191–198
Farooq M, Hussain M, Wakeel A, Kadambot HMS (2015) Salt stress in maize: effects, resistance mechanisms, and management. A review. Agron Sustain Dev 35:461–481
Fernández-Aparicio M, Andolfi A, Evidente A, Rubiales D (2006) Orobanche species specific responses to *Trigonella foenum-graecum* root exudates. Workshop parasitic plant management in sustainable agriculture final meeting of COST849. 23–24 November 2006, ITQB Oeiras-Lisbon, Portugal, p 10
Finch-Savage WE, Gray D, Dickson GM (1991) The combined effects of osmotic priming with plant growth regulator and fungicide soaks on the seed quality of five bedding plant species. Seed Sci Technol 19:495–503
Foti S, Cosentino SL, Patanè C, D'Agosta GM (2002) Effects of osmo-conditioning upon seed germination of sorghum (*Sorghum bicolor* (L.) Moench) under low temperatures. Seed Sci Technol 30:521–533
França MB, Panek AD, Eleutherio ECA (2007) Oxidative stress and its effects during dehydration. Comp Biochem Physiol A: Mole Integr Phys 146(4):621–631
Fu J, Sun Y, Chu X, Xu Y, Hu T (2014) Exogenous 5-aminolevulenic acid promotes seed germination in *Elymus nutans* against oxidative damage induced by cold stress. PLoS One 9(9):e107152
Gadallah MMA (1999) Effects of proline and GlyBet on *Vicia faba* responses to salt stress. Plant Biol 42:249–257
Gallardo K, Job C, Groot SPC, Puype M, Demol H, Vandekerckhove J, Job D (2001) Proteomic analysis of *Arabidopsis* seed germination and priming. Plant Physiol 126:835–848
Gautam S, Singh PK (2009) Salicylic acid induced salinity tolerance in corn under NaCl stress. Acta Physiol Plant 31:1185–1190
Ghassemi-Golezani K, Khomari S, Dalil B, Hosseinzadeh-Mahootchy A, Chadordooz-Jeddi A (2010) Effects of seed aging on field performance of winter oilseed rape. J Food Agric Environ 8:175–178
Ghiyasi M, Zardoshty MR, Mogadam AF, Tajbakhsh M, Amirnia R (2008) Effect of osmo-priming on germination and seedling growth of corn (*Zea mays* L.) seeds. Res J Biol Sci 3:779–782
Golldack D, Li C, Mohan H, Probst N (2013) Gibberellins and abscisic acid signal cross talk: living and developing under unfavorable conditions. Plant Cell Rep 32:1007–1016
Gong H, Zhu X, Chen K, Wang S, Zhang C (2005) Silicon alleviates oxidative damage of wheat plants in pots under drought. J Plant Sci 169:313–321
Gopie-shkhanna V, Kannabiran K (2007) Larvicidal effect of *Hemidesmus indicus, Gymnema sylvestre*, and *Eclipta prostrata* against *Culex quinquifaciatus* mosquito larva. Afr J Biotech 6:307–311
Graeber K, Linkies A, Muller K, Wunchova A, Rott A, Leubner-Metzger G (2010) Cross-species approaches to seed dormancy and germination: conservation and biodiversity of ABA-regulated mechanisms and the Brassicaceae DOG1 genes. Plant Mol Biol 73:67–87
Guan YJ, Hu J, Wang XJ, Shao CX (2009) Seed priming with chitosan improves maize germination and seedling growth in relation to physiological changes under low temperature stress. J Zhejiang Univ Sci B 10(6):427–433
Gupta A, Dadlani M, Arun Kumar MB, Roy M, Naseem M, Choudhary VK, Maiti RK (2008) Seed priming: the aftermath. Int J Agric Environ Biotechnol 1:199–209

Gutiérrez RMP, Mitchell S, Solis RV (2008) *Psidium guajava*: A review of its traditional uses, phytochemistry and pharmacology. J Ethnopharmacol 117:1–27

Haigh AM, Barlow EWR (1987) Germination and priming of tomato, carrot, onion and sorghum seeds in a range of osmotica. J Am Soc Hortic Sci 112:202–208

Hameed A, Sheikh MA, Jamil A, Maqsood S, Basra A (2013) Seed priming with sodium silicate enhances seed germination and seedling growth in wheat (*Triticum aestivum* L.) under water deficit stress induced by polyethylene glycol. Pak J Life Soc Sci 11(1):19–24

Hameed M, Sheikh MA, Hameed A, Farooq T, Basra SMA, Jamil A (2014) Chitosan seed priming improves seed germination and seedling growth in wheat (*Triticum aestivum* L.) under osmotic stress induced by polyethylene glycol. Philipp Agric Sci 97(3):294–299

Hamid M, Ashraf MY, Ur-Rahman K, Arshad M (2008) Influence of salicylic acid seed priming on growth and some biochemical attributes in wheat grown under saline conditions. Pak J Bot 40(1):361–367

Hamidi H, Safarnejad A (2010) Effect of drought stress on alfalfa cultivars (*Medicago sativa* L.) in germination stage. Am Eurasian J Agric Environ Sci 8(6):705–709

Hamilton EW, Heckathorn SA (2001) Mitochondrial adaptations to NaCl. Complex I is protected by anti-oxidants and small heat shock proteins, whereas complex II is protected by proline and betaine. Plant Physiol 126:1266–1274

Harris D, Joshi AJ, Khan PA, Gothkar P, Sodhi PS (1999) On-farm seed priming in semi-arid agriculture: development and evaluation in maize, rice and chickpea in India using participatory methods. Exp Agric 35:15–29

Harris PA, Hollington RA, Khattak RA (2002) On-farm seed priming: a key Technology for Improving the livelihoods of resource poor farmers on saline lands. In: Ahmad R, Malik KA (eds) Prospects for saline agriculture. Kluwer Academic Publishers, Boston, pp 423–431

Hasanuzzaman M, Nahar K, Alam MM, Roychowdhury R, Fujita M (2013) Physiological, biochemical, and molecular mechanisms of heat stress tolerance in plants. Inter J Mol Sci 14(5):9643–9684

Hassanpouraghdam MB, Emarat PJ, Farsad AN (2009) The effect of osmo-priming on germination and seedling growth of *Brassica napus* L. under salinity conditions. J Food Agric Environ 7(2):620–622

Hauvermale AL, Ariizumi T, Steber CM (2012) Gibberellin signaling: a theme and variations on della repression. Plant Physiol 160:83–92

Hayashi K (2012) The interaction and integration of auxin signaling components. Plant Cell Physiol 53:965–975

Hayat S, Fariduddin Q, Ali B, Ahmad A (2005) Effect of salicylic acid on growth and enzyme activities of wheat seedlings. Acta Agron Hung 53:433–437

Hayat Q, Hayat S, Irfan M, Ahmad A (2010) Effect of exogenous salicylic acid under changing environment: a review. Environ Exp Bot 68:14–25

Hellmann H, Funk D, Rentsch D, Frommer WB (2000) Hyper-sensitivity of an *Arabidopsis* sugar signaling mutant toward exogenous proline application. Plant Physiol 122:357–368

Hentrich M, Boettcher C, Duchting P (2013) The jasmonic acid signaling pathway is linked to auxin homeostasis through the modulation of YUCCA8 and YUCCA9 gene expression. Plant J 74:626–637

Hernandez-Ruiz J, Cano A, Arnao MB (2004) Melatonin: a growth-stimulating compound present in lupine tissues. Planta 220:140–144

Heydecker W, Coolbear T (1977) Seed treatments for improved performance-survey and attempted prognosis. Seed Sci Technol 5:353–425

Heydecker W, Higgins J, Gulliver RL (1973) Accelerated germination by osmotic seed treatment. Nature 5427:42–44

Heydecker W, Higgins J, Turner YJ (1975) Invigoration of seeds? Seed Sci Technol 3:881–888

Heyl A, Riefler M, Romanov G, Schmulling T (2012) Properties, functions and evolution of cytokinin receptors. Europ J Cell Biol 91:246–256

Hodge S, Thompson GA, Powell G (2005) Application of DL-β-aminobutyric acid (BABA) as a root drench to legumes inhibits the growth and reproduction of the pea aphid *Acyrthosiphon pisum* (Hemiptera: Aphididae). Bull Entomol Res 95:449–455

Hoque MA, Okuma E, Banu MNA, Nakamura Y, Shimoishi Y, Murata Y (2007) Exogenous proline mitigates the detrimental effects of salt stress more than the betaine by increasing antioxidant enzyme activities. J Plant Physiol 164:553–561

Horvath E, Szalai G, Janda T (2007) Induction of abiotic stress tolerance by salicylic acid signalling. J Plant Growth Regul 26:290–300

Hotta Y, Tanaka T, Takaoka H, Takeuchi Y, Konnai M (1997a) Promotive effects of 5-aminolevulinic acid on the yield of several crops. Plant Growth Regul 22:109–114

Hotta Y, Tanaka T, Takaoka H, Takeuchi Y, Konnai M (1997b) New physiological effects of 5-aminolevulinic acid in plants: the increase of photosynthesis, chlorophyll content, and plant growth. Biosci Biotechnol Biochem 61:2025–2028

Hu J, Zhu ZY, Song WJ, Wang JC, Hu WM (2005) Effects of sand priming on germination and field performance in direct-sown rice (*Oryza sativa* L.). Seed Sci Technol 33:243–248

Hu L, Hu T, Zhang X, Pang H, Fu J (2012) Exogenous GlyBet ameliorates the adverse effect of salt stress on perennial ryegrass. J Am Soc Hortic Sci 137(1):38–46

Hutchkinson CE, Kileber JJ (2002) Cytokinin signaling in *Arabidopsis*. Plant Cell 14:547–559

Jahari M, Pireivatlou N, Qasimov MH (2010) Effect of soil water stress on yield and praline content of four wheat lines. Afr J Biotech 9(1):036–040

Jakab G, Ton J, Flors V, Zimmerli L, Metraux JP, Mauch-Mani B (2005) Enhancing *Arabidopsis* salt and drought stress tolerance by chemical priming for its abscisic acid responses. Plant Physiol 139(1):267–274

Janas KM, Posmyk MM (2013) Melatonin, an under estimated natural substance with great potential for agricultural application. Acta Physiol Plant 35(12):3285–3292

Jiang X, Li H, Song X (2016) Seed priming with melatonin effects on seed germination and seedling growth in maize under salinity stress. Pak J Bot 48(4):1345–1352

Jisha KC, Puthur JT (2014) Halopriming of seeds imparts tolerance to NaCl and PEG induced stress in *Vigna radiata* (L.)Wilczek varieties. Physiol Mol Biol Plants 20(3):303–312

Jisha KC, Puthur JT (2016) Seed priming with beta-amino butyric acid improves abiotic stress. Tolerance Rice Seedlings. Rice Sci 23(5):242–254

Jisha KC, Vijayakumari K, Puthur JT (2013) Seed priming for abiotic stress tolerance: an overview. Acta Physiol Plant 35(5):1381–1396

Kabiri R, Farahbakhsh H, Nasibi F (2012) Effect of drought and its interaction with salicylic acid on black cumin germination and seedling growth. World Appl Sci J 18:520–527

Kamaraj A, Padmavathi S (2012) Effect of seed enhancement treatment using leaf extract on physiological and morphological characteristics of green gram (*Vigna radiate* l.) seed adt 3. Int J Curr Res 4:110–114

Kanto U, Jutamanee K, Osotsapar Y, Chai-arree W, Jattupornpong S (2015) Promotive effect of priming with 5-aminolevulinic acid on seed germination capacity, seedling growth and antioxidant enzyme activity in rice subjected to accelerated ageing treatment. Plant Prod Sci 18(4):443–454

Kaya MD, Day S (2008) Relationship between seed size and NaCl on germination, seed vigor and early seedling growth of sunflower (*Helianthus annuus* L.). Afr J Agric Res 3(11):787–791

Kaya MD, Okcu G, Atak M, Cıkıhı Y, Kolsarıcı O (2006) Seed treatments to overcome salt and drought stress during germination in sunflower (*Helianthus annuus* L.). Eur J Agron 24:291–295

Kester ST, Geneve RL, Houtz RL (1997) Priming and accelerated ageing effect Lisoaspartylmethyltransferase activity in tomato (*Lycopersicon esculentum* L.) seed. J Exp Bot 48:943–949

Keunen E, Schellingen K, Vangronsveld J, Cuypers A (2016) Ethylene and metal stress: small molecule, big impact. Front Plant Sci 7:23

Khaliliaqdam N, Mir-Mahmoodi T (2013) Response of barley to hormonal seed priming. Agric Sci Dev 2(12):128–131

Khan AA (1991) Preplant physiological seed conditioning. Hortic Rev 13:131–181
Khan AA, Tao AL, Knypl JS, Borkowska B, Powell LE (1978) Osmotic conditioning of seeds: physiological and biochemical changes. Acta Hortic 83:267–278
Khan HA, Ayub CM, Pervez MA, Bilal RM, Shahid MA, Ziaf K (2009) Effect of seed priming with NaCl on salinity tolerance of hot pepper (*Capsicum annuum* L.) at seedling stage. Soil Environ 28:81–87
Khatami SR, Sedghi M, Sharifi RS (2015) Influence of priming on the physiological traits of corn seed germination under drought stress. Ann West Univ Timişoara Ser Biol 1:1–6
Khedr AA, Abbas MA, Abdel Wahid AA, Quick PW, Abogadallah GM (2003) Proline induces the expression of salt-stress-responsive proteins and may improve the adaptation of *Pancratium maritimum* L. to salt stress. J Exp Bot 54:2553–2562
Kibinza S, Bazin J, Bailly C, Farrant JM, Corbineau F, El-Maarouf-Bouteau H (2011) Catalase is a key enzyme in seed recovery from ageing during priming. Plant Sci 181:309–315
Kishor PBK, Hong Z, Miao GH, Hu CAA, Verma DPS (1995) Overexpression of 1-pyrroline-5-carboxylate synthetase increase proline production and confer osmotolerance in transgenic plants. Plant Physiol 108:1387–1394
Korkmaz A, Korkmaz Y, Demirkiran AZ (2010) Enhancing chilling stress tolerance of pepper seedlings by exogenous application of 5-aminolevulinic acid. Environ Exp Bot 67:495–501
Kumar M, Sirhindi G, Bhardwaj R, Kumar S, Jain G (2010) Effect of exogenous H_2O_2 on antioxidant enzymes *Brassica juncea* L. seedlings in relation to 24-epibrassinolide under chilling stress. Ind J Biochem Biophysics 47:378–382
Lantieri S, Saracco F, Kraak HL, Bino RJ (1994) The effects of priming on nuclear replication activity and germination of pepper (*Capsicum annuum* L.) and tomato (*Lycopersicon esculentum* Mill.) seeds. Seed Sci Res 4:81–87
Lara TS, Lira JMS, Rodrigues AC, Rakocevic M, Alvarenga AA (2014) Potassium nitrate priming affects the activity of nitrate reductase and antioxidant enzymes in tomato germination. J Agric Sci 6(2):72–80
Lei XY, Zhu RY, Zhang GY, Dai YR (2004) Attenuation of cold-induced apoptosis by exogenous melatonin in carrot suspension cells: the possible involvement of polyamines. J Pineal Res 36:126–131
Li DM, Zhang J, Sun WJ, Li Q, Dai AH, Bai JG (2011) 5-Aminolevulinic acid pretreatment mitigates drought stress of cucumber leaves through altering antioxidant enzyme activity. Sci Hortic 130:820–828
Li C, Wang P, Wei Z, Liang D, Liu C, Yin L, Jia D, Fu M, Ma F (2012) The mitigation effects of exogenous melatonin on salinity-induced stress in *Malus hupehensis*. J Pineal Res 53:298–306
Liang YC, Chen Q, Liu Q, Zhang WH, Ding RX (2003) Exogenous silicon (Si) increases antioxidant enzyme activity and reduces lipid peroxidation in roots of salt-stressed barley (*Hordeum vulgare L.*). J Plant Physiol 160:1157–1164
Liang Y, Sun W, Zhu YG, Christie P (2007) Mechanisms of silicon-mediated alleviation of abiotic stresses in higher plants: a review. Environ Pollut 147:422–428
Liu D, Pei ZF, Naeem MS, Ming DF, Liu HB, Khan F, Zhou WJ (2011) 5-Aminolevulinic acid activates antioxidative defence system and seedling growth in *Brassica napus* L. under water-deficit stress. J Agron Crop Sci 197:284–295
Logan BA, Kornyeyev D, Hardison J, Holaday AS (2006) The role of antioxidant enzymes in photoprotection. Photosynth Res 88:119–132
Lopez CML, Takahashi H, Yamazaki S (2002) Plant water relations of kidney bean plants treated with NaCl and foliarly applied glycinebetaine. J Agron Crop Sci 188:73–80
Ma Q, Wang W, Li Y, Li D, Zou Q (2006) Alleviation of photoinhibition in drought-stresses wheat (*Triticum aestivum*) by foliar-applied GlyBet. Plant Physiol 163:165–175
MacDonald H (1997) Auxin perception and signal transduction. Physiol Plant 100:423–430
Mansour MMF (1998) Protection of plasma membrane of onion epidermal cells by GlyBet and proline against NaCl stress. Plant Physiol Biochem 36(10):767–772

Massarat N, Siadat A, Sharafizade M, Habibi B (2014) The effect of halo priming and hydro priming ongermination and growth of maize seedling hybrid SC704 in drought and salinity stress condition. J Crop Physiol 5(19):49–59
Matilla AJ, Matilla-Vazquez MA (2008) Involvement of ethylene in seed physiology. Plant Sci 175:87–97
Matysik J, Bhalu BA, Mohanty P (2002) Molecular mechanism of quenching of reactive oxygen species by proline under stress in plants. Curr Sci 82:525–532
Mauromicale G, Cavallaro V (1996) Effects of seed osmopriming on germination of three herbage grasses at low temperatures. Seed Sci Technol 24:331–338
Mauromicale G, Cavallaro V (1997) A comparative study of the effects of different compounds on priming of tomato seed germination under suboptimal temperature. Seed Sci Technol 25:99–408
McDonald MB (2000) Seed priming. In: Black M, Bewley JD (eds) Seed technology and its biological basis. Sheffield Academic Press, Sheffield, pp 287–325
Meng JF, Xu TF, Wang ZZ, Fang YL, Xi ZM, Zhang ZW (2014) The ameliorative effects of exogenous melatonin on grape cuttings under water-deficient stress: antioxidant metabolites, leaf anatomy, and chloroplast morphology. J Pineal Res 57(2):200–212
Mexal J, Fisher JT, Osteryoung J, Reid CP (1975) Oxygen availability in polyethylene glycol solutions and its implications in plant-water relations. Plant Physiol 55:20–24
Miao BH, Han XG, Zhang WH (2010) The ameliorative effect of silicon on soybean seedlings grown in potassium-deficient medium. Ann Bot 105:967–973
Miransari M, Smith DL (2014) Plant hormones and seed germination. Environ Exp Bot 99:110–121
Morant AV, Jorgensen K, Jorgensen K, Paquette SM, Sanchez-Perez R, Moller BL, Bak S (2008) Beta-glucosidases as detonators of plant chemical defense. Phytochemistry 69:1795–1813
Mouradi M., Bouizgaren A, Farissi M, Makoudi B, Kabbadj A, Very Anne-Alienor, Sentenac H, Qaddoury A, Ghoulam C (2016) Osmopriming improves seeds germination, growth, antioxidant responses and membrane stability during early stage of Moroccan alfalfa populations under water deficit. Chilean J Agric Res 76(3). https://doi.org/10.4067/S0718-58392016000300002
Munns, Tester M (2008) Mechanisms of salinity tolerance. Annu Rev Plant Biol 59:651–681
Murata N, Takahashi S, Nishiyama Y, Allakhverdiev SI (2007) Photoinhibition of photosystem II under environmental stress. Biochem Biophys Acta 1767:414–421
Naeem M, Bhatti I, Ahmad RH, Ashraf YM (2004) Effect of some growth hormones (GA_3, IAA and kinetin) on the morphology and early or delayed initiation of bud of lentil (*Lensculinaris* Medik). Pak J Bot 36:801–809
Narwal SS (1994) Allelopathy in crop production. Scientific Publishers, Jodhpur, p 288
Nascimento WM (2003) Muskmelon seed germination and seedling development in response to seed priming. Sci Agric 60:71–75
Nascimento WM (2004) Condicionamento osmótico de sementes de hortaliças: potencialidades e implicações. Hortic Bras 16:106–109
Nouman W, Siddiqui MT, Basra SMA, Afzal I, Rehman HU (2012) Enhancement of emergence potential and stand establishment of *Moringa oleifera* Lam. by seed priming. Turk J Agric For 36:227–235
Oka Y, Cohen Y, Spiegel Y (1999) Local and systemic induced resistance to the root-knot nematode in tomato by DL-β-amino*n*- butyric acid. Phytopathology 89:1138–1143
Omami EN (2005) Response of amaranth to salinity stress. Submitted in partial fulfillment of the requirements for the degree Ph.D. Horticulture in the Department of Plant Production and Soil Science Faculty of Natural and Agricultural Sciences University of Pretoria
Orabi SA, Dawood MG, Saleem SR (2015) Comparative study between the physiological role of hydrogen peroxide and salicylic acid in alleviating low temperature stress on tomato plants grown under sand-ponic culture. Sci Agric 9(1):49–59
Osborne DJ (1983) Biochemical control of system operating in the early hours of germination. Can J Bot 61:3568–3577

Ozbingol N, Corbineau F, Groot SPC, Bino RJ, Côme D (1999) Activation of the cell cycle in tomato (*Lycopersicon esculentum* Mill.) seeds during osmoconditioning as related to temperature and oxygen. Ann Bot 84:245–251

Paredes SD, Korkmaz A, Manchester LC, Tan DX, Reiter RJ (2009) Phytomelatonin: a review. J Exp Bot 60(1):57–69

Parera CA, Cantliffe DJ (1994) Presowing seed priming. Hortic Rev 16:109–141

Passam HC, Kakouriotis D (1994) The effects of osmoconditioning on the germination, emergence and early plant growth of cucumber under saline conditions. Sci Horticult 57:233–240

Patanè C, Cavallaro V, D'Agosta G, Cosentino SL (2008) Plant emergence of PEG-osmoprimed seeds under suboptimal temperatures in two cultivars of sweet sorghum differing in seed tannin content. J Agron Crop Sci 194:304–309

Pei ZF, Ming DF, Liu D, Wan GL, Geng XX, Gong HJ, Zhou WJ (2010) Silicon improves the tolerance to water-deficit stress induced by polyethylene glycol in wheat (*Triticum aestivum* L.) seedlings. J Plant Growth Regul 29(1):106–115

Peleg Z, Blumwald E (2011) Hormone balance and abiotic stress tolerance in crop plants. Curr Opin Plant Biol 14:290–295

Perry DA, Harrison JG (1974) Studies on the sensitivity of monogerm sugar beet germination to water. Ann Appl Biol 77:51–60

Phuwiwat W, Wichittrakarn W, Laosinwattana C, Teerarak M (2012) Inhibitory effects of *Melia azedarach* L. leaf extract on seed germination and seedling growth of two weed species. Pak J Weed Sci Res 18(special issue):485–492

Pill WG (1995) Low water potential and presowing germination treatments to improve seed quality. In: Basra S (ed) Seed quality. Basic mechanisms and agricultural implications. Food Products Press, London, pp 319–360

Pill WG, Collins CM, Goldberger B, Gregory N (2009) Responses of nonprime or primed seeds of "Marketmore 76" cucumber (*Cucumis sativus* L.) slurry coated with *Trichoderma* species to planting in growth media infested with *Pythium aphanidermatum*. Sci Hortic 121:54–62

Popko J, Hänsch R, Mendel R, Polle A, Teichmann T (2010) The role of abscisic acid and auxin in the response of poplar to abiotic stress. Plant Biol 12:242–258

Posmyk MM, Kuran H, Marciniak K, Janas KM (2008) Presowing seed treatment with melatonin protects red cabbage seedlings against toxic copper ion concentrations. J Pineal Res 45:24–31

Posmyk MM, Bałabusta M, Wieczorek M, Sliwinska E, Janas KM (2009) Melatonin applied to cucumber (*Cucumis sativus* L.) seeds improves germination during chilling stress. J Pineal Res 46:214–223

Pospíšilová J, Synková H, Rululcová J (2000) Cytokinins and water stress. Biol Plant 43:321–328

Pouramir-Dashtmian F, Khajeh-Hosseini M, Esfahani M (2014) Improving rice seedling physiological and biochemical processes under low temperature by seed priming with salicylic acid. Research article. Int J Plant Anim Environ Sci 4(2):565

Prabha D, Negi S, Kumari P, Negi YK, Chauhan JS (2016) Effect of seed priming with some plant leaf extract on seedling growth characteristics and root rot disease in tomato. Int J Agric System (IJAS) 4(1):46–51

Qureshi KM, Chughtai S, Qureshi US, Abbasi NA (2013) Impact of exogenous application of salt and growth regulators on growth and yield of strawberry. Pak J Bot 45:1179–1186

Rae-Hyun K, Song HG (2007) Effects of application of *Rhodopseudomonas* sp. on seed germination and growth of tomato under axenic conditions. J Microbiol Biotechnol 17:1805–1810

Rahman MS, Miyake H, Takeoka Y (2002) Effects of exogenous GlyBet on growth and ultrastructure of salt-stressed rice seedlings (*Oryza sativa* L.). Plant Prod Sci 5:33–44

Rajpar I, Wright D (2000) Effects of sowing methods on survival, ion update and yield of wheat *(Triticum aestuvum L.)* in-sodic soils. J Agric Sci Camb 134:269–278

Rajpar I, Kanif YM, Memon AA (2006) Effect of seed priming on growth and yield of wheat (*Triticum aestivum* L.) under non-saline conditions. Int J Agric Res 1:259–264

Rangaswamy A, Purushothaman S, Devasenapaty P (1993) Seed hardening in relation to seedling quality characters of the crops. Madras Agric J 80:535–537

Raphael E (2012) Phytochemical constituents of some leaves extract of *Aloe vera* and *Azadirachta indica* plant species. Global Adv Res J Environ Sci Toxicol 1:14–17
Rathinavel K, Dharmalingam C (1999) Seed hardening to augment the productivity of cotton cv. LRA 5116 (*Gossypium hirsutum* L.). Madras Agric J 86:68–72
Raza MAS, Saleem MF, Shah GM, Khan IH, Raza A (2014) Exogenous application of glycinebetaine and potassium for improving water relations and grain yield of wheat under drought. J Soil Sci Plant Nutr 14:348–364
Renugadevi J, Vijayageetha V (2006) Organic seed fortification in cluster bean (*Cyamopsis tetragonoloba* L.) TAUB. International conference on indigenous vegetables and Legumes. Prospects for fighting poverty, hunger and malnutrition
Renugadevi J, Natarajan N, Srimathi P (2008) Efficacy of botanicals in improving seeds and seedling quality characteristics of cluster bean. Legum Res 31:164–168
Rowse HR (1995) Drum priming, a non-osmotic method of priming seeds. Seed Sci Technol 24:281–294
Ruan S, Xue QZ (2002) Effects of chitosan coating on seed germination and salt-tolerance of seedlings in hybrid rice (*Oryza sativa* L.). Acta Agron Sin 28(6):803–808
Sadeghi H, Khazaei F, Yari L, Sheidaei S (2011) Effect of seed osmopriming on seed germination behavior and vigor of soybean (*Glycine max* L.). ARPN J Agric Biol Sci 6(1):39–43
Sairam RK, Rao KV, Srivastava GC (2002) Differential response of wheat genotypes to long-term salinity stress in relation to oxidative stress, antioxidant activity and osmolyte concentration. Plant Sci 163:1037–1046
Sajedi N, Ardakani M, Madani H, Naderi A, Miransari M (2011) The effects of selenium and other micronutrients on the antioxidant activities and yield of corn (*Zea mays* L.) under drought stress. Physiol Mol Biol Plants 17:215–222
Sakamoto A, Murata N (2000) Genetic engineering of GlyBet synthesis in plants: current status and implications for enhancement of stress tolerance. J Exp Bot 51:81–88
Sakhabutdinova AR, Fatkhutdinova DR, Bezrukova MV, Shakirova FM (2003) Salicylic acid prevents the damaging action of stress factors on wheat plants. Bulg J Plant Physiol 54(special issue):314–319
Sakr MT, Arafa AA (2009) Effect of some antioxidants on canola plants grown under soil salt stress condition. Pak J Biol Sci 12:582–588
Satish S, Mohana DC, Ranhavendra MP, Raveesha KA (2007) Antifungal activity of some plant extracts against important seed borne pathogens of *Aspergillus* sp. J Agric Technol 3:109–119
Sedghi M, Nemati A, Esmailepour B (2010) Effect of seed priming on germination and seedling growth of two medicinal plants under salinity. Emir J Food Agric 22:130–139
Shah SH (2007) Effect of kinetin spray on growth and productivity of black cumin plants. Russ J Plant Physiol 54:702–705
Shao CX, Hu J, Song WJ, Hu WM (2005) Effects of seed priming with chitosan solutions of different acidity on seed germination and physiological characteristics of maize seedling. J Zhejiang Univ (Agric Life Sci) 31(6):705–708
Sharifzadeh M, Naderi A, Siadat SA, Sakinejad T, Lak S (2013) Effects of salicylic acid pretreatment on germination of wheat under drought stress. J Agric Sci 5(3):179–199
Sheikh SK, AL-Malki FM. (2011) Growth and chlorophyll responses of bean plants to the chitosan applications. Eur J Sci Res 50(1):124–134
Simiroff N, Cumbes QJ (1989) Hydroxyl radical scavenging activity of compatible solutes. Phytochemistry 28:1057–1060
Singh BG, Rao G (1993) Effect of chemical soaking of sunflower (*Helianthus annuus* L.) seed on vigour index. Indian J Agric Sci 63:232–233
Sivakumar P, Sharmila P, Saradhi PP (2000) Proline alleviates salt-stress-induced enhancement in Ribulose-1, 5-bisphosphate oxygenase activity. Biochem Biophys Res Commun 279:512–515
Sivritepe HO, Dourado AM (1995) The effect of priming treatments on the viability and accumulation of chromosomal damage in aged pea seeds. Ann Bot 75:165–171
Sivritepe N, Sivritepe HO, Eris A (2003) The effects of NaCl priming on salt tolerance in melon seedlings grown under saline conditions. Sci Horticult 97:229–237

Skopelitis DS, Paranychianakis NV, Paschalidis KA, Pliakonis ED, Delis ID, Yakoumakis DI, Kouvarakis A, Papadakis AK, Stephanou EG, Roubelakis-Angelakis KA (2006) Abiotic stress generates ROS that signal expression of anionic glutamate dehydrogenases to form glutamate for proline synthesis in tobacco and grapevine. Plant Cell 18:2767–2781

Sneideris LC, Gavassi MA, Campos ML, Damião VA, Carvalho RF (2015) Effects of hormonal priming on seed germination of pigeon pea under cadmium stress. An Acad Bras Cienc 87(3):1847–1852

Soccio M, Laus M, Spera G, Trono D, Pompa M, Flagella Z, Pastore D (2010) Mitochondrial proline oxidation is affected by hyperosmotic stress in durum wheat seedlings. Ann Appl Biol 157:1–11

Soeda Y, Konings MCJM, Vorst O, Van Houwelingen AMML, Stoopen GM, Maliepaard CA, Kodde J, Bino RJ, SPC G, van der Geest AHM (2005) Gene expression programs during *Brassica oleracea* seed maturation, osmopriming, and germination are indicators of progression of the germination process and the stress tolerance level. Plant Physiol 137:354–368

Soltani E, Akram-Ghaderi F, Maemar H (2008) The effect of priming on germination components and seedling growth of cotton seeds under drought. J Agric Sci Nat Res 14(5):13–17

Staroske N, Conrad U, Kumlehn J, Hensel G, Radchuk R, Erban A, Kopka J, Weschke W, Weber S (2016) Increasing abscisic acid levels by immunomodulation in barley grains induces precocious maturation without changing grain composition. J Exp Bot 67(9):2675–2687

Suchada B, Sarobol E, Meechoui S, Sooksathan I (2007) Drought recovery and grain yield potential of rice after chitosan application. Kasetsart J (Nat Sci) 41:1–6

Sung FJM, Chang YH (1993) Biochemical activities associated with priming of sweetcorn seed to improve vigor. Seed Sci Technol 21:97–105

Szafrańska K, Glińska S, Janas KM (2012) Changes in the nature of phenolic deposits after re-warming as a result of melatonin pre-sowing treatment of *vigna radiata* seeds. J Plant Physiol 169:34–40

Szafrańska K, Szewczyk R, Janas KM (2014) Involvement of melatonin applied to *Vigna radiata* L. seeds in plant response to chilling stress. Cent Eur J Biol 9:1117–1126

Taie HAA, Abdelhamid MT, Dawood MG, Nassar MG (2013) Pre-sowing seed treatment with proline improves some physiological, biochemical and anatomical attributes of faba bean plants under sea water stress. J Appl Sci Res 9(4):2853–2867

Tan DX, Manchester LC, Terron MP, Flores LJ, Reiter RJ (2007) One molecule, many derivatives: a never-ending interaction of melatonin with reactive oxygen and nitrogen species? J Pineal Res 42(1):28–42

Tan DX, Hardeland R, Manchester LC, Paredes SD, Korkmaz A, Sainz RM et al (2010) The changing biological roles of melatonin during evolution: from an antioxidant to signals of darkness, sexual selection and fitness. Biol Rev 85:607–623

Taylor AG, Harman GE, Nielsen PA (1994) Biological seed treatments using *Trichoderma harzianum* for horticultural crops. HortTechnology 4:105–109

Taylor AG, Beresniewicz MM, Goffinet MC (1997) Semipermeable layer in seeds. In: Ellis RH, Black M, Murdoch AJ, Hong TD (eds) Basic and applied aspects of seeds biology. Kluwer Academic Publ, Dordrecht, pp 429–436

Taylor AG, Allen PS, Bennet MA, Bradford KJ, Burris JS, Misra MK (1998) Seed enhancements. Seed Sci Res 8:245–256

Thornton JM, Collins ARS, Powell AA (1993) The effect of aerated hydration on DNA synthesis in embryos of *Brassica oleracea* L. Seed Sci Res 3:195–199

Tiong SH, Looi CY, Hazni H, Arya A, Paydar M, Wong WF, Cheah SC, Mustafa MR, Awang K (2013) Antidiabetic and antioxidant properties of alkaloids from *Catharanthus roseus* (L.) G. Don. Molecule 18:9770–9784

Tiryaki I, Buyukcingil Y (2009) Seed priming combined with plant hormones: influence on germination and seedling emergence of sorghum at low temperature. Seed Sci Technol 37:303–315

Tiryaki I, Korkmaz A, Ozbay N, Nas MN (2004) Priming in the presence in the plant growth regulators hastens germination and seedling emergence of dormant annual ryegrass (*Lilium multiflorum* Lam.) seeds. Asian J Plant Sci 3:655–659

Van Tassel DL, Roberts N, Lewy A, O'Neill SD (2001) Melatonin in plant organs. J Pineal Res 31(1):8–15

Vanderhoef LN, Dute RR (1981) Auxin-regulated wall loosening and sustained growth in elongation. Plant Physiol 67:146–149

Varier A, Vari AK, Dadlani M (2010) The subcellular basis of seed priming. Curr Sci India 99:450–456

Vendruscolo ACG, Schuster I, Pileggi M, Scapim CA, Molinari HBC, Marur CJ, Vieira LGC (2007) Stress-induced synthesis of proline confers tolerance to water deficit in transgenic wheat. J Plant Physiol 164(10):1367–1370

Voegele A, Linkies A, Muller K, Leubner-Metzger G (2011) Members of the GIBBERELLIN receptor gene family GID1 (GIBBERELLIN INSENSITIVE DWARF1) play distinct roles during *Lepidium sativum* and *Arabidopsis thaliana* seed germination. J Exp Bot 155:1851–1870

Voetberg GS, Sharp RE (1991) Growth of the maize primary root at low water potentials. III. Role of increased proline deposition in osmotic adjustment. Plant Physiol 96:1125–1130

Wahid A, Gelani S, Ashraf M, Foolad MR (2007a) Heat tolerance in plants: an overview. Environ Exp Bot 61:199–223

Wahid A, Perveen M, Gelani S, Basra SMA (2007b) Pretreatment of seed with H_2O_2 improves salt tolerance of wheat seedlings by alleviation of oxidative damage and expression of stress proteins. J Plant Physiol 164:283–294

Wahid A, Noreen A, Basra SMA, Gelani S, Farooq M (2008) Priming induced metabolic changes in sunflower (*Helianthus annuus*) achenes improves germination and seedling growth. Bot Stud 49:343–350

Wang Y, Mopper S, Hasentein KH (2001) Effects of salinity on endogenous ABA, IAA, JA, and SA in *Iris hexagona*. J Chem Ecol 27:327–342

Wang LJ, Jiang WB, Huang BJ (2004) Promotion of 5-aminolevulinic acid on photosynthesis of melon (*Cucumis melo*) seedlings under low light and chilling stress conditions. Physiol Plant 121:258–264

Wang AX, Wang XF, Ren YF, Gong XM, Bewley JD (2005) Endo-b-mannanase and b-mannosidase activities in rice grains during and following germination, and the influence of gibberellin and abscisic acid. Seed Sci Res 15:219–227

Wang X, Wei Z, Liu D, Zhao G (2011) Effects of NaCl and silicon on activities of antioxidative enzymes in roots, shoots and leaves of alfalfa. Afr J Biotechnol 10:545–549

Warren JM, Bennett MA (1999) Bio-osmopriming tomato (*Lycopersicon esculentum* Mill.) seeds for improved stand establishment. Seed Sci Technol 27:489–499

Weeda S, Zhang N, Zhao X, Ndip G, Guo Y, Buck GA, Fu C, Ren S (2014) *Arabidopsis* transcriptome analysis reveals key roles of melatonin in plant defense systems. PLoS One 9:e93462

Wei W, Qing-Tian L, Ya-Nan C, Reiter RJ, Xiao-Min Y, Dan-Hua Z, Zhang W (2015) Melatonin enhances plant growth and abiotic stress tolerance in soybean plants. J Exp Bot 66:695–707

Welbaum GE, Bradford KJ (1990) Water relations of seed development and germination in muskmelon (*Cucumis melo* L.). IV. Characteristics of the perisperm during seed development. Plant Physiol 92:1038–1045

Welbaum GE, Shen Z, Oluoch MO, Jett LM (1998) The evolution and effects of priming vegetable seeds. Seed Technol 20:209–235

Worrall D, Holrod GH, Moore JP, Glowacz M, Croft P, Taylor JE, Paul ND, Roberts MR (2012) Treating seeds with activators of plant defence generates long-lasting priming of resistance to pests and pathogens. New Phytol 193:770–778

Xin Z, Browse J (2000) Cold comfort farm: the acclimation of plants to freezing temperatures. Plant Cell Environ 23:893–902

Yamaguchi S (2008) Gibberellin metabolism and its regulation. Annu Rev Plant Biol 59:225–251

Yi Z, Zhang M, Ling B, Xu D, Ye J (2005) Inhibitory effects of *Lantana camara* and its contained phenolic compounds in *Eichhornia crassipes* growth. J Appl Ecol 17:1637–1640

Yildirim E, Karlidag H, Dursun A (2011) Salt tolerance of physalis during germination and seedling growth. Pak J Bot 43(6):2673–2676

Zapata PJ, Serrano M, Pretel MT, Amorós A, Botella MA (2004) Polyamines and ethylene changes during germination of different plant species under salinity. Plant Sci 167:781–788

Zhang S, Hu J, Zhang Y, Xie XJ, Knapp A (2007) Seed priming with brassinolide improves lucerne (*Medicago sativa* L.) seed germination and seedling growth in relation to physiological changes under salinity stress. Aust J Agric Res 58:811–815

Zhang M, Wang Z, Yuan L, Yin C, Cheng J, Wang L, Huang J, Zhang H (2012) Osmopriming improves tomato seed vigor under aging and salinity stress. Afr J Biotechnol 11:6305–6311

Zhang CP, Li YC, Yuan FG, Hu SJ, Liu HY, He P (2013a) Role of 5-aminolevulinic acid in the salinity stress response of the seeds and seedlings of the medicinal plant *Cassia obtusifolia* L. Bot Stud 54:18

Zhang N, Zhao B, Zhang HJ, Weeda S, Yang C, Yang ZC, Ren S, Guo YD (2013b) Melatonin promotes water stress tolerance, lateral root formation, and seed germination in cucumber (*Cucumis sativus* L.). J Pineal Res 54:15–23

Zhang HJ, Zhang N, Yang RC, Wang L, Sun QQ, Li DB, Cao YY, Weeda S, Zhao B, Ren S, Guo YD (2014) Melatonin promotes seed germination under high salinity by regulating antioxidant systems, ABA and GA_4 interaction in cucumber (*Cucumis sativus* L.). J Pineal Res 57:269–279

Zhong YP, Wang B, Yan JH, Cheng LJ, Yao LM, Xiao L, Wu TL (2014) DL-β-amino butyric acid induced resistance in soybean against *Aphis glycines* Matsumura (Hemiptera: Aphididae). PLoS One 9(1):1–11

Zhou YG, Yang YD, Qi YG, Zhang ZM, Wang XJ, Hu XJ (2002) Effects of chitosan on some physiological activity in germinating seed of peanut. J Peanut Sci 31(1):22–25

Zhou YH, Yu JQ, Mao WH, Huang LF, Song XS, Nogues S (2006) Genotypic variation of Rubisco expression, photosynthetic electron flow and antioxidant metabolism in the chloroplasts of chill-exposed cucumber plants. Plant Cell Physiol 47:192–199

Zimmerli L, Hou BH, Tsai CH, Jakab G, Mauch-Mani B, Omerville S (2008) The xenobiotic beta-amino butyric acid enhances *Arabidopsis* thermo-tolerance. Plant J 53(1):144–156

Part II

Case Studies on Priming

Seed Priming: A Low-Cost Technology for Resource-Poor Farmers in Improving Pulse Productivity

11

Malay K. Bhowmick

Abstract

Pulses continue to become an integral component of sustainable crop production system for their ability towards biological nitrogen fixation, low water requirement, comparatively shorter duration and capacity to withstand abnormal weather conditions. Average pulse productivity in India often becomes low at famers' fields due to a number of constraints. Moreover, these crops are mainly grown under energy-starved conditions by small and marginal farmers, who do not have sufficient resources with regard to fine seedbed preparation before sowing as well as post-sowing management practices along with appropriate plant protection measures. Hence, poor seed germination, slow and insufficient seedling emergence and inappropriate stand establishment are not uncommon in stress-prone areas. Seed priming is a simple, inexpensive, highly effective and risk-averting tool for improving plant acclimatization under both biotic and abiotic stresses, besides ensuring uniform seed germination, rapid emergence, better stand establishment, improved crop growth, and higher productivity of pulses. The possible technological options for seed priming in pulses include hydro-priming, osmo-priming, bio-priming, nutri-priming, solid matrix priming, hormo-priming, halo-priming, nano-priming and ultra-priming. The present chapter highlights different aspects, techniques and importance of seed priming with particular reference to pulse crops.

Keywords

Crop establishment · Pulse productivity · Seed priming

M. K. Bhowmick (✉)
Rice Research Station (Government of West Bengal), Chinsurah, West Bengal, India

© Springer Nature Singapore Pte Ltd. 2018

A. Rakshit, H. B. Singh (eds.), *Advances in Seed Priming*,
https://doi.org/10.1007/978-981-13-0032-5_11

11.1 Introduction

Belonging to the Fabaceae family, pulses represent the world's third largest group of plant life. Humans have cultivated pulse crops since the dawn of farming as one of the first plants in the world to be domesticated. Pulses are one of the most important sources of high-quality dietary protein, especially for a large part of the population who are vegetarian. These crops are hardier than most other crops, require less quantity of water and withstand severe weather extremes like droughts and floods, where other crops usually fail, thereby playing an important role in mitigating, adapting and reducing the adverse effects of climate change. Besides, their unique attributes, especially their ability to biologically fix nitrogen, also have a direct and positive impact on soil biodiversity. Being legumes, they become an ideal and integral component of conservation agriculture (Siddique et al. 2012). As recently reported by the Food and Agriculture Organization of the United Nations (FAO) on World Soil Day, 'climate smart' pulses are essential to global food security by delivering high-nutrition protein to people and critical nutrients to soil (FAO 2016a).

Globally, India becomes a home to about 24% of undernourished people. About 15.2% of people in the country are undernourished. This signifies the importance of pulses in food and nutrition security for Indian population. However, India is the largest producer, importer and consumer of pulses, accounting for 25% of global production from 35% of global area under pulses (Ahlawat et al. 2016; FAO 2016b). More than a dozen pulses crops are grown in different parts of the country. Among them, chickpea (Bengal gram: *Cicer arietinum*), pigeon pea (redgram, *arhar*, *tur*: *Cajanus cajan*), mung bean (green gram, *moong*: *Vigna radiata*), urdbean (black gram, *biri*, mash: *Vigna mungo*), lentil (masur: *Lens culinaris* subsp. *culinaris*) and field pea (*mator*: *Pisum sativum*) are the most common ones. Other pulses of regional or local importance include *Lathyrus* (grass pea, chickling pea, *khesari*: *Lathyrus sativus*), rajmash (French bean, kidney bean, common bean: *Phaseolus vulgaris*), moth bean (moth, Turkish gram: *Vigna aconitifolia*), horse gram (*kulthi*: *Macrotyloma uniflorum*), cowpea (lobia, *barbati*: *Vigna unguiculata*), soybean (*soya bean*: *Glycine max*), broad bean (faba bean, *bakla*: *Vicia faba*) and ricebean (climbing mountain bean, red bean: *Vigna umbellata*). Still the on-farm productivity of pulses as a whole in India is less than half of the productivity levels in the USA and Canada, because of the fact that pulse crops are mostly confined to marginal areas under stress-prone environments, getting subjected to a number of constraints. Moreover, these crops are mainly grown by small and marginal farmers, who do not have sufficient resources, unlike progressive farmers, in investing much towards fine seedbed preparation since sowing along with agronomic management practices including plant protection measures till crop harvest.

Since pulse seeds tend to absorb water vapour in humid atmosphere, the decline in germination and viability of seeds becomes a major problem upon ageing of seeds, especially under hot and humid climatic situations. In many agricultural areas, the major cause of poor stand establishment and low crop yield is due to prevalence of unfavourable environmental conditions for seed germination and seedling emergence, especially during *kharif* season. Poor crop establishment is a major constraint for pulse production, particularly in drought-prone environment, exhibiting irregular (asynchronous) emergence trends that can extend over a long period of time (Naseem et al. 1997; Rahmianna et al. 2000; Nayban et al. 2017). Such an unreliable, irregular and uneven germination as well as delayed emergence is also of common occurrence during pre-*kharif* season. In case of *rabi* season, pulses are mostly grown on the residual soil moisture in rice fallows as *utera* (*paira* or relay) crop (Bhowmick et al. 2005, 2010, 2011, 2014a; Gupta and Bhowmick 2005, 2012a, b, 2013; Mondal and Ghosh 2005; Biswas and Bhowmick 2015). Low productivity especially under *utera* system is a major problem associated with *rabi* pulses (Bhowmick et al. 2005). Even there is a limited scope for agronomic manipulation under rice-*utera* system, there is a huge potential for increasing cropping intensity in considerable areas that remain idle after *kharif* (*aman*) rice (Rautaray 2008).

There are many reports towards improving seed germination as well as crop performance in the field, involving a wide range of techniques, which may be categorized into physiological (seed priming, coating and pelleting), physical (magnetic, radiation and plasma) and biological (seed enhancement) aspects (Afzal et al. 2016). Of these, seed priming (interchangeable with on-farm seed priming or invigoration) is a simple, effective and low-cost technology towards improving seed germination, seedling emergence, stand establishment, crop growth, nodulation and productivity of pulses (Ali et al. 2005; Gupta and Bhowmick 2012a; Sujatha et al. 2013; Ghosh et al. 2016; Bhowmick et al. 2017). Not only that, it can also be an appropriate tool for resource-poor small and marginal farmers towards better managing climate risk crop husbandry, besides guaranteeing the full expression of crop yield potential (Padgham 2009). The present chapter is intended to illustrate different aspects, techniques and importance of seed priming with particular reference to pulses.

11.2 Basis of Seed Priming

The theory of seed priming was proposed by Heydecker in 1973 (Heydecker 1973). Seed priming is a controlled hydration process, used widely for enhancing seed performance by improving the rate and uniformity of germination and decreasing seed sensitivity to external factors (Corbineau and Côme 2006; Sharma et al. 2015; Afzal et al. 2016). Once sown in the fields, seeds are to spend a significant period of time for absorbing water from the soil. By reducing this time to a minimum, seeds can be made to germinate, and the germinating seeds to emerge more quickly and uniformly in the way, known as seed priming. This technique is based on the progress of germination in three phases: imbibition (phase I), transition (the true germination process) or germination *stricto* sensu (phase II), and growth (phase III). In particular, water uptake follows this triphasic pattern with an initial rapid imbibition phase (phase I), followed by a lag period (phase II) referred to as germination *stricto* sensu and finally a second water uptake phase (phase III) associated with radicle growth (Côme 1980; Bewley 1997; Hadas 2004). There is a rapid initial water uptake due to the seed's low water potential during imbibition that results in the resumption of respiratory activity and protein synthesis using extant messenger ribonucleic acid (mRNA). Most important one is the phase II, which is associated with a slow increase in seed water content along with initiation of physiological activities related to germination (Varier et al. 2010). Various cellular and biochemical events including mitochondria repair and synthesis, protein synthesis relying on the translation of new RNA, changes in soluble sugars, etc. are carried out in phase II (Bray 1995; Bewley 1997). The basis of seed priming is to allow a controlled water uptake by the seeds up to the end of phase II, before the radicle protrudes from the seed coat. Since most seeds are desiccation tolerant up to this stage, the germination process can be arrested by drying. Phase II is more sensitive to external factors than phase III (Côme and Thévenot 1982). Therefore, seeds that have passed through this phase in the priming process germinate in a wider range of environmental factors than non-primed seeds (Corbineau and Côme 2006). In phase III, the process of germination is completed, culminating in radicle emergence (Bennett et al. 2013). Varier et al. (2010) described the sub cellular basis of seed priming in detail.

11.3 Different Techniques of Seed Priming

Seed priming can be accomplished through a number of shotgun techniques such as hydro-priming, osmo-priming, bio-priming, nutri-priming, solid matrix priming, hormonal priming using plant growth regulators, halo-priming, etc. (Bose 2014; Singh et al. 2015a; Nayban et al. 2017). These techniques have been briefed hereunder.

11.3.1 Hydro-priming

As proposed by Harris (1992), it is a very low-cost technique, designated as on-farm seed priming, that involves soaking of seeds in water for a certain period of time prior to sowing (Pill and Necker 2001). It may or may not be followed by air-drying of the seeds (Nawaz et al. 2013). This pre-sowing seed treatment, known as hydro-priming, allows the seeds to imbibe water and go through the first phase of germination in which pregermination metabolic activities get started, whilst the latter two phases of germination are inhibited (Pill and Necker 2001). Another kind of hydro-priming is known as 'on-farm steeping' that is practiced by incubating seeds (cereals, legumes) for a limited time in warm water (Sivasubramaniam et al. 2011). A drum is used for this purpose, and the water can also be applied as humid air or as water vapour, and hence, it is also called as drum priming (Rowse 1996; Taylor et al. 1998). Seed priming in water has been found to decrease time between sowing and emergence and to improve seedling vigour (Harris 1996; Parera and Cantliffe 1994). Besides, it can reduce the effect of salinity on the morphological parameters of the plants (Rafiq et al. 2006). Any factor that facilitates rapid germination may contribute to the establishment of a successful crop. Although soaking the seeds in water and drying before sowing is the easiest way to achieve hydration, a major disadvantage is that it may result in uneven hydration and non-uniform germination (Nawaz et al. 2013). According to Swarnkar et al. (2012), seed priming (soaking seeds overnight in water) helps germination and gives seedlings a head start. After priming, the seeds need to dry just enough to be sown easily. For example, soaking chickpea before sowing costs practically nothing, but such a simple practice can raise the seed yield of chickpea by over 40% in many places (Swarnkar et al. 2012). Significant increase in seed yield of *Lathyrus* has been reported in sowing of sprouted seeds under rice-*utera* system in rice fallows (Bhowmick et al. 2012, 2014a, b; Bhowmick 2013). Yucel (2012) studied on priming of lentil seeds by soaking in 1% KH_2PO_4 and 1% KNO_3 for 12 h, comparing with hydro-priming in distilled water (24 h) at 24 °C and control (untreated seeds) when primed seeds were subjected to germination under six different constant temperatures (5, 10, 15, 20, 25 and 30 ± 0.5 °C). Amongst these treatments, seed priming with water for 12 h at 15 °C recorded the highest germination percentage along with the least mean germination time and synchrony. In general, soaking seeds of pulses in water for 4–8 h and subsequent sowing of these primed seeds at 8–10 days before harvest of rice proved to be a simple and effective practice to improve seed germination, crop growth, plant stand and seed yield by 10–20% under relay cropping system (Urkurkar 2012; Bhowmick et al. 2010, 2013, 2014a).

11.3.2 Osmo-priming

Osmotic priming or osmo-priming (also known as osmotic conditioning or osmo-conditioning) is a commercially and widely used standard technique, employed for improving seed germination and vigour, requiring only small quantities of seeds (Halmer 2004). It is a pre-sowing treatment in an osmotic solution that allows seeds to imbibe water to proceed to the first stage of germination but prevent radicle protrusion through the seed coat (Heydecker et al. 1973). It involves controlled imbibition of seeds to start the initial events of germination followed by seed drying up to its original weight. Osmo-priming has many advantages including rapid and uniform germination and emergence, improved seedling growth and better stand establishment under any environmental and soil conditions (Chiu and Sung 2002; Nawaz et al. 2013; Singh et al. 2015a; Afzal et al. 2016). In this technique, seeds are soaked for a certain period of time in well aerated solutions with a low water potential and later washed and air-dried. The low water potential of the solutions can be achieved by adding osmotica like sugar, polyethylene glycol, i.e. PEG [$H\text{-}(O\text{-}CH_2\text{-}CH_2)_n\text{-}OH$], glycerol ($C_3H_8O_3$), sorbitol ($C_6H_{14}O_6$), mannitol ($C_6H_{14}O_6$) or salts like potassium chloride (KCl). Some more examples of the osmotica that can be used include potassium nitrate (KNO_3), monopotassium phosphate or potassium dihydrogen orthophosphate or potassium dihydrogen phosphate (KH_2PO_4), dipotassium phosphate or dipotassium hydrogen orthophosphate (K_2HPO_4), tripotassium phosphate (K_3PO_4), calcium chloride ($CaCl_2$), zinc sulphate ($ZnSO_4$), borax or sodium tetraborate decahydrate ($Na_2B_4O_7 \cdot 10H_2O$), magnesium chloride ($MgCl_2$), manganese sulphate ($MgSO_4$), sodium chloride (NaCl), sodium sulphate (Na_2SO_4), and different organic compounds, viz. agrosan, cycocel, citric, furamic, succinic, malic acids, purines, pyrimidines, caffeine, uracil, xanthine and uridine diphosphate (De Chandra

1999; Sivasubramaniam et al. 2011). Sivasubramaniam et al. (2011) made an indicative list of different osmotica used for seed priming in food legumes.

Bhowmick et al. (2012, 2014b, c) made a study at Pulses and Oilseeds Research Sub-station, Beldanga, Murshidabad, West Bengal during *rabi* season to evaluate different levels of seed priming in *Lathyrus* raised under rice-*utera* system. The treatments included seed soaking in water (6 h), seed soaking in 2% KH_2PO_4 solution (6 h), use of sprouted seeds and control (no soaking). Amongst these treatments, use of sprouted seeds and 2% KH_2PO_4 soaked seeds significantly recorded higher number of pods plant^{-1}, resulting in higher seed yield. Higher plant height, more number of branches plant^{-1}, seeds pod^{-1} as well as 100-seed weight were also registered under these treatments which ultimately exhibited 17.64–24.61% yield advantages over no soaking plots. Comparatively better performance of crop plants under all the seed priming treatments except 'no soaking' could be attributed to their good establishment as well as tolerance to soil moisture stress, which might be explained due to a number of physico-chemical changes within the cytoplasm including greater hydration of colloids, higher viscosity and elasticity of the protoplasm, etc. (Solaimalai and Subburamu 2004). Values of all the growth and yield attributes along with seed yield were, however, found to be the lowest when non-soaked (non-primed) seeds were sown (Bhowmick et al. 2014b, c). Similar findings were also reported earlier in *Lathyrus* (Bhowmick 2013; Bhowmick et al. 2017), lentil (Gupta and Bhowmick 2005; Bhowmick 2010) and chickpea (Bhowmick et al. 2010, 2013) under rice-*utera* system.

Singh et al. (2014) studied on osmo-priming in cowpea. Three levels of osmo-priming duration by soaking in 1% KNO_3 salt for 6, 8 and 10 h were compared with hydro-primed control (10 h). The results showed that osmo-priming with KNO_3 for different durations were superior to unprimed treatment in terms of seed germination, emergence, plant height and dry matter accumulation in cowpea. Use of primed seeds (both osmo-priming and hydro-priming) was found to improve the performance of cowpea crop. Moreover, osmo-priming with KNO_3 salt (soaked in 1% KNO_3 salt solution and dried before sowing) for 6 h resulted greater seed germination and seedling height than hydro-priming.

11.3.3 Bio-priming

Bio-priming is a process of biological seed treatment that refers to a combination of seed hydration (physiological aspect of disease control) and inoculation (biological aspect of disease control and nutrient uptake) of seed with beneficial organism to protect seed. It integrates both the biological (seed inoculation) and physiological (seed hydration) aspects. It is an ecological approach using either bacteria or selected fungal antagonists against the soil- and seed-borne pathogens (Sivasubramaniam et al. 2011; Afzal et al. 2016; Rakshit et al. 2015). Bio-priming techniques involve the addition of beneficial rhizosphere microorganisms in the priming process, either as a method for efficient delivery to the crop or to control pathogen proliferation during priming itself (Sharma et al. 2015; Meena et al. 2016, 2017).

Bio-priming with plant growth promoting rhizobacteria (PGPR) has been an alternative to mineral fertilizers to increase crop yield by their ability to produce and/or change the concentration of plant hormones, asymbiotic N_2 fixation and solubilization of mineral phosphate and other nutrients. Some most prominent bioagents are *Trichoderma*, *Pseudomonas*, *Glomus*, *Bacillus*, *Agrobacterium* and *Gliocladium*. Inoculation of seeds with these beneficial microorganisms in combination with priming (bio-priming) potentially can promote rapid and more uniform seed germination and plant growth, improve the versatility of crop performance and stabilize the efficacy of biological agents in the present fragile agricultural ecosystem by reducing dependency on chemical inputs (Sivasubramaniam et al. 2011; Rakshit et al. 2014; Sharma et al. 2015) and by ameliorating a wide variety of biotic, abiotic and physiological stresses to seed and seedlings (Mastouri et al. 2010; Rajendra Prasad et al. 2017). Seed bio-priming process has been discussed in detail (Sivasubramaniam et al. 2011; Rakshit et al. 2014). According to Vishwas et al. (2017), seed priming with *Rhizobium* + *Pseudomonas* at 10% for 12 h recorded significantly higher percent and speed of germination, and seedling vigour, and was followed by seed priming with *Rhizobium* at 10% for 12 h in chickpea. Rakshit et al. (2014), however, made an indicative list of bioagents used for different food legumes.

11.3.4 Nutri-priming

Nutrient priming or nutri-priming means soaking of seeds in nutrient solution of a specific concentration, instead of pure water, for a certain period of time or duration prior to sowing (Shivay et al. 2016). Seed priming with nutrients (macro or micro) can increase seed nutrient content and improve seed quality for better germination, seedling establishment, plant growth, nutrient uptake and water use efficiency of several crop species. Rakshit et al. (2013) discussed in detail about the usefulness of micronutrient seed priming in integrated nutrient management. In micronutrient seed priming, micronutrients are used as osmotica (Imran et al. 2004; Singh 2007). For example, seed priming with 0.02% and 0.04% sodium molybdate dihydrate ($Na_2MoO_4.2H_2O$) for 5 h has been reported to improve productivity of mung bean (Umair et al. 2011).

Swarnkar et al. (2012) reported improvement in germination due to water soaking of chickpea seeds. Adding *Rhizobium* to the soaking water further helps the seedlings form nitrogen-fixing nodules that can capture nitrogen from the air. This is especially true for acidic soils, often encountered in rice fallows, where many legumes become relatively unproductive because of limited nodulation owing to poor availability of molybdenum (Mo). Large yield increases in chickpea are possible at farmers' fields following seed priming with tiny amounts of Mo because it is an essential micronutrient for chickpea. Priming seeds with both Mo and *Rhizobium* can raise seed yield by a third compared to priming with just water. According to Swarnkar et al. (2012), 0.5 g sodium molybdate (Na_2MoO_4) and 5.0 g *Rhizobium* inoculum can be added in 1 L of water required for each kilogram of chickpea seed.

Alkaline soils are often deficient in zinc (Zn). In moderately Zn-deficient soils, zinc priming is an effective tool, whereas it may not fulfil the zinc requirement of the plant under severe deficiency. For example, seed priming alone is not sufficient to fulfil the requirement of kidney bean (Harris et al. 2008). However, seed priming with Zn is helpful in improving crop emergence, stand establishment, plant growth, seed yield and nutrient concentration (Shivay et al. 2016). Seed priming of chickpea seeds in a 0.05% solution of zinc sulphate heptahydrate ($ZnSO_4.7H_2O$) has been found quite effective to exhibit 19% higher seed yield and 29% more Zn concentration in seeds over that of non-primed seeds (Harris et al. 2008).

Dey et al. (2014) studied on chickpea with four levels of seed priming (control, hydro-priming, 0.05% zinc sulphate priming and 0.1% ammonium molybdate priming) under rainfed condition. In case of priming treatments, seeds were fully immersed in priming solution at room temperature for 8 h. All seeds were rinsed thoroughly with tap water to remove excess salts from seed coat and lightly hand-dried using blotting paper. Afterwards, primed seeds were allowed to dry back to their original moisture content under shade for 3 days and in sun for 1 day. Significantly higher values of major growth and yield attributes (plant height, branches plant^{-1}, pod number plant^{-1} and pod weight plant^{-1}) were recorded in plots having seed primed with 0.05% $ZnSO_4.7H_2O$ and was followed by seed priming with 0.1% ammonium molybdate [$(NH_4)_2MoO_4$] and simple water (hydro-priming). The positive effect of seed priming on plant growth might be due to earlier emergence of seedlings (Passam and Kakouriotis 1994), leading to uniform crop stand and more vigorous plants. Moussavi-Nik et al. (1997) reported the stimulatory effect of Zn on crop germination and establishment as well as the final yield, especially in Zn-deficient soils. Use of zinc sulphate might play the role in protein synthesis, cell membrane function and cell elongation (Cakmak 2000). Being an important element for nitrogenase enzyme, Mo helped in nitrogen fixation as well as crop productivity when seeds were primed with 0.1% ammonium molybdate (Dey et al. 2014).

11.3.5 Solid Matrix Priming

Solid matrix priming (SMP) or matriconditioning is a process in which seeds are mixed with a solid carrier material (e.g. ground leonardite shale, agro-lig, vermiculite, expanded calcined clay, synthetic calcium silicates, etc.) and water in known proportion (Khan 1992; Parera and Cantliffe 1992; Harman and Nelson 1994; Rogis et al. 2004; Hacısalihoglu 2007; Ermiş et al. 2016). This mixture allows the seeds to imbibe and attain the threshold moisture content but prevents radicle emergence (Taylor et al. 1988). Both the holding capacity and density of solid carriers and the amount of carrier relative to seed and water used for optimum conditioning have differed greatly (Khan et al. 1992). Bioprotectants and/or chemical pesticides may be used in conjugation with SMP. These materials may be added first to seeds as slurry, followed by the addition of solid particulate and water (Taylor and Harman 1990). SMP can increase germination at suboptimal temperatures and increase anti-oxidative enzyme activity and levels of antioxidants in treated seeds (Wang et al. 2003; Kepczynska et al. 2007). Chitosan is a large cationic polysaccharide mainly obtained from waste materials from seafood processing. As an abundant and comparatively cheap organic compound in China, chitosan can be used for seed priming (Zhou et al. 2002). For example, SMP with 0.1% chitosan at 5% moisture level and 0.2% chitosan at 10–20% moisture level can enhance seed germination and seedling invigoration that synchronize with emergence of radicle and salinity stress tolerance up to the level of 6 dS m^{-1} in mung bean (Sen and Mandal 2016).

11.3.6 Hormo-priming

Hormonal priming or hormo-priming is the pre-seed treatment with different hormones, viz. salicylic acid (SA), abscisic acid (ABA), ascorbate, kinetin, gibberellic acid (GA_3), etc., which promote the growth and development of seedlings (Nawaz et al. 2013; Singh et al. 2015a). Plant growth hormones or their derivatives contained by several products are indole-3-butyric acids (IBA), an auxin and kinetin type of cytokinin (Afzal et al. 2016). Cytokinins are known to play a vital role in all phases of plant development starting from seed germination up to senescence. These cytokinins can be used as priming agents, which are mainly involved in the breakdown of dormancy of some seeds (Arteca 1996). For example, GA_3 treated soybean seeds recorded better field performance due to its stimulation effect in the formation of enzymes which are important in the early phase of germination and emergence (Riedell et al. 1985; Maske et al. 1997; Assefa et al. 2010). Bassi et al. (2011) reported that priming with 50 ppm GA_3 for 2 h enhanced emergence, germination percentage and speed of germination as compared to non-primed seed lots in soybean.

Various naturally occurring growth promoting substances such as moringa (*Moringa oleifera*) leaf extract, chitosan, sorghum (*Sorghum bicolor*) water extract and seaweed extract are also commonly used for seed priming (Afzal et al. 2016). Seed soaked with chitosan increased the energy of germination, germination percentage, lipase activity and GA_3 and indole acetic acid (IAA) levels in peanut (Zhou et al. 2002).

11.3.7 Halo-priming

Halo-priming refers to soaking of seeds in solution of inorganic salts such as sodium chloride (NaCl), potassium nitrate (KNO_3), calcium chloride ($CaCl_2$), calcium sulphate ($CaSO_4$), etc. A number of studies have shown a significant improvement in seed germination, seedling emergence and establishment and final crop yield in salt-affected soils in response to halo-priming (Nawaz et al. 2013; Singh et al. 2015a). Halo-priming not only promotes seed germination but also stimulates subsequent growth, thereby enhancing final yield of certain food legumes (Eleiwa 1989; Sallam 1999). In an evaluation study on the response of seeds primed with NaCl solution at different salinity levels (0, 3, 6 and 9 dS m^{-1}) in relation to early growth stage, seed priming with NaCl proved to be better treatment as compared to non-primed seeds (Khan et al. 2009). Seed treatment with $CaCl_2$ or KNO_3 generally exhibited improvement in proteins, free amino acids and soluble sugars during germination under salt stress in pigeon pea (Jyotsna and Srivastava 1998). Beneficial effects of KCl have been reported in redgram (Rajandran 1982) and green gram (Basha 1982). Major yield attributes (pods plant^{-1} and seeds plant^{-1}) along with seed yield plant^{-1} in soybean were significantly enhanced by seed priming particularly with KNO_3. On contrary, seedling emergence percentage and germination time were significantly adversely affected by seed priming (KH_2PO_4 and KNO_3) as compared to non-primed seeds (Golezani et al. 2011). Pradhan et al. (2017) reported higher germination percentage, energy of emergence, seedling length, seedling dry weight and seed vigour index when seeds of black gram were soaked in 1% KCl solution (12 h), compared with control (unsoaked seed), and it was followed by priming in 1% KNO_3 solution (12 h).

11.3.8 Nano-priming

It is a new method of seed priming with nanoparticles for increasing germination percentage and improving seedling vigour. Maroufi et al. (2011) reported significant effect of nano-priming with the spray of titanium dioxide (TiO_2) nanoparticles (0.02%) in terms of improved seed germination percentage, seedling dry weight and seedling vigour in mung bean. However, comprehensive studies on nano-priming in pulses have not been elucidated.

11.3.9 Ultra-priming

Ultrasonic priming or ultra-priming involves the use of ultrasound energy for achieving more complete and faster seed germination as well as higher crop yields. Ultrasound energy (or simply ultrasound) is a type of mechanical energy called sound characterized by vibrating or moving particles within a medium. Ultrasound is distinguished by vibrations with a frequency higher than 20,000 Hz, which is not audible to humans. These ultrasounds only impose mechanical pressure on seeds to break the dormancy; therefore, no chemicals would contaminate the seeds (Nazari and Eteghadipour 2017). Water is used as a medium for treating the seeds by ultrasounds (Nazari et al. 2014). Seeds of the desired species are placed into an ultrasonic wave emitting apparatus containing water (Sharififar et al. 2015). Ultrasonication enforces the uptake of water, nutrients and/or other beneficial substances into the seed grains. Furthermore, the ultrasonic waves help in activating enzymatic and other biological reactions within the plant cell, resulting in a faster and more uniform germination. Even weak seeds/grains can be activated so that the overall seed performance gets improved (Hielscher 2017). The ultrasound technology has been successfully used in many mass transfer processes in food, such as in drying, extraction, osmotic dehydration, desalting and hydration (Miano et al. 2016). For example, use of ultrasound in mung bean seed priming was found to reduce ~25% of the hydration process time, causing an acceleration of seed germination (Miano et al. 2016). Ultrasonic priming is a simple low-cost method to improve seed germination percentage, seed vigour index and root and shoot length of seedlings.

11.4 Importance of Seed Priming in Pulse Cultivation

In seed priming, seeds are partially hydrated to allow metabolic events to occur without actual germination and then redried to permit routine handling. Such a simple, inexpensive and economically viable technology offers a handful of benefits in pulse cultivation by resource-poor farmers. These are discussed hereunder.

11.4.1 Enhancement in Seed Germination/Seedling Emergence

Soil moisture is the most serious constraint for cultivating *rabi* pulses in rice fallows. Because, growing of *rabi* pulses solely depends upon the availability of soil moisture in the field after rice harvest when availability of soil moisture becomes low, followed by a fast decline in water table with the advancement of *rabi* season, resulting into mid- and terminal drought at flowering and pod filling stages, thereby adversely affecting pulse productivity in rice fallows (Ali et al. 2014; Ghosh et al. 2016). Hence, timely sowing of *rabi* pulses is very crucial in view of the moisture deficit during critical periods under rice fallow conditions (Kumar et al. 2012; Ali et al. 2014). Germination is an important aspect of seedling establishment under these conditions. Rice fallows affect seed germination, seedling emergence and crop establishment due to disruption of soil structure, soil water deficit, poor aeration and mechanical impedance of the seed zone. Of these, soil hardiness is the major limiting factor, followed by low organic matter content in the soil. Soil hardiness in the puddle rice fields deteriorates the hydraulic properties of the soil, which adversely affects the distribution of soil moisture as well as root growth of deep-rooted pulse crops. Such a hostile environment creates potential threat to microbial activity, nutrient availability, root growth (mostly confined to top soil layer) and water and nutrient uptake, which together make the subsoil resources unutilized in rice fallows. Even under relay (*paira*) cropping, plant population often becomes low due to low seed rate, poor contact of seed with soil, seed rotting as well as dryness of soil in patches (Ali et al. 2014; Ghosh et al. 2016).

Seed priming techniques have been found to improve germination characteristics including germination process, germination rate, germination uniformity (synchronized germination), germination percentage, etc., owing simply to less imbibition time (Brocklehurst and Dearman 2008; McDonald 2000; Taylor et al. 1998) and build-up of germination-enhancing metabolites (Basra et al. 2005; Farooq et al. 2006), thereby obtaining sufficient emergence followed by vigorous seedling growth at early stage, adequate crop stand (Arif et al. 2005; Ali et al. 2007; Diniz et al. 2009; Tiwari and Shivhare 2016; Nayban et al. 2017) as well as seed yield under abiotic stress conditions (Farooq et al. 2006, 2009; Harris 1996, 2006; Iqbal and Ashraf 2007; Kaya et al. 2006; Khan et al. 2008; Patade et al. 2011; Saglam et al. 2010; Ansari et al. 2012; Dey et al. 2014). Simple overnight soaking of seeds can hasten seed germination and establishment under relay cropping (Ali et al. 2014; Singh et al. 2016). Pre-sowing soaking of seeds with KH_2PO_4, Na_2HPO_4, etc. has also earlier been reported to improve seed germination, seedling vigour and root growth early in the season, resulting in good establishment, better drought tolerance and more yield of crop plants (Solaimalai and Subburamu 2004). Osmo-priming not only improves seed germination but also enhances general crop performance under nonsaline or saline conditions.

11.4.2 Improvement in Root and Shoot Length of Seedling

Seed priming ensures rapid seed germination and seedling emergence, helping in faster emergence, producing deep roots before the upper layers of the soil are dried and crusted and finally leading to better crop establishment and higher crop yield (Ashraf and Foolad 2005). Healthy plants with well-developed root system can more effectively mobilize limiting nutrients from the soil and can also better withstand adverse conditions (Nayban et al. 2017). Seed priming (osmo-conditioning) with KH_2PO_4 (200 mM) and mannitol (20 g L^{-1}) has been found to significantly increase the root as well as shoot length of mung bean seedlings possibly due to priming-induced nuclear replication in root tips of fresh seedlings (Umair et al. 2013). Vishwas et al. (2017) reported significantly higher root and shoot length of chickpea seedling due to seed priming with *Rhizobium* + *Pseudomonas* at 10% for 12 h. Likewise, Sajjan et al. (2017) reported an improvement in different growth parameters including root and shoot length of seedling due to seed priming with mesquite (*Prosopis juliflora*) leaf extracts (2%) for 1 h in pigeon pea.

11.4.3 Improvement in Nodulation

Seed treatment with beneficial microbes is becoming increasingly important (Howell et al. 1997). Treatment of leguminous seeds with *Rhizobium* spp. is well known for many years for nitrogen fixation. *Azospirillum* and other nitrogen-fixing bacteria are also in use. There are reports on improvement in nodulation, besides imparting a good control of seed-borne diseases with the use of microbiological products for seed priming in pulses. SMP, osmo-priming and hydro-priming methods have all been employed to increase beneficial microbial populations on the seed (Sharma et al. 2015), helping better root nodulation in pulses. Dugesar et al. (2017) evaluated four different methods of seed priming (hydro-priming with distilled water; osmo-priming with 20% PEG 6000; halo-priming with 1% NaCl; halo-priming with 1% $CaCl_2$; organic priming with 5% tulsi leaf extract; organic priming with 5% curi leaf extract) for 12 h in black gram. They reported significant differences among priming treatments with the control (no priming). The highest values of field emergence (%) and plant nodulation characters were observed under osmo-priming, whilst the highest nodulation was observed under the seeds primed with

PEG and $CaCl_2$. Mishra et al. (2017) reported better root nodulation in pigeon pea due to osmo-priming with PEG (20%), compared with control.

11.4.4 Increased Nutrient Uptake and Improved Crop Nutrition

There is an opportunity to improve crop nutrition through seed priming with both macro and micronutrients (Harris 2006). Seed priming is an easiest way of supplying starter nutrient(s) for seedling establishment and early growth and to integrate mineral nutrients into food legumes (agronomic biofortification). In particular, micronutrient deficiency (hidden hunger) is a global issue in food crops, causing reduced yields and nutritional quality of the produce (Dey et al. 2014). Micronutrient application through seed priming not only improves germination, stand establishment, plant vigour and yield but also micronutrient concentration (enrichment) in the grain (Singh et al. 2015b). For example, priming chickpea seeds in solutions of optimal concentration of micronutrients resulted in an increase in Zn content from about 40–60 (unprimed) to 500–800 mg kg^{-1} (Zn-primed). Priming lentil seeds exhibited similar results. There was no difference in Zn content of seeds primed for 8 h, compared to those primed for 12 h in solutions of similar nutrient concentration. However, an increase in concentration of Zn in the priming solution increases the amount of nutrient taken into both chickpea and lentil seeds (Johnson et al. 2005). Nutrient priming has been found more economical and convenient as compared to soil application in low fertile soils (Slaton et al. 2001). Umair et al. (2013) reported an improved nutrient uptake and crop nutrition through seed priming in mung bean.

11.4.5 Increased Crop Yield

Since pulses are commonly grown under energy-starved conditions with sub-optimal agronomic management, seed priming can be effective for legumes. There are many reports that yields of a number of legumes can be increased considerably by priming seeds before sowing (Parera and Cantliffe 1994; Musa et al. 2001; Harris et al. 1999, 2004; Rashid et al. 2004; Grant et al. 2005; Gupta and Bhowmick 2005; Khan et al. 2008; Bhowmick 2010; Umair et al. 2011, 2013; Bhowmick et al. 2013). However, yield increases due to priming are not always only due to increased stand density but also may be due to a combination of population density and improved individual plant performance (Musa et al. 2001; Rashid et al. 2004; Harris 2006). According to Afzal et al. (2016), hydro-primed seeds produced healthy seedlings, which resulted in uniform crop stand, drought resistance, early maturity and somewhat improved yield.

11.4.6 Increased Pest Resistance

Seed priming can be a viable option (Rakshit et al. 2013) for improving plant acclimatization not only under abiotic stresses but also it is quite effective for the management of biotic stresses (insect pests, diseases and weeds), besides sustaining agricultural production without endangering natural resource base (Rakshit et al. 2014; Nayban et al. 2017). There are several field reports that primed crops suffer less from insect pests and diseases (Harris 2006). Musa et al. (2001) reported that seed priming significantly reduced the seed-borne diseases like collar rot (*Sclerotium rolfsii*) and *Fusarium* wilt in chickpea. Likewise, the primed chickpea crop suffered less damage from pod borer damage in India (Musa et al. 2001; Harris 2006). Seed priming in water for 8 h has been reported to significantly minimize the incidence of viral disease caused by *Mungbean yellow mosaic virus* in mung bean (Rashid et al. 2004). Rakshit et al. (2014) reported that seed bio-priming with bioagent (*Trichoderma harzianum* and/or *Pseudomonas fluorescens*) at the rate of 10 g kg^{-1} seed might provide protection against seed- and soil-borne plant pathogens and improve seed germination and seedling growth. Seed priming with zinc salts is used to increase growth and disease resistance of seedlings (Afzal et al. 2016). Seed priming ensures good crop establishment, which can increase the competitiveness of crop plants, enabling them to suppress the weeds also (Nayban et al. 2017).

11.4.7 Seasonal Climate Risk Management

Greater uncertainties under changing climatic situation often combined with the use of poor quality seeds and pulse cultivation in marginal lands lead to slow the germination and emergence, causing patchy stand as well as multiple and delayed replanting of seeds. Seed priming is a simple, effective and climate-resilient technology for addressing this problem. It has been found that primed seeds usually emerge faster, produce more vigorous seedlings with better-developed root systems, reach flowering and maturity earlier and result in higher yield than non-primed seeds, which is important for avoiding or escaping terminal drought and heat stress (Uddin et al. 2005; Padgham 2009). When crop planting gets delayed due to adverse climatic conditions such as low temperature or high rainfall at sowing and high temperature at reproductive stage, seed priming may be beneficial in avoiding detrimental conditions by earlier crop maturity without compromising yield. In fact, earlier and vigorous crop stand usually captures more resources of water and nutrients through better root system, owing to larger leaf area and duration with enhanced photoassimilation that subsequently contributes towards better yield (Afzal et al. 2016). Seed priming with beneficial microorganisms promotes plant growth and increases abiotic stress tolerance in arid or semi-arid areas.

11.5 Conclusion

Pulses continue to occupy a prominent place from the view point of food and nutritional security and also become well suited to any cropping system owing to their ability towards biological nitrogen fixation, low water requirement, comparatively shorter duration and capacity to withstand weather abnormalities. Since these crops are mainly raised in the hands of small and marginal farmers, their productivity levels often become low due to suboptimal agronomic management practices. Seed priming is an easily applicable, practicable, low-cost, risk-averting and economically viable technology for the resource-poor farmers towards ensuring uniform seed germination, rapid emergence, successful crop establishment and better yield performance of pulse crops under stress-prone conditions, including adverse soil temperature, variable soil moisture, soil hardiness, etc. Several research accomplishments clearly show that priming pulse seeds before sowing can increase the rate and extent of emergence, improve seedling vigour, produce deeper roots with better nodulation and nutrient acquisition, hasten flowering and maturity, impart pest resistance and ultimately enhance crop yield in most of the cases. But the longevity and viability of primed seeds during storage may be reduced. There may have certain other instances, where priming does not benefit farmers, but negative effects of seed priming are uncommon. Further studies are, however, needed for developing and fine-tuning different alternative seed priming options and their effective combinations, based on easily available natural resources, crop seasons, growing situations and prevailing soil and environmental conditions. Moreover, location-specific selection of suitable pulse crops as well as priming modalities (appropriate priming material, its solution concentration and/or soaking duration, as applicable) need to be standardized and validated through on-farm testing before making any solid recommendation to the resource-poor farmers.

References

Afzal I, Rehman HU, Naveed M, Basra SMA (2016) Recent advances in seed enhancements. In: New challenges in seed biology – basic and translational research driving seed technology. Intech Open, pp 47–74. https://doi.org/10.5772/64791

Ahlawat IPS, Sharma P, Singh U (2016) Production, demand and import of pulses in India. Indian J Agron 61(4th IAC Special Issue):S33–S41

Ali MO, Sarkar A, Rahman MM, Gahoonia TS, Uddin MK (2005) Improvement of lentil yield through seed priming in Bangladesh. J Lentil Res 2:54–59

Ali S, Arif M, Gul R, Khan A, Shah SS, Ali I (2007) Improving maize seed emergence and early seedling growth through water soaking. Sci Khyber 19:173–177

Ali M, Ghosh PK, Hazra KK (2014) Resource conservation technologies in rice fallow. In: Ghosh PK, Kumar N, Venkatesh MS, Hazra KK, Nadarajan N (eds) Resource conservation technology in pulses. Scientific Publishers (India), Jodhpur, pp 83–89

Ansari O, Choghazardi HR, Sharif Zadeh F, Nazarli H (2012) Seed reserve utilization and seedling growth of treated seeds of mountain ray (*Secale montanum*) as affected by drought stress. Cercetări Agronomice Moldova 2(150):43–48

Arif M, Ali S, Shah A, Javed N, Rashid A (2005) Seed priming maize for improving emergence and seedling growth. Sarhad J Agric 21:539–543

Arteca RN (1996) Brassinosteroids. In: Davies PJ (ed) Plant hormones, physiology, biochemistry and molecular biology. Huwer Academic Publishers, New York, pp 206–213

Ashraf M, Foolad MR (2005) Pre-sowing seed treatment – a shotgun approach to improve germination, plant growth, and crop yield under saline and non-saline conditions. Adv Agron 88:223–276

Assefa MK, Hunje R, Koti RV (2010) Enhancement of seed quality in soybean following priming treatment. Karnataka J Agric Sci 23:787–789

Basha M (1982) Effect of chemical fortification of seeds on storability, field performance and quality of the resultant seeds in greengram. M.Sc. thesis, Tamil Nadu Agricultural University, Coimbatore, Tamil Nadu, India

Basra SMA, Farooq M, Tabassum R (2005) Physiological and biochemical aspects of seed vigor enhancement treatments in fine rice (*Oryza sativa* L.). Seed Sci Technol 33:623–628

Bassi G, Sharma S, Gill BS (2011) Pre-sowing seed treatment and quality invigoration in soybean {*Glycine max* (L) Merrill}. Seed Res 31:81–84

Bennett MA, Grassbaugh EM, Evans AE (2013) Vegetable crop seed vigor and seedling performance. Acta Hortic 975:172–179

Bewley JD (1997) Seed germination and dormancy. Plant Cell 9:1055–1066

Bhowmick MK (2010) Effect of planting time and seed priming on growth and yield of lentil under rice-*utera* system. J Food Legum 23:152–153

Bhowmick MK (2013) Seed priming and foliar nutrition towards enhancing Lathyrus productivity under rice-*utera* system. In: Proceedings of the Centenary Session, 100th Indian Science Congress, Section of Agriculture and Forestry Sciences, 3–7 Jan 2013, Kolkata, West Bengal, India, p 219

Bhowmick MK, Aich A, Aich SS, Shrivastava MP, Gupta S, Man GC (2005) Crop diversification through *paira* (*utera*) cropping with *rabi* pulses. SATSA Mukhapatra – Ann Tech Issue 9:43–60

Bhowmick MK, Biswas PK, Sen P, Bhattacharyya P (2010) Studies on seed priming, row spacing and foliar nutrition in chickpea under rainfed conditions in West Bengal, India, In: Proceedings of the international seminar on Climate Change and Environmental Challenges of 21st century, 7–9 Dec 2010, Institute of Environmental Science. University of Rajshahi, Rajshahi, pp 79–80

Bhowmick MK, Dhara MC, Islam S, Biswas PK, Duary B (2011) Maximization of lentil productivity under rice-*utera* system in West Bengal, Abstracts of Research Papers. The Eighteenth West Bengal State Science and Technology Congress, 28 Feb–01 Mar 2011. Ramakrishna Mission Residential College, Narendrapur, pp 204–205

Bhowmick MK, Gupta S, Bhattacharyya P (2012) Effect of seed priming and foliar nutrition on growth and productivity of lathyrus under rice-*utera* system in West Bengal, Abstracts. International conference on recent trends in *Lathyrus sativus* research, 8–9 Nov 2012. National Institute of Nutrition (ICMR), Hyderabad, p 39

Bhowmick MK, Duary B, Biswas PK, Rakshit A, Adhikari B (2013) Seed priming, row spacing and foliar nutrition in relation to growth and yield of chickpea under rainfed condition. SATSA Mukhapatra – Ann Tech Issue 17:114–119

Bhowmick MK, Duary B, Biswas PK (2014a) Promoting grasspea (*Lathyrus sativus* L.) cultivation in rice-fallows of West Bengal. In: Chattopadhyay PK (ed) Some empirical aspects of economic growth and diversification in India's emerging economy. New Delhi Publishers, New Delhi, pp 367–373

Bhowmick MK, Dhara MC, Duary B, Biswas PK, Bhattacharyya P (2014b) Improvement of lathyrus productivity through seed priming and foliar nutrition under rice-*utera* system. In: Proceedings of international seminar on "Integrating agriculture and allied research: prioritizing future potentials for secure livelihoods", 6–9 Nov 2014, Crop and Weed Science Society, Bidhan Chandra Krishi Viswavidyalaya, Nadia, West Bengal, India, p 97

Bhowmick MK, Dhara MC, Duary B, Biswas PK, Bhattacharyya P (2014c) Improvement of *Lathyrus* productivity through seed priming and foliar nutrition under rice-*utera* system. J Crop Weed 10(2):277–280

Bhowmick MK, Duary B, Dana I, Biswas PK, Dhara MC (2017) Low cost technologies for improving productivity of *Lathyrus* under rice-*utera* system in West Bengal. In: Moktan A (ed) Contemporary issues in agriculture, environment and rural development in India. New Delhi Publishers, New Delhi, pp 113–118

Biswas PK, Bhowmick MK (2015) Crop diversification through pulses in the northeastern plain zone of India. In: Ghosh M, Sarkar D, Roy BC (eds) Diversification of agriculture in eastern India, India studies in business and economics. Springer, New Delhi, pp 132–139

Bose B (2014) Role of seed priming in abiotic stress management in plants, Book of Abstracts. National Seminar on "Agriculture and Bio-security in Changing Scenario", 1–3 Feb 2014. Department of ASEPAN, PSB, Visva-Bharati, Sriniketan, pp v–vii

Bray CM (1995) Biochemical processes during osmopriming of seeds. In: Kigel J, Galili G (eds) Seed development and germination. Marcel Dekker, New York, pp 767–789

Brocklehurst PA, Dearman J (2008) Interaction between seed priming treatments and nine seed lots of carrot, celery and onion. II. Seedling emergence and plant growth. Ann Appl Biol 102:583–593

Cakmak I (2000) Possible roles of zinc in protecting plant cells from damage by reactive oxygen species. New Phytol 146:185–205

Chiu KY, Sung JM (2002) Effect of priming temperature on storability of primed sweet corn seed. Crop Sci 42:1996–2003

Côme D (1980) Problems of embryonal dormancy as exemplified by apple embryo. Israel J Bot 29:145–156

Côme D, Thévenot C (1982) Environmental control of embryo dormancy and germination. In: Khan AA (ed) The physiology and biochemistry of seed development, dormancy and germination. Elsevier Biomedical Press, Amsterdam, pp 271–298

Corbineau F, Côme D (2006) Priming: a technique for improving seed quality. Seed Testing International. ISTA News Bulletin No. 132:38–40

De Chandra G (1999) Fundamentals of agronomy. Oxford and IBH Publishing Company, New Delhi

Dey S, Pramanik K, Mukherjee S, Poddar S, Barik K (2014) Effect of seed priming and straw mulching on growth and productivity of chickpea (*Cicer arietinum* L.) under rainfed condition. SATSA Mukhapatra – Ann Tech Issue 18:126–131

Diniz KA, Silva PA, Oliveira JA, Evangelista JRE (2009) Sweet pepper seed responses to inoculation with microorganisms and coating with micronutrients, amino acids and plant growth regulators. Sci Agric 66(3):293–297

Dugesar V, Chaurasia AK, Bara BM, Kant K (2017) Effect of different priming methods on nodulation in blackgram (*Vigna mungo* L.) seeds. Int J Curr Microbiol App Sci 6(7):1137–1143

Eleiwa ME (1989) Effect of prolonged seed soaking on the organic and mineral components of immature pods of soybeans. Egypt J Bot 32:149–160

Ermiş S, Kara F, Özden E, Demir I (2016) Solid matrix priming of cabbage seed lots: repair of ageing and increasing seed quality. J Agric Sci 22:588–595

FAO (2016a) 'Climate smart' pulses are key to global food security: FAO. Food and Agriculture Organization of the United Nations, Rome. http://foodretailnetwork.asia/2016/12/climate-smart-pulses-are-key-to-global-food-security-fao-2/. Accessed 7 Jan 2017

FAO (2016b) Pulses Nutritious seeds for a sustainable future. International Year of Pulses, www.fao.org/pulses-2016. Food and Agriculture Organization of the United Nations (FAO), 189 p

Farooq M, Basra SMA, Khalid A, Tabassum R, Mehmood T (2006) Nutrient homeostasis, reserves metabolism and seedling vigor as affected by seed priming in coarse rice. Can J Bot 84:1196–1202

Farooq M, Basra SMA, Wahid A, Khaliq A, Kobayashi N (2009) Rice seed invigoration. In: Lichtfouse E (ed) Sustainable agriculture reviews. Springer, Dordrecht, pp 137–175

Ghosh PK, Hazra KK, Nath CP, Das A, Acharya CL (2016) Scope, constraints and challenges of intensifying rice (*Oryza sativa*) fallows through pulses. Indian J Agron 61(4th IAC Special Issue):S122–S128

Golezani KG, Jafari SF, Kolvanagh JS (2011) Seed priming and field performance of soybean (*Glycine max* L.) in response to water limitation. Not Bot Horti Agrobo 39(2):186–189

Grant C, Bittman S, Montreal M, Plenchette C, Morel C (2005) Soil and fertilizer phosphorus: effect of plant P supply and mycorrhizal development. Can J Plant Sci 85:3–14

Gupta S, Bhowmick MK (2005) Scope of growing Lathyrus and lentil in relay cropping systems after rice in West Bengal, India. Lathyrus Lathyrism Newsl 4:28–33

Gupta S, Bhowmick MK (2012a) Enhancing lentil productivity under rice-*utera* system in West Bengal, India. J Lentil Res 5:23–32

Gupta S, Bhowmick MK (2012b) Strategies for maximizing lathyrus productivity under rice-utera system in West Bengal, Abstracts, International conference on recent trends in *Lathyrus sativus* research, 8–9 Nov 2012. National Institute of Nutrition (ICMR), Hyderabad, p 25

Gupta S, Bhowmick MK (2013) Achieving higher seed yield of lathyrus in rice-fallows of West Bengal. SATSA Mukhapatra – Ann Tech Issue 17:65–76

Hacısalihoglu G (2007) Germination characteristics of three warm-season turfgrasses subjected to matriconditioning and aging. Hort Technol 17(4):480–485

Hadas A (2004) Seedbed preparation – the soil physical environment of germinating seeds. In: Benech-Arnold RL, Sánchez RA (eds) Handbook of seed physiology-applications to agriculture. Food Products Press® and The Haworth Reference Press, imprints of The Haworth Press, New York, pp 3–49

Halmer P (2004) Methods to improve seed performance in the field. In: Benech-Arnold RL, Sánchez RA (eds) Handbook of seed physiology-applications to agriculture. Food Products Press® and The Haworth Reference Press, imprints of The Haworth Press, New York, pp 125–166

Harman GE, Nelson EB (1994) Mechanisms of protection of seed and seedlings by biological seed treatments: implications for practical disease control, in seed treatment: progress and prospects, mono, vol 57. BCPC, Thornton Health, Surrey, pp 283–292

Harris D (1992) Staying in control of rainfed crops. In: Proceedings of the first annual scientific conference of the SADCC/ODA land and water management programme, Gaborone, Botswana, pp 257–262

Harris D (1996) The effects of manure, genotype, seed priming, depth and date of sowing on the emergence and early growth of *Sorghum bicolor* (L.) Moench in semi-arid Botswana. Soil Tillage Res 40:73–88

Harris D (2006) Development and testing of 'on-farm' seed priming. Adv Agron 90:129–178

Harris D, Joshi A, Khan PA, Gothkar P, Sodhi PS (1999) On-farm seed priming in semiarid agriculture: development and evaluation in maize, rice and chickpea in India using participatory methods. Exp Agric 35:15–29

Harris D, Rashid A, Hollington PA, Ali S, Rafiq M (2004) Mungbean production in the North West Frontier Province of Pakistan and the influence of on-farm seed priming. In: Improving income and nutrition by incorporating mungbean in cereal fallows in the Indo-Gangetic plains of South Asia. Proceedings. Final workshop and planning meeting, Punjab Agricultural University, Ludhiana, Punjab, India, 27–31 May 2004, Asian Vegetable Research and Development Center (AVRDC), South Asia Vegetable Research Network, Great Britain, Dept for International Development, The World Vegetable Center, 2004 – Mung bean. AVRDC Publication No. 06-682, 342 pp

Harris D, Rashid A, Miraj G, Arif M, Yunas M (2008) 'On-farm' seed priming with zinc in chickpea and wheat in Pakistan. Plant Soil 306:3–10

Heydecker W (1973) Germination of an idea: the priming of seeds. School of Agriculture Research, University of Nottingham, Nottingham, pp 50–67

Heydecker WJ, Higgins J, Gulliver K (1973) Accelerated germination by osmotic seed treatment. Nature 246:42–46

Hielscher (2017) Ultrasound technology. https://www.hielscher.com/ultrasonic-priming-of-seeds-how-sonication-improves-crop-germination.htm. Accessed 12 Apr 2017

Howell CR, DeVay JE, Garber RH, Batson WE (1997) Field control of cotton seedling disease with *Trichoderma virens* in combination with fungicide seed treatment. J Cotton Sci 1:15–20

Imran M, Neumann G, Ròmheld V (2004) Nutrient seed priming improves germination rate and seedling growth under submergence stress at low temperature. International Research on Food Security, Natural Resource Management and Rural Development, Cuvillier Verlag, Gòttingen

Iqbal M, Ashraf M (2007) Seed treatment with auxins modulates growth and ion partitioning in salt-stressed wheat plants. J Integr Plant Biol 49:1003–1015

Johnson SE, Lauren JG, Welch RM, Duxbury JM (2005) A comparison of the effects of micronutrient seed priming and soil fertilization on the mineral nutrition of chickpea (*Cicer arietinum*), lentil (*Lens culinaris*), rice (*Oryza sativa*) and wheat (*Triticum aestivum*) in Nepal. Exp Agric 41:427–448

Jyotsna V, Srivastava AK (1998) Physiological basis of salt stress resistance in pigeon pea (*Cajanus cajan* L.)-II. Pre-sowing seed soaking treatment in regulating early seeding metabolism during seed germination. Pl Physiol Biochem 25:89–94

Kaya MD, Okçu G, Atak M, Cikili Y, Kolsarici O (2006) Seed treatments to overcome salt and drought stress during germination in sunflower (*Helianthus annuus* L.). Eur J Agron 24:291–295

Kepczynska E, Piekna-Grochala J, Kepczynski J (2007) Seed germination of two tomato cultivars following matrix-conditioning under optimal and stress temperatures. Seed Sci Technol 35:749–753

Khan AA (1992) Preplant physiological seed conditioning. Hortic Rev 13:131–181

Khan AA, Maguire JD, Abawi GS, Ilyas S (1992) Matriconditioning of vegetable seeds to improve stand establishment in early field plantings. J Am Soc Hortic Sci 117(1):41–47

Khan A, Khalil SK, Khan AZ, Marwat KB, Afzal A (2008) The role of seed priming in semi-arid area for mungbean phenology and yield. Pak J Bot 40(6):2471–2480

Khan HA, Ayub CM, Pervez MA, Bilal RM, Shahid MA, Ziaf K (2009) Effect of seed priming with NaCl on salinity tolerance of hot pepper (*Capsicum annuum* L.) at seedling stage. Soil Environ 28:81–87

Kumar N, Singh MK, Ghosh PK, Venkatesh MS, Hazra KK, Nadarajan N (2012) Resource conservation technology in pulse based cropping systems. Publication No. 1/2012. ICAR-Indian Institute of Pulses Research, Kanpur, 32

Maroufi K, Farahani HA, Moradi O (2011) Evaluation of nano priming on germination percentage in greengram (*Vigna radiata* L.). Adv Environ Biol 5(11):3659–3663

Maske VG, Dotale RD, Sorte PN, Chore CN (1997) Germination, root and shoot studies in soybean as influenced by GA_3 and NAA. J Soils Crops 7:147–149

Mastouri F, Bjorkman T, Harman GE (2010) Seed treatment with *Trichoderma harzianum* alleviates biotic, abiotic, and physiological stresses in germinating seeds and seedlings. Phytopathology 100(11):1213–1221

McDonald MB (2000) Seed priming. In: Black M, Bewley JD (eds) Seed technology and its biological basis. Sheffield Academic Press, Sheffield, pp 287–325

Meena SK, Rakshit A, Meena VS (2016) Effect of seed bio-priming and N doses under varied soil type on nitrogen use efficiency (NUE) of wheat (*Triticum aestivum* L.) under greenhouse conditions. Biocatal Agric Biotechnol 6:68–75

Meena SK, Rakshit A, Singh HB, Meena VS (2017) Effect of nitrogen levels and seed bio-priming on root infection, growth and yield attributes of wheat in varied soil type. Biocatal Agric Biotechnol 7:172–178

Miano AC, Pereira JDC, Castanha N, Júnior MDDM, Augusto PED (2016) Enhancing mung bean hydration using the ultrasound technology: description of mechanisms and impact on its germination and main components. Sci Rep Nat 6:388996. https://doi.org/10.1038/srep38996

Mishra SN, Chaurasia AK, Tripathi V, Kumar B (2017) Effects of seed priming methods on growth and nodulation characters in pigeonpea (*Cajanus cajan* L.) seeds. J Pharmacogn Phytochem 6(4):620–623

Mondal SS, Ghosh A (2005) Integrated nutrient management on the productivity and nutrient uptake of crops in rice-lathyrus (as *utera*)-sesame cropping system under rainfed lowland ecosystem. J Crop Weed 1:12–16

Moussavi-Nik M, Rengel Z, Pearson JN, Hollamby G (1997) Dynamics of nutrient remobilization from seed of weed genotypes during imbibition, germination and early seedling growth. Plant Soil 197:271–280

Musa AM, Harris D, Johansen C, Kumar J (2001) Short duration chickpea to replace fellow after *aman* rice: the role of on-farm seed priming in the high barind tract of Bangladesh. Exp Agric 37:509–521

Naseem SB, Khan AH, Islam M, Mollah U, Ali MA (1997) Effect of seeding methods and varying surface soil moisture on the stand establishment of mungbean (*Vigna radiata* L.). Bang J Sci Indus Res 32:295–301

Nawaz J, Hussain M, Jabbar A, Nadeem GA, Sajid M, Subtain MU, Shabbir I (2013) Seed priming: a technique. Int J Agric Crop Sci 6(20):1373–1381

Nayban G, Mandal AK, De BK (2017) Seed priming: a low-cost climate-resilient tool for improving germination, growth and productivity of mungbean. SATSA Mukhapatra-Ann Tech Issue 21:162–172

Nazari M, Eteghadipour M (2017) Impacts of ultrasonic waves on seeds: a mini-review. Agric Res Tech Open Access J 6(3):5. https://doi.org/10.19080/ARTOAJ.2017.06.555688

Nazari M, Sharififar A, Asghari HR (2014) *Medicago scutellata* seed dormancy breaking by ultrasonic waves. Plant Breed Seed Sci 69(1):15–24

Padgham J (2009) Agricultural development under a changing climate: opportunities and challenges for adaptation. Joint departmental discussion paper- Issue 1. Agriculture and Rural Development & Environment Departments. The International Bank for Reconstruction and Development, The World Bank, 1818 H Street, NW Washington, DC 20433, 169

Parera CA, Cantliffe DJ (1992) Enhanced emergence seedling vigor in shrunken-2 sweet corn via seed disinfection and solid matrix priming. J Am Hortic Sci 117:400–403

Parera CA, Cantliffe DJ (1994) Presowing seed priming. Hortic Rev 16:109–141

Passam HC, Kakouriotis D (1994) The effect of osmoconditioning on the germination, emergence and early plant growth of cucumber under saline conditions. Sci Hortic 57:233–240

Patade VY, Maya K, Zakwan A (2011) Seed priming mediated germination improvement and tolerance to subsequent exposure to cold and salt stress in capsicum. Res J Seed Sci 4(3):125–136

Pill WG, Necker AD (2001) The effects of seed treatments on germination and establishment of Kentucky bluegrass (*Poa pratense* L.). Seed Sci Technol 29:65–72

Pradhan V, Rai PK, Bara BM, Srivastava DK (2017) Influence of halopriming and organic priming on germination and seed vigour in blackgram (*Vigna mungo* L.) seeds. J Pharmacogn Phytochem 6(4):537–540

Rafiq S, Iqbal T, Hameed A, Rafiq ZA, Rafiq N (2006) Morphobiochemical analysis of salinity stress response of wheat. Pak J Bot 38:1759–1767

Rahmianna AA, Adisarwanto T, Kirchhof G, So HB (2000) Crop establishment of legumes in rainfed lowland rice-based cropping system. Soil Tillage Res 56:67–82

Rajandran P (1982) Effect of chemical infusion of seeds on storability, field performance and quality of the resultant seed in redgram. M. Sc. Thesis, Tamil Nadu Agricultural University, Coimbatore, Tamil Nadu, India

Rajendra Prasad S, Kamble UR, Sripathy KV, Bhaskar KU, Singh DP (2017) Seed bio-priming for biotic and abiotic stress management. In: Singh D, Singh H, Prabha R (eds) Microbial inoculants in sustainable agricultural productivity. Springer, New Delhi, pp 211–228. https://doi.org/10.1007/978-81-322-2647-5_12

Rakshit A, Pal S, Rai S, Rai A, Bhowmick MK, Singh HB (2013) Micronutrient seed priming: a potential tool in integrated nutrient management. SATSA Mukhapatra-Ann Tech Issue 17:77–89

Rakshit A, Pal S, Meena S, Manjhee B, Preetipriya RS, Rai A, Bhowmick MK, Singh HB (2014) Seed bio-priming: a potential tool in integrated resource management. SATSA Mukhapatra-Ann Tech Issue 18:94–103

Rakshit A, Singh HB, Abhijit S (2015) XXIII, 417 p. 56 illus., 38 illus. in color. Springer-Verlag GmbH, Berlin. ISBN:978-81-322-2169-2

Rashid A, Harris D, Hollington P, Ali S (2004) On-farm seed priming reduces yield losses of mungbean (*Vigna radiata*) associated with mungbean yellow mosaic virus in NWFP of Pakistan. Crop Prot 23:1119–1124

Rautaray SK (2008) Productivity and economics of rice based *utera* crops for lower Assam. J Food Legum 21:51–52

Riedell WE, Khoo U, Inglett GE (1985) Effect of bio regulators on soybean leaf structure and chlorophyll retention. In: Plant growth regulation, Lake Alfred Florida. Lake Alfred, pp 204–212

Rogis C, Gibson LR, Knapp AD, Horton R (2004) Can solid matrix priming with GA_3 break seed dormancy in eastern gamagrass? J Range Manag 57:656–660

Rowse HR (1996) Drum priming – a non-osmotic method of priming seeds. Seed Sci Technol 24:281–294

Saglam S, Day S, Kaya G, Gurbuz A (2010) Hydro-priming increases germination of lentil (*Lens culinaris* Medik) under water stress. Notulae Scientia Biologicae 2(2):103–106
Sajjan AS, Dhanelappagol MS, Jolli RB (2017) Seed quality enhancement through seed priming in pigeonpea [*Cajanus cajan* (L.) Millsp.]. Legum Res 40(1):173–177
Sallam HA (1999) Effect of some seed-soaking treatments on growth and chemical components on faba bean plants under saline conditions. Ann Agric Sci 44:159–171
Sen SK, Mandal P (2016) Solid matrix priming with chitosan enhances seed germination and seedling invigoration in mung bean under salinity stress. J Cent Eur Agric 17(3):749–762
Sharififar A, Nazari M, Asghari HR (2015) Effect of ultrasonic waves on seed germination of *Atriplex lentiformis*, *Cuminum cyminum*, and *Zygophyllum eurypterum*. J Appl Res Med Aromat Plants 2(3):102–104
Sharma KK, Singh US, Sharma P, Kumar A, Sharma L (2015) Seed treatments for sustainable agriculture-a review. J Appl Nat Sci 7(1):521–539
Shivay YS, Singh U, Prasad R, Kaur R (2016) Agronomic interventions for micronutrient biofortification of pulses. Indian J Agron 61(4th IAC Special Issue):S161–S172
Siddique KHM, Johansen C, Turner NC, Jeuffroy M-H, Hashem A, Sakar D, Gan Y, Alghamdi SS (2012) Innovations in agronomy for food legumes. A review. Agron Sustain Dev 32(1.), Springer/EDP Sciences/INRA):45–64. https://doi.org/10.1007/s13593-011-0021-5
Singh MV (2007) Efficiency of seed treatment for ameliorating zinc deficiency in crops. In: Zinc crops 2007, improving crop production and human health, May 24–26, 2007, Istanbul, Turkey
Singh A, Dahiru R, Musa M, Haliru BS (2014) Effect of osmo-priming duration on germination, emergence, and early growth of cowpea (*Vigna unguiculata* (L.) Walp.) in the Sudan Savanna Nigeria. Int J Agron 2014:4. 10.1155/2014/841238
Singh H, Jassal RK, Kang JS, Sandhu SS, Kang H, Grewal K (2015a) Seed priming techniques in field crops – a review. Agric Rev 36(4):251–264
Singh U, Praharaj CS, Shivay YS, Kumar L, Singh SS (2015b) Ferti-fortification: an agronomic approach for micronutrient enrichment of pulses. In: Dixit GP, Singh J, Singh NP (eds) Pulses: challenges and opportunities under changing climatic scenario, In: Proceedings of the national conference on "Pulses: challenges and opportunities under changing climatic scenario", 29 Sept–01 Oct 2014, JNKVV, Jabalpur, Madhya Pradesh. Indian Society of Pulses Research and Development. ICAR-Indian Institute of Pulses Research, Kanpur, pp 208–222
Singh NP, Praharaj CS, Sandhu JS (2016) Utilizing untapped potential of rice fallow of East and North-east India through pulse production. Indian J Genet 76(4):388–398. https://doi.org/10.5958/0975-6906.2016.00058.4
Sivasubramaniam K, Geetha R, Sujatha K, Raja K, Sripunitha A, Selvarani R (2011) Seed priming: triumphs and tribulations. Madras Agric J 98(7–9):197–209
Slaton NA, Wilson CR Jr, Natanatngiro S, Norman JR, Boothe LD (2001) Evaluation of Zn seed treatments for rice. Agron J 93:152–157
Solaimalai A, Subburamu K (2004) Seed hardening for field crops – a review. Agric Rev 25:129–140
Sujatha K, Sivasubramaniam K, Padma J, Selvarani K (2013) Seed hardening. Int J Agric Sci 9(1):392–412
Swarnkar VK, Khajanji SN, Upadhyay N (2012) Effective utilization of rice fallows and rice bunds for cultivation of *rabi* pulses. In: Training manual, model training course on "Production techniques in *rabi* pulses", 9–16 Jan 2012, Directorate of Extension Services, Indira Gandhi Krishi Vishwavidhyalaya, Raipur, Chhattisgarh, India, pp 62–67
Taylor AG, Harman GE (1990) Concepts and technologies of selected seed treatments. Phytopathology 28:312–339
Taylor AG, Klein DE, Whitlow TH (1988) SMP: solid matrix priming of seeds. Sci Hortic 37(1–2):1–11
Taylor AG, Allen PS, Bennett MA, Bradford KJ, Burns JS, Misra MK (1998) Seed enhancements. Seed Sci Res 8:245–256
Tiwari AK, Shivhare AK (2016) Pulses in India: retrospect and prospects. Publication no. DPD/Pub.1/Vol. 2/2016. Government of India, Ministry of Agriculture & Farmers Welfare

(Department of Agriculture, Cooperation and Farmers Welfare), Directorate of Pulses Development, Vindhyachal Bhavan, Bhopal, 316

Uddin MJ, Ali MO, Rahman MM (2005) Prospects of chickpea in rice-based cropping systems in Bangladesh. In: Pande S, Stevenson PC, Neupane RK, Grzywacz D (eds) Policy and strategy for increasing income and food security through improved crop management of chickpea in rice fallows in Asia, Summary of a NARC-ICRISAT-NRI workshop, Nov 17–18, 2004, Kathmandu, Nepal. International Crops Research Institute for the Semi-Arid Tropics, Patancheru, pp 35–46

Umair A, Ali S, Hayat R, Ansar M, Tareen MJ (2011) Evaluation of seed priming in mungbean (*Vigna radiata*) for yield, nodulation and biological nitrogen fixation under rainfed conditions. Afr J Biotechnol 10(79):18122–18129

Umair A, Ali S, Sarwar M, Bashir K, Tareen MJ, Malik MA (2013) Assessment of some priming techniques in mungbean (*Vigna radiata*): a greenhouse study. Pak J Agric Res 26(4):265–274

Urkurkar JS (2012) Technology transfer strategies for enhancing rabi pulse production in Chhattisgarh. In: Training manual, model training course on "Production techniques in rabi pulses", Jan 09–16, 2012, Directorate of Extension Services, Indira Gandhi Krishi Vishwavidhyalaya, Raipur, Chhattisgarh, India, pp 11–13

Varier A, Vari AK, Dadlani M (2010) The subcellular basis of seed priming. Curr Sci 99(4):450–456

Vishwas S, Chaurasia AK, Bara BM, Debnath A, Parihar NN, Brunda K, Saxena R (2017) Effect of priming on germination and seedling establishment of chickpea (*Cicer arietinum* L.) seeds. J Pharmacogn Phytochem 6(4):72–74

Wang HY, Chen CL, Sung JM (2003) Both warm water soaking matriconditioning treatments enhance anti-oxidation of bitter gourd seeds germinated at suboptimal temperature. Seed Sci Technol 31:47–56

Yucel DO (2012) The effect of different priming treatments and germination temperatures on germination performance of lentil (*Lens culinaris* Medik) seeds. ARPN J Agric Biol Sci 7(12):977–981

Zhou YG, Yang YD, Qi YG, Zhang ZM, Wang XJ, Hu XJ (2002) Effects of chitosan on some physiological activity in germinating seed of peanut. J Peanut Sci 31:22–25

Studies on Seed Priming in Pepper (*Capsicum annuum* L.)

12

Nusret Ozbay

Abstract

Plants are regularly exposed to various adverse environmental conditions such as drought, salinity, chilling, and high temperatures. These abiotic stresses adversely affect the plant growth and productivity and, in extreme cases, cause plant death. One feature expected of high-quality seeds is to be able to germinate under adverse growing conditions. Rapid germination and emergence are essential for successful crop establishment, for which seed priming could play an important role. Seed priming is an effective technology to enhance rapid and uniform emergence and to achieve high vigor, leading to better stand establishment and yield. Over the past 50 years, there has been extensive literature published on seed priming and related pre-sowing seed treatments. These have been particularly promising for high-value small-seeded vegetable crops such as pepper where rapid, uniform germination is at a premium either in the cell transplant production or where accurate and uniform plant population density is required from direct sowing. The aim of this chapter is to review some of the current ideas and recent work on physiological priming treatments designed to enhance germination and emergence performances of pepper seeds.

Keywords

Pepper · Seed priming · Germination · Emergence · Seedling growth and establishment · Vigor · Environmental stress

N. Ozbay (✉)
Department of Horticulture, Agricultural Faculty, Bingol University, Bingol, Turkey

© Springer Nature Singapore Pte Ltd. 2018

A. Rakshit, H. B. Singh (eds.), *Advances in Seed Priming*,
https://doi.org/10.1007/978-981-13-0032-5_12

12.1 Introduction

Pepper is one of the most important vegetable crops in the world. Peppers are grown worldwide because of their adaptation to different agroclimatic regions and their wide variety of shapes, sizes, colors, and pungencies of the fruit (Qin et al. 2014). Bell pepper or sweet pepper belongs to the genus *Capsicum*, a member of the Solanaceae family that includes tomatoes, potatoes, and eggplants. Pepper is a cultivar group of the species *Capsicum annuum* L. Cultivars of the plant produce fruits in different colors, including red, yellow, and orange, but more exotic colors include purple, white, and lime green. The fruit is also frequently consumed in its unripe form, when the fruit is still green. In the United States, Canada, the United Kingdom, Ireland, and some other parts of the world, in addition to the terms "bell pepper" and "sweet pepper," the fruit is often referred to simply as a "pepper," whereas in many Commonwealth of Nations countries, such as Australia, India, Malaysia, and New Zealand, they are called "capsicum."

Peppers are native to Mexico, Central America, and northern South America. Pepper seeds were later carried to Spain in 1493 and from there spread to other European, African, and Asian countries (Andrews 1993; Abdel-Kader and El-Mougy 2014). According to FAO statistical database, in 2014, pepper is cultivated on 3,625,452 ha in the world, with an annual production of 36,143,113 metric tons (MT) representing an annual yield of 9.97 MT ha^{-1}. Today, China is the world's largest pepper producer, followed by Mexico, Turkey, and Indonesia (FAO 2017).

Peppers grow better in evenly moist and evenly warm soil and can be grown year round in frost-free conditions. If moisture levels or temperature levels fluctuate, too much growers will have problems related to some diseases and yield. It has also been reported that pepper seed germination and emergence are slow and nonuniform under normal as well as stressful conditions (Belletti and Quagliotti 1988; Chartzoulakis and Klapaki 2000; Demir and Okcu 2004; Khan et al. 2009a). Low ability of seed germination during seed production and rapid deterioration during storage of pepper seed are also the major problems that affect seed quality (Siri et al. 2013). Rapid germination and emergence are essential for successful crop establishment. Various seed treatments have been suggested to improve seed germination and emergence, and one of them is priming. Seed priming is an effective technology to enhance rapid and uniform emergence and to achieve high vigor, leading to better stand establishment and yield (Singh et al. 2015; Ibrahim 2016).

During priming, the seeds are partially hydrated in such a way that pre-germination metabolic activities start; however, radicle protrusion is prevented followed by drying of seeds to the original moisture content (Sharma et al. 2015). Seed is primed to obtain faster and more uniform germination resulting in a stronger crop stand. Yadav et al. (2011) reported that priming treatments improved germination percentage as well as rate of seed germination in pepper under normal as well as the stressed conditions. Maiti et al. (2009) studied the effect of priming on seedling vigor and productivity of pepper, during post-rainy seasons demonstrating that priming improved germination and seedling development and yield pepper. Another primary benefit of priming is the extension of the temperature range at which a seed can germinate (Valdes and Bradford 1987; Ellis and Butcher 1988). This has particular importance for crops such as pepper, in which seedlings are produced in the early spring at low temperatures for planting in the open field and in late summer at higher temperatures for glasshouse production. It has been reported that priming improved seed germination at suboptimal temperatures (Pandita et al. 2007; Korkmaz and Korkmaz 2009). Nascimento et al. (2011) reported that priming improved the physiological quality of hot pepper seeds by improving germination at high temperature (35 °C).

12.2 Seed Priming Methods

In seed technology, different pre-sowing treatments have been utilized to reduce the time from sowing until germination or emergence and to shorten the interval between the first and last seed germination. One of the most important pre-sowing treatments is known as seed priming. The principle of priming is based on the fact that it is possible to hydrate seed in some ways at a moisture level sufficient to initiate the early events of germination but not sufficient to permit radical protrusion through the seed coat (Sivritepe and Sivritepe 2016). Seed priming enhances seed performance by rapid and uniform germination, more uniform emergence, normal and vigorous seedlings, greater tolerance to environmental stress, and reduced dormancy in many species (Khan 1992; Desai et al. 1997; Cantliffe 2003). Additionally, this physiological treatment provides invigoration treatment to partially aged seed lots. Excellent review studies have considered the importance of seed priming and the factors that affect its success (Khan 1992; Parera and Cantliffe 1994; Bray 1995; Pill 1995; Taylor et al. 1998; Welbaum et al. 1998; McDonald 2000). Beneficial effects from priming have been reported for several vegetable seeds including pepper (Gomes et al. 2012). Improvement of germination in pepper plant by priming with water and NaCl has been reported (Smith and Cobb 1991a). The results of Kaya et al. (2010) showed that priming pepper seeds can increase germination percentage and accelerate germination at suboptimal temperatures. Priming pepper seeds in PEG and KNO^3 solutions significantly increased emergence percentage and decreased mean time to emergence of pepper seedlings (El-Shatoury 2010). Various priming treatments have been developed to enhance seed germination, seedling growth, and yield in most of the crops under normal and stress conditions and also to avoid biotic and abiotic stresses during germination and emergence phases (Bradford 1986; McDonald 2000). These priming methods can be listed as follows (Ashraf and Foolad 2005; Lutts et al. 2016; Sivritepe and Sivritepe 2016): hydropriming, osmopriming, solid matrix priming, halopriming, drum priming, hormopriming, thermopriming, biopriming, nutripriming, and organic priming.

12.2.1 Hydropriming

It is a simple and low-cost priming technique in which seeds are soaked in water (aerated distilled water is preferred) for a certain period of time to a point where pre-germination metabolic activities start without actual germination and subsequently dried back to storage moisture contents before sowing (Singh et al. 2015). This process is especially useful in economically disadvantaged, arid crop-growing areas. Hydropriming has been considered as a simple and cost-effective strategy for improving germination and emergence of pepper. For example, Demir and Okcu (2004) reported that aerated hydration treatment of pepper and eggplant seeds significantly increased final germination of both species over a wide range of temperature (18–35 °C). Hydropriming with normal water or warm water improved the radicle protrusion of pepper seeds significantly as compared to the control treatment (Yadav et al. 2011). Similarly, hydropriming of pepper seeds for 24 h enhanced the germination and emergence percentage, ensured early germination and seedling emergence, increased uniformity of emergence, and increased seedling vigor compared to unprimed seeds (Uche et al. 2016).

12.2.2 Osmotic Priming

This method is originally developed and described by Heydecker et al. (1973, 1975). Osmotic priming technique, also called osmopriming or osmoconditioning, is based on the controlled hydration of seeds to a level that allows pre-germination metabolic activity but inhibits radicular emergence. It is achieved by immersing seeds in an aqueous solution of a chemically inert but osmotically active compound such as polyethylene glycol (PEG) for a specific period at a specific temperature (Heydecker et al. 1975; Khan 1992). At the end of the process, seeds are rinsed before further processing. Taylor and Harman (1990) and Gray (1994) provide access to the very large literature on osmotic priming. Some of the compounds that are used for osmotic priming include sugars, polyethylene glycol (PEG), glycerol, sorbitol, or mannitol. The improvement of germination and emergence performance of seeds of many plant species through a pre-sowing treatment with a chemically inert osmoticum have been reported by many researchers (Bennett et al. 1992). The effect of seed priming on field emergence of pepper seeds has ranged from no improvement to some advancement in mean germination time (Yaklitch and Orzolek 1977; Bradford et al. 1990). Aljaro and Wyneken (1985) reported that osmoconditioning of sweet pepper seed did not affect the germination percentage but shortened germination time and gave more uniform germination. Pepper seeds primed in mannitol had improved final percent germination and produced larger seedlings than nonprimed seeds (Georghiou et al. 1987; Passam et al. 1989). Kikuti et al. (2005) reported that sweet pepper seeds primed with PEG 6000 performed better in all the vigor tests assessed, except the germination test. Demirkaya (2006) studied osmotic conditioning with polyethylene glycol to enhance the germination percentage and to reduce the mean germination time of pepper seeds. He found that osmoconditioned seed enhanced germination percentage and reduced the mean germination time. Cortez-Baheza et al. (2007) indicate osmopriming treatments using PEG in combination with GA_3 or KNO_3 have shown to be good approaches to revigorate pepper seeds for commercial purposes. Siri et al. (2013) reported that seed germination was improved in osmoprimed pepper seeds primed in a PEG 6000 solution with the osmotic potential of −1.5 MPa for 6 days.

12.2.3 Halopriming

In halopriming, the seeds are immersed in different inorganic salt solutions (NaCl, KNO_3 $CaCl_2$, $CaSO_4$, etc.) which facilitate the process of seed germination and subsequent seedling emergence even under adverse environmental conditions. Halopriming is a simple and cheap agro-technique and therefore found suitable to be recommended to the farmers owing to better synchrony of emergence and crop stand under various conditions of environment (Sedghi et al. 2010). A number of studies have shown a significant improvement in seed germination, seedling emergence and establishment, and final crop yield under unfavorable conditions such as suboptimal temperature and salinity as a result of seed halopriming. For example, Smith and Cobb (1991a) reported that priming with water and NaCl improved germination in pepper plant. Carter (1994) initiated a work to examine the effects of NaCl priming on total germination and the rate of germination under suboptimal temperature conditions. Seeds of "Tam Veracruz" pepper were primed at 23 °C for 5 days in double-distilled H_2O, or 0.2, 0.4, or 0.6 M NaCl solution (equivalent to an osmotic potential of −0.89 MPa, −1.77 MPa, and −2.66 MPa, respectively). He suggested that priming "Tam Veracruz" pepper seeds for 5 days at 23 °C in 0.2 M NaCl increased the germination rate at temperatures from 15 to 23 °C. A previous study on pepper (Amjad et al. 2007) showed that halopriming improved seed germination, seedling emergence, and growth under saline and drought conditions. Khan et al. (2009a) observed that priming of seeds using NaCl improved seedling vigor and establishment under salt stress conditions. It has been reported that halopriming with salt (NaCl) improved the rate of pepper seed germination (Yadav et al. 2011). In a study comparing hydropriming and halopriming with dehydration treatments for physiological enhancement of pepper seeds, both hydro- and halopriming treatments with and without dehydration conditions caused increases in normal germination percentage and germination index of pepper seeds compared with the control (Sivritepe and Senturk 2011). Maiti et al. (2013) reported that halopriming in 3% KNO_3 solution for 40 h at normal room temperature increased speed of emergence, seedling vigor index, root length and shoot length over hydropriming, and control in pepper. They also reported that, at field level, the halopriming treatment increased yield compared to the control and hydropriming treatments. Dutta et al. (2015) also reported that pepper seeds primed in 1% KNO_3 recorded the highest germination percentage as compared to nonprimed control.

12.2.4 Drum Priming

Drum priming is a special seed priming method in which seeds are hydrated to predetermine humidity content by placing them inside a horizontal rotating drum. Drum priming has the capability to accurately hydrate seeds without concern for spent solution or solid carrier disposal (Rowse 1996). In this novel method, the supply of water to the seeds is controlled by physical rather than osmotic means or solid matrix materials, limiting the amount of water introduced to the seed. Specially designed apparatus enables monitoring of the seed weight, precise regulation of time, and water amount during hydration process, what ultimately results in an appropriate and uniform moisture level of the seeds (Warren and Bennett 1997). The method enables the priming of much larger quantities of seed than previously been practical and avoids the environmental problems of disposal of the used priming solution. Unlike other priming methods, it does not require an oxygen enrichment of the atmosphere because of the nature of the process. Drum priming systems are among the most common methods of seed priming for commercial treatment. Da Silva et al. (2015) combined drum priming with 24-epibrassinolide (24-EpiBL) for enhancement of bell pepper seed performance. Several advantages were verified in the drum priming technique with added 24-EpiBL compared with the traditional procedure (water alone). Drum priming with 24-epibrassinolide showed positive effect on germination time and seedling growth of bell pepper.

12.2.5 Solid Matrix Priming

Solid matrix priming (SPM), also known as matriconditioning, is a process in which seeds are mixed with a solid material and water in known proportions and then incubated for a given duration at constant temperature. Solid matrix priming utilizes carriers like vermiculite, diatomaceous earth, or another highly water-absorbent polymer (Celite® or Micro-cel®) possessing characteristics such as high water holding capacities, low osmotic potentials, and low bulk density (Taylor et al. 1988, 1998; Khan et al. 1990; Khan 1992; McDonald 2000). In order to improve the control of imbibition, pure water may be replaced by an osmotic solution, as in osmotic priming (Khan 1991). After incubation the extraneous solid material is sieved off. Solid matrix priming is similar to osmotic priming, allowing the seed to imbibe and attain threshold moisture content and pre-germination metabolic activities but preventing radicle emergence. However, it has the advantages of allowing aeration, incorporation of biological agents to combat soilborne pathogens, and improved ease of handling (Taylor et al. 1988; Harman et al. 1989; Paparella et al. 2015). This technique has been used to improve seed germination and rate in many species including peppers. Kubik et al. (1988) compared SPM with commercially osmoprimed tomato and pepper seeds under growth chamber conditions. They used expanded calcine clay as a solid material. SMP-treated seed performed equally as well as, or better than, commercially osmoprimed seed. Ilyas et al. (2002) reported that matriconditioning improved quality and protein level of medium vigor hot pepper seed. Kang et al. (2003) compared osmotic with solid matrix priming to determine the more effective treatment for improving seed germination in pepper and tomato and reported that solid matrix primed pepper seeds germinate faster than osmotic primed seeds at all temperatures (15, 20, 25, and 30 °C). On the other hand, they also reported that early growth was not significantly influenced by osmotic priming or SMP treatment of pepper and tomato seeds. Hacisalihoglu and White (2006) conducted an experiment to determine the optimum duration, temperature, and ratio (seed/carrier/water) of matriconditioning "long red cayenne" pepper seeds for improved germination percentage and mean germination time. The results showed that matriconditioning, which is carried out in 453 g glass jars by mixing seeds, water, and synthetic calcium silicate at 25–30 °C for up to 5 days, increased final germination percentage to 96.7%, compared with 82% in the control seeds. Furthermore, matriconditioning decreased mean germination time tenfold, compared with a nonprimed control. Pandita et al. (2007) found that solid matrix priming improved germination of hot pepper seed by 10–16% depending on temperature, and this effect enhanced when SMP was followed by halopriming and osmopriming. QianQian et al. (2009) indicated that priming treatment with vermiculite could improve the vigor of hot pepper seeds and the salt tolerance of hot pepper seedlings.

12.2.6 Biopriming

Biopriming seed treatment is gaining importance in seed germination and management of plant pathogens as another alternative to chemical fungicides in recent years. The biopriming procedure first described by Callan et al. (1990) for biological control of *Pythium* preemergence damping-off of sh2 sweet corn integrated imbibition at an optimal temperature with protection by a biocontrol agent. Biopriming is a process of biological seed treatment that refers to combination of seed hydration (physiological aspect of disease control) and inoculation (biological aspect of disease control) of seed with beneficial organisms to protect seed (Rakshit et al. 2015). The seeds are removed before radicle emergence (Callan et al. 1990). Because seeds can fail to germinate due to a number of reasons, both biological and physiological seed treatments used in combination seem to provide the best seed protection (Bennett et al. 1992). The leakage of seed exudates during biopriming may supply nutrients and energy for biocontrol agents (Wright et al. 2003). This favorable environment contributed to colonization and proliferation of biocontrol agents over the seed surface to facilitate water uptake and nutrients during biopriming. Biopriming with different beneficial microbes may not only enhance seed quality but also boost seedling vigor and ability to withstand abiotic and biotic stressors and thus offer an innovative crop protection toll for the sustainable improvement of crop yield (Rakshit et al. 2015). Biopriming is recently used as an alternative method for controlling many seed- and soilborne pathogens (Reddy 2012). For example, combined effect of *Pseudomonas fluorescens* and *Trichoderma harzianum* as seed biopriming resulted in significant growth of pepper seedlings (Kumar et al. 2010).

In another work on bell pepper, Da Silva and Filho (2012) compared the efficiency of biostimulant priming (10 mL of Stimulate® in 100 mL of distilled water) to hydropriming in bell pepper seeds as evaluated by SVIS® analysis and recommended vigor tests. They indicated that hydroprimed and biostimulant primed seeds performed better than control. An investigation was carried out using "PKM 1" chili (*Capsicum annuum* L.) seed to standardize biopriming with the biocontrol agents *Trichoderma viride* or *Pseudomonas fluorescens* in order to improve seed germination and seedling vigor (Ananthi et al. 2014). They have suggested that seed biopriming with 60% (w/v) preparations of *T. viride* or *P. fluorescens* for 3 h or 12 h, respectively, can therefore be adopted to improve seed germination and seedling vigor in chili. Ilyas et al. (2015) reported that biopriming with the biofungicide and clove oil 0.06% or 0.1% was an effective seed treatment to improve the vigor and relative speed of germination while reducing the percentage of *Colletotrichum capsici*, a seed-borne pathogen causing anthracnose disease in hot pepper seeds (Ilyas et al. 2015). However, it still provides inconsistent results. Copeland and McDonald (1995) indicate that, in many cases, the biological treatment of seeds is not persistent in soil, seed, or plant, under natural conditions. Therefore, a deeper research concerning the methodology of application, doses, storage of coated seeds, responses from different species to the materials, types of products, and polymers should be performed to standardize.

12.2.7 Priming with Plant Growth Regulators and Other Organic Sources

Soaking or treating seeds in optimal concentrations of plant growth regulators (PGRs) improves germination, stand establishment, growth, and yield of crop plants under both normal and stress conditions. Carter and Stevens (1998) determined that when primed with gibberellic acid (GA3), pepper seeds had higher germination (91%) as compared to nonprimed seeds at high temperature (40 °C). The direct benefits of seed priming with PGR were reported by Mendoza et al. (2002), who demonstrated that that imbibing pepper seeds in 0.1 mM salicylic acid prevented seedling from subsequent chilling-induced damage. Yogananda et al. (2004) found that bell pepper seeds invigorated with GA_3 (200 ppm) recorded higher germination, root and shoot length seedling dry weight, rate of germination, and seedling vigor indices over control. A laboratory study was performed by Khan et al. (2012) to explore the benefits of seed priming with polyamines on seed germination and seedling growth of hot pepper. Hot pepper seeds were primed in aerated solution of putrescine, spermine, or spermidine (25, 50, 75, and 100 mM) for 48 h at 25 ± 2 °C. Results showed that polyamine priming resulted in earlier and synchronized germination via improving final germination percentage, time to 50% germination, mean germination time, germination energy, germination speed, and germination index compared with control. Seed performance of various crops including pepper can also be improved by inclusion of plant growth regulators and hormones during priming and other pre-sowing treatments (Lee et al. 1998;

Korkmaz and Korkmaz 2009). For example, priming with PEG and GA_{4+7} increased early germination in pepper especially at low temperature (15 °C), 3.8 days as compared 4.7 days in nonprimed seeds (Watkins and Cantliffe 1983). Andreoli and Khan (1999) reported that matriconditioning pepper seeds at 25 °C in the presence of gibberellic acid hastened seed germination and improved stand establishment of pepper. Korkmaz (2005) reported that inclusion of acetyl salicylic acid and methyl jasmonate into the priming solution improved low-temperature germination and emergence of sweet pepper.

Several new compounds are also being tested for their potential use as priming agents. For example, priming pepper seeds in 3% KNO_3 with the presence of 5-aminolevulenic acid (ALA) improved final germination percentage (FGP) and germination rate at 15 °C compared to nonprimed seeds (Korkmaz and Korkmaz 2009). Ozbay and Susluoglu (2016a) reported that sweet pepper were treated in 3% KNO_3, 2% KH_2PO_4, and 10% PEG solutions containing 0, 25, 50, and 100 mg L-1 prohexadione-calcium (Pro-Ca) in darkness at 25 °C for 3 days. Priming pepper seeds, in the presence or absence of plant growth regulator, improved final germination percentage, mean time to germination, and germination index, final emergence percentage, and mean time to emergence compared to nonprimed seeds. The highest final germination percentage and the lowest mean time to germination were obtained from KH_2PO_4 + 25 mg L-1 Pro-Ca treatment.

Priming with some organic agents could also be used in pepper seeds for physiological enhancement. In an earlier study conducted by Sivritepe and Sivritepe (2008), possibility of using seaweed extract (as an organic material) in priming technique was examined in pepper seeds. The seeds (cv. California Wonder) were subjected to priming treatments performed by the use of 1:1, 1:5, 1:10, 1:25, 1:50, 1:100, 1:250, 1:500, and 1:1000 dilutions of seaweed extract (Maxicrop) and also H_2O, for 1, 2, and 3 days at 20 °C. They concluded that seaweed extract could be used as an osmotic agent in organic priming of pepper seeds for physiological enhancement. They also commented that seaweed extracts caused physiological enhancements and increases in performance of seeds due to their hormonal components such as cytokinins and amino acid contents and also their hygroscopic characteristics. Demirkaya (2010) demonstrated that seaweed extract could be used in osmotic conditioning treatments of onion seeds as well as pepper seeds. Similarly, Sivritepe et al. (2015) reported that organic priming with continuously aerated seaweed extract (1000 ppm) for 2 days at 20 °C and dehydration treatments gave the best results in terms of physiological enhancement of pepper seeds. An experiment was conducted to improve the germination and quality of seedling through seed priming of bell pepper cv. California Wonder using 16 botanicals and animal by-products/wastes for 12 and 24 h. The best germination and quality of seedlings were obtained through pre-sowing seed priming treatments of *Melia azedarach* leaf extract 10% followed by Eucalyptus leaf extract 10%, garlic clove extract 5%, cow urine 5%, and cow dung extract 5% at seed soaking duration of 24 h (Mehta et al. 2010). Dutta et al. (2015) carried out an experiment to find out efficacy of organic (200 g/Kg seed neem leaf powder) and inorganic priming (3% KNO_3 and 2% KH_2PO_4) on germination and seedling vigor of bird's eye chili (*Capsicum frutescens* L.) seeds. Among all priming treatments, 1% KNO3 (39.68% higher) recorded the highest germination percentage as compared to nonprimed control. In a recent study conducted to investigate effect of organic priming with marigold herbal tea on seed quality in Aji pepper (*Capsicum baccatum* var. *pendulum* Willd.), the herbal teas obtained from flowers of the *Tagetes patula* and *Tagetes erecta* species were used as an organic priming solution. It has been concluded that organic priming effectively increased germination, emergence, and the seedling fresh weight at different maturity levels of pepper (Mavi 2016). On the other hand, Teksan and Kavak (2016) reported that organic priming treatments with marigold and rose flower herbal teas did not significantly affect germination and emergence of pepper seeds compared to the control. Demirkaya (2016) found that organic priming treatments with seaweed extracts and methyl jasmonate improved the germination percentages and reduced the mean germination times of pepper seeds at low temperatures (15 and 20 °C).

12.3 Factors Affecting Priming Success

Priming efficiency is affected by many factors such as plant species (even cultivars within a species), priming technique, priming media, its concentration, water potential, oxygen availability, priming duration, temperature, seed condition, and storage conditions (Parera and Cantliffe 1994; Bray 1995). Optimal priming conditions may vary among cultivars and even seed lots of a given species. Bradford et al. (1990) indicated that there was a strong interaction between pepper seed lots and priming response, with the slowly germinating lots exhibiting the greatest benefit from priming. It has been reported that the germination response of two pepper varieties, viz., California Wonder and Yolo Wonder, was investigated under the osmotic potential of −5, −10, and −15 bar created by aqueous solution of PEG 6000 and KNO_3. Variety California Wonder appeared to be more responsive to preconditioning treatments as compared to the variety Yolo Wonder (Thakur et al. 1997). Priming in salt solutions leads to faster germination than priming in mannitol or PEG (Passam et al. 1989). Similar results were also observed for pepper seeds primed in KNO_3 and $KNO_3 + K_3PO_4$ solutions (O'Sullivan and Bouw 1984). $KN0_3$-primed jalapeno pepper seeds resulted in significantly earlier germination and accelerated vegetative seedling development, whereas priming in PEG appeared to retard vegetative seedling development (Rivas et al. 1984). In a study conducted by Jones and Sanders (1987), pepper seeds were soaked in water, a 1% KNO_3 + 1% K_2HPO_4 solution or a 1.5% KNO_3 + 1.5% K_2HPO_4 solution at 21 °C for either 72 h or 96 h. Seeds were air-dried and germinated at 15 °C, 20 °C, or 25 °C. All soaking treatments hastened germination and resulted in more uniform germination. Soaking seed in water or a 1% KNO_3 + 1% K_2HPO_4 solution gave similar results and was slightly better in promoting germination than the 1.5% KNO_3 + 1.5% K_2HPO_4 solution. Hydropriming (48 h) for tomato and sand matrix priming (80% water holding capacity, 3 days) for eggplant and chili were established as best methods of priming treatment capable of improving seed vigor (Venkatasubramanian and Umarani 2007).

12.4 Effect of Priming on Seed Germination and Seedling Establishment

High yield of the vegetable crops can only be obtained with high stand establishment of seedlings so that they can compete with the environment and can produce best plantation, ultimately increased yield and quality (Grassbaugh and Bennett 1998; Cantliffe 2003). The nonuniformity of pepper seed germination is one of the main problems faced by growers, since these seeds usually show low germination rate and low vigor (Silva et al. 2012). High germination and uniform stand establishment for chili pepper production are essential to maintaining profitable yields. However, pepper has non-starchy endosperm, and this offered a mechanical barrier to the growing embryo resulting in poor germination and emergence (Andreoli and Khan 1999). Nonuniform crop emergence results in plants of variable size and competitive ability. Subsequent management practices may be less effective on fields with nonuniform emergence. If stand establishment is poor, yields and quality of once-over machine-harvested crops will be poor. An increasing number of investigators are becoming interested in seed biology with the objective of understanding and controlling the many aspects of seed germination and seedling establishment (Bradford 1995). Seed priming is an alternative to achieve this objective. A number of processes stimulating germination are activated by seed priming and persist after the dehydration of the seed (Asgedom and Becker 2001). Therefore, upon sowing, the primed seeds can rapidly imbibe and restore the seed metabolism, resulting in an increased germination rate, decreased nonuniform germination, and better seedling development (Rowse 1995). Many evidences have shown seed priming could improve germination and early seedling growth under normal and stress conditions compared to plants grown from untreated seed (Bradford 1986; Chen et al. 2012). There are several investigations reporting either no effect or negative or beneficial responses of seed priming for pepper. Ghate and Phatak (1982) reported a significant decrease in germination rate when pepper seeds were primed with K_2HPO_4 plus $(NH4)_2HPO_4$ solution. It has been also reported that seed priming has not proved beneficial effect for tabasco pepper (*Capsicum frutescens* L.) field stand establishment (Sundstrom et al. 1987). In a study carried out by Jones and Sanders (1987), pepper seeds were primed in water, a 1% KNO_3 + 1% K_2HPO_4 solution, or a 1.5% KNO_3 + 1.5% K_2HPO_4 solution at 21 °C for either 72 h or 96 h. Seeds were air-dried and germinated at 15 °C, 20 °C, or 25 °C. All priming treatments hastened germination and resulted in more uniform germination. They also reported that priming seed in water or a 1% KNO_3 + 1% K_2HPO_4 solution gave similar results and was slightly better in promoting germination than the 1.5% KNO_3 + 1.5% K_2HPO_4 solution. Sundstrom and Edwards (1989) reported that an increased rate of germination was observed when jalapeno (*Capsicum annuum* L.) and tabasco (*Capsicum frutescens* L.) peppers were primed in a 3.0% or 2.75% KNO_3 solution, respectively. In bell pepper, priming of seeds enhanced the rate of germination (Bradford et al. 1990; Khan et al. 1992). Cooksey et al. (1994) compared non-treated seed, primed seed, and transplants for effects on stand establishment, plant morphology, and yield of paprika pepper. They concluded that non-treated seed was satisfactory for stand establishment, although primed seed had the potential to provide greater initial stands. They also concluded that transplanting is not recommended for stand establishment of paprika pepper intended for mechanical harvest. Lanteri et al. (1994) revealed that priming of pepper seeds in −1.1, −1.3, and −1.5 MPa PEG solution for 14 days at 25 °C reduced the mean time to germination. Lee et al. (1997) found that the rate of germination and improvement of seedling stands were also accelerated as a result of seed priming in pepper. Pepper seeds osmoprimed with PEG 6000 and KNO_3 germinated more rapidly than the control or water presoaked ones. In addition to enhanced germinability, osmopriming treatment was found to significantly improve emergence index and vigor index (Thakur et al. 1997). It was also observed that aerated hydration treatment of pepper seeds produced larger seedlings and better stands in the field or greenhouse compared to untreated (Demir and Okcu 2004). It was noticed by Yogananda et al. (2004) that bell pepper seeds invigorated with GA_3 (200 ppm) or KNO_3 (1.0%) recorded higher germination, root and shoot length, seeding dry weight, rate of germination, and seedling vigor index over control. Korkmaz (2005) for sweet pepper reported that priming treatments generally improved the germination synchrony. In capsicum, Pandita et al. (2007) observed that osmo- and solid matrix priming improved seed germination over nonprimed seeds at 15 °C, 20 °C, and 25 °C temperatures. They also observed that priming also reduced mean days to germination significantly over control. It was reported by QianQian et al. (2009) that after priming treatment of the seeds with vermiculite, the germination rate, the germination energy, the germination index and vigor index of hot pepper seeds, and the fresh and dry weights of seedlings were significantly higher than those of control. Yadav et al. (2011) noticed that germination percentage of primed pepper seeds was increased to as compared to control and also tolerated cold and salt stress for 10 days with 100% survival, whereas control seedling could not survive. Ameri et al. (2011) conducted a laboratory experiment on seed priming of pepper "California Wonder" with 1% NaCl, 1% $CaCl_2$, 3% KNO_3, 3% $FeSO_4$, and control. All treatments resulted in a higher germination rates compared to the control. They also indicated that seed priming with $FeSO_4$ was the best treatment resulting in the maximum radicle dry weight, germination percentage, and germination rate with values of 0.126%, 70%, and 5.07%, respectively, while in the control values were 0.063%, 36.81%, and 0.83%, respectively. In a work investigating effects of osmotic conditioning and humidification applications on emergence percentage and mean emergence time of pepper seeds (Demirkaya 2012), the seeds were osmoprimed with PEG-6000 and −1.0 MPa for 1, 2, and 3 days and hydroprimed for 1, 2, and 3 days. The results of the study indicated that humidification and priming applications with PEG-6000 positively affected mean emergence time and emergence percentages in pepper seeds.

12.5 Effect of Priming on Performances of Pepper Seeds at Low Temperature

Low temperature is one of the most important environmental factors affecting pepper seed germination or seedling emergence. Most of the pepper cultivars are extremely sensitive to chilling stress particularly during emergence and early stages of seedling development. However, different varieties of the same species can tolerate low temperature differently. Peppers are very sensitive to germination temperatures, tolerating 16 °C on the lower end, with an optimum range between 18 and 35 °C and a mean optimum of 29.5 °C (Nonnecke 1996). The rate of germination and emergence of pepper seeds is markedly reduced at a temperature ranging from 15 to 20 °C. Pepper plants experience chilling injury with prolonged temps of 0–10 ° C. Methods to bypass the slow germination of pepper seed at low temperatures have been studied by researchers. Seed priming is an effective and practical technique that enables the seed to germinate and emerge faster even at suboptimal temperatures and also known to reduce the imbibitional damage associated with planting seeds in cold soils (Bennett and Waters 1987; Jisha et al. 2013). The effect of priming on low-temperature performance of pepper seeds has ranged from no improvement to some advancement in germination and emergence percentages and rates. For example, imbibition at 30 °C for 48 h in water or in aerated KNO_3 solutions for 6 or 8 days enhanced the germination at 15 °C when the seeds were not redried after treatment (Sachs et al. 1980). It has been reported that field emergence percentages of pepper at 20 °C generally were unaffected by priming (Bradford et al. 1990). Carter (1994) initiated to examine the effects of NaCl priming on total germination and the rate of germination under suboptimal temperature conditions. Seeds of "Tam Veracruz" chili were primed at 23 °C for 5 days in double-distilled H_2O or 0.2, 0.4, or 0.6 M NaCl solution (equivalent to an osmotic potential of −0.89 MPa, −1.77 MPa, and − 2.66 MPa, respectively). He suggested that priming "Tam Veracruz" chili seed for 5 days at 23 °C in 0.2 M NaCl increased the germination rate at temperatures from 15 to 23 °C. Matriconditioning pepper seeds at 15 °C for 7 days or at 25 °C for 4 days reduced the time needed for germination on filter paper at 15 °C, and conditioning at 25 °C was more effective than conditioning at 15 °C in reducing the germination time. The priming treatment also improved the performance of pepper seeds in early field plantings at suboptimal temperatures (averaged over 10 days after planting) ranging from 12 to 18 °C (Khan et al. 1995). Increasing evidence suggests that benzoic acid derivatives such as salicylic acid (SA) or ASA regulate stress tolerance in plants (Lopez-Delgado et al. 1998). Mendoza et al. (2002) reported that priming pepper seeds in 0.1 mM SA prevented seedlings from subsequent chilling-induced damage. The incorporation of plant growth regulators into the priming solution has been shown to be effective to enhance germination and seedling emergence of crops under adverse condition. For example, incorporation of acetyl salicylic acid (ASA) or methyl jasmonate (MeJA) into the priming solution improved sweet pepper germination and emergence at low temperature (Korkmaz 2005). Priming sweet pepper seeds in KNO_3 supplemented with 0.1 mM acetyl salicylic acid resulted in 91% germination and 85% emergence at 15 °C which are significantly higher than the germination and emergence of nonprimed and seeds primed in KNO_3 only (Korkmaz 2005). In the same study, it has been also reported that the primed seeds stored for 1 month at 4 °C still exhibited improved germination performance at 15 °C. Similarly, Korkmaz and Korkmaz (2009) indicate that priming seeds in 25 ppm and 50 ppm ALA (5-aminolevulinic acid) incorporated into the KNO_3 solution improved low-temperature performance of red pepper seeds. They also indicated that primed seeds stored for 1 month at 4 °C or 25 °C still exhibited improved germination and emergence performance at 15 °C. Chemical seed priming treatments imparted tolerance to subsequent cold (4 °C) or salt stress (NaCl, 200 mM) exposure at later growth stages. Among the treatments, seed priming with thiourea (TU, 1.3 mM) was effective in imparting cold as well as salt stress tolerance (Yadav et al. 2011). In a recent work carried out by Sharma et al. (2015), the effect of seed priming using PEG-6000, gibberellic acid, KH_2PO_4, Na_2HPO_4, distilled water, and cow urine on seedling vigor of bell pepper seeds was studied. The seeds were primed at 20 °C for 24 and 48 h. They reported that although all the priming treatments significantly improved seed performance over control, seed priming with 100 ppm GA_3 for 48 h was more effective in improving seed performances under low temperature than the other treatments. The addition of prohexadione-Ca into the priming solution significantly improved sweet pepper germination and emergence at low temperatures compared to unprimed seeds and the seeds primed in KNO_3, KH_2PO_4, and PEG only (Ozbay and Susluoglu 2016a). Samarah et al. (2016) compared hydropriming and natural compounds that could be used as seed treatments to increase seed germination rate and improve cold tolerance of bell pepper. They reported that treatments with nanochitin, chitosan, acetic acid, or hydropriming improved low-temperature germination in soil by 17–39% compared with untreated seeds, being hydropriming or nanochitin more effective in reducing mean time to germination than chitosan or acetic acid treatments. Result of the study conducted by Korkmaz et al. (2017) revealed that seed application of 1 μM or 5 μM melatonin significantly improved pepper seed germination and seedling emergence at chilling temperatures compared to seeds not treated with melatonin.

12.6 Effect of Priming on Performances of Pepper Seeds at High Temperature

High temperature is also one of the most important environmental factors affecting seed germination and seedling emergence of many vegetable crops including pepper. Pepper seeds for fall production are sown when summer greenhouse temperatures can reach 40–45 °C (Vavrina 1994). This range is far above the optimum temperature (29 °C) for pepper germination (Maynard and Hochmuth 1997). At supraoptimal temperatures, pepper seeds may enter into the state of thermoinhibition. The inability to germinate at higher than optimal temperatures is attributed to a condition called thermoinhibition or thermodormancy (Carter and Varina 2001). Most of the pepper cultivars are extremely sensitive to high temperature during germination and emergence and also later stages of plant development such as flowering and fruit set (Aloni et al. 2001). However, different varieties of the same species can tolerate high temperature differently as in low temperature mentioned previously. Cultivar differences in final germination pronounced at 35 °C. Therefore, growers should consider this, when choosing cultivars for fall transplants. However, no cultivar germinated well at temperatures higher than 35 °C (Carter 2000). Priming can also help alleviate high-temperature inhibition of germination and improve seedling emergence of pepper and other species (Ellis and Butcher 1988; Cantliffe 1991). For example, pepper seed germination was improved under supraoptimal conditions if the seeds were first soaked in solutions of ethephon or gibberellic acid (GA_3) (Carter and Stevens 1998). Nascimento et al. (2011) reported that osmoconditioning improves the physiological quality of hot pepper seeds by improving germination at high temperature (35 °C). In a study conducted by Silva et al. (2012), hot pepper "Mari" seeds were osmoconditioned in aerated solution of polyethylene glycol (PEG 6000) for 7 days and subjected to germination test at 15, 20, 25, 30, and 35 °C. The total seed germination decreased with increasing temperature. The osmotic conditioning was effective in improving seed germination at all temperatures, especially at high temperatures. Ozbay and Susluoglu (2016b) reported that priming pepper seeds in the presence or absence of plant growth regulators improved final germination and emergence percentage, mean emergence time, emergence index, and emergence rate (E_{50}) at high temperature (35 °C) compared to control seeds.

12.7 Priming to Overcome Salt Stress

Among the abiotic stresses, salinity is a major limiting factor in the crop productivity all over the world. Almost 20% of cultivated area of the world and half of the world's irrigated lands are stressed by the salinity (Chinnusamy et al. 2005). Soil salinity, if not properly managed, often results in poor stand establishment, reduced plant growth, and reduced yield of many horticultural crops such as peppers (Flynn et al. 2003; Niu et al. 2010). When exposed to high salt concentration, growth of pepper, which is not a salt-tolerant crop, can be negatively affected, and yield and fruit quality are diminished (Ibn Maaouia-Houimli et al. 2011). Salinity may reduce crop yield by upsetting water and nutritional balance of the plant (Khan et al. 2007). The salinity delays or prevents the seed germination through various factors, such as a reduction in water availability by high osmotic potential and toxicity of Na and Cl ions, changes in the mobilization of stored reserves, and affecting the structural organization of proteins (Kumar 1995; Ibrahim 2016). Salt concentrations affected percentage and rate of germination as well as the opening of cotyledonary leaves in *Capsicum annuum* L. cv. California Wonder. The 100 mM or higher salt concentrations reduced the rate of radicle protrusion significantly (Yadav et al. 2011). Similarly, Chartzoulakis and Klapaki (2000) reported the significant reduction in the germination as well as growth of two bell pepper hybrids on exposure to 100 and 150 mM NaCl. Furthermore, pepper yield is reduced up to 14% for every increase in unit of salinity above its threshold (Rhoades et al. 1992). To alleviate this problem, a number of studies were conducted with the aim of removing the inhibitory effect of salt stress on plant growth. The beneficial effect of seed priming has been used for improving germination, emergence, and stand establishment of peppers under salt conditions (Hassen et al. 2014). Plants derived from primed seed have a higher adaptation capacity to salinity originating in the osmoregulation process (Levitt 1980). In studies conducted by Smith and Cobb (1991a), Amjad et al. (2007), and Yadav et al. (2011) with pepper, it was concluded that seed priming improves seed germination, seedling emergence, and growth under saline conditions. Priming of seeds with NaCl was found to be effective in alleviating the adverse effects of salt stress on pepper plants. Khan et al. (2009a) observed that priming of seeds using 1 mM NaCl improved seedling vigor and establishment under salt stress conditions. It has been reported that hormonal priming with salicylic acid (0.8 mM) and acetylsalicylic acid (0.2 mM) showed significantly better results over the control by improvement in time taken to 50% emergence, final emergence percentage, root and shoot length, seedling fresh and dry weight, and seedling vigor under normal as well as saline conditions. It has also been reported that acetylsalicylic acid exhibited superiority over salicylic acid (Khan et al. 2009b). Pepper seeds treated with 10 mM glyciebetaine (GB), an organic osmolyte accumulated in variety of plants in response to abiotic stress, for 24 h in darkness improved germination and synchrony of germination under salt stress. Glyciebetaine decreases the melonoldehyde (MDA) and increases the proline content and SOD activity in seeds. The enhanced tolerance to salt stress may be due to reduced lipid peroxidation and elevated SOD enzyme activity (Korkmaz and Sirikci 2011). Osmopriming with $CaCl_2$, KCl, and NaCl improved germination rate, chlorophyll content, proline, and protein accumulation in pepper under salt stress conditions (Hassen et al. 2014). Hassen et al. (2017) also conducted a study to assess the effects of priming with concentrations of KCl, NaCl, and CaCl2 on morphological and biochemical parameters of the pepper cvs. Beldi, Baklouti, and Anaheim chili. They reported that seedlings developed from primed seed had improved biomass, water content, carotenoid content, soluble sugar, polyphenols, and soluble proteins at salt concentrations of 6 $g{\cdot}L^{-1}$.

12.8 Reversal of Seed Deterioration by Priming

Another area of practical concern of priming is the interaction between priming treatments and seed deterioration. Seed deterioration may be defined as the loss of seed viability and vigor due aging effects and adverse environmental factors particularly higher temperature, relative air humidity, and oxygen/carbon dioxide ratio (McDonald 1999; Jyoti and Malik 2013). Seed deterioration is associated with various cellular, metabolic, and chemical alterations including lipid peroxidation, membrane disruption, DNA damage, impairment of RNA, and protein synthesis and causes several detrimental effects on seed (Jyoti and Malik 2013). Seed storage causes a decrease in the protein content which may be related to oxidation of the amino acids due to the increase in the respiratory activity and advance in the deterioration process of the stored seeds (Manonmani et al. 2014). Poor storage conditions may accelerate seed deterioration in primed and unprimed seeds (Georghiou et al. 1987). It is a serious problem particularly in developing countries where seeds are stored in places usually without a proper control of air humidity, temperature, and O_2/CO_2 concentration (Khan et al. 2016). As seed deterioration increases, seed performance progressively decreases. Plants that have originated from deteriorate seed can also reduce growth rate (Kapoor et al. 2010). Pepper is a warm climate crop, and seeds are sensitive to storage conditions and lose germination potential within a short period of time, particularly under adverse storage conditions (Priestly 1986). Decline in germination potential due to seed aging may result in low seedling emergence and stand establishment (Ermis et al. 2016). Accelerated aging of sweet pepper seeds at 42 °C and 100% R.H. for 0, 5, 10, 15, 20, 25, and 30 days resulted in decreased germinability in terms of radical emergence (%) that was also differed significantly among different aging duration (Kaewnaree et al. 2011). According to Li et al. (2005), the main mechanism for aging of pepper seeds is associated with increased peroxidation of lipid membranes. The aging of pepper seeds, during long-term storage, deteriorated their vital status which was expressed in changes in their moisture content, decreasing of their sowing qualities, and development of weaker seedlings with higher water content (Panayotov and Aladjadjiyan 2014). Priming can reverse some of the aging-induced deteriorative factors and thus improve seed performance (Taylor et al. 1998). The beneficial effects of priming are associated with the repair and building up of nucleic acid, increased synthesis of proteins, as well as the repair of both mitochondria and membranes (McDonald 1999, 2000). Under invigoration, metabolic repair processes in deteriorated seeds occur before onset of seed germination process (Srinivasan et al. 2009). Many seed priming treatments have been used to reduce the damage of aging and invigorate their performance in many crops including pepper (Farooq et al. 2009; Ermis et al. 2016). In pepper, osmoconditioning was found to be also an effective treatment for protecting seeds stored under high temperatures by increasing longevity of seeds. However, osmoconditioning after storage did not seem to have any significant effect on seed viability, though it enhanced the germination rate (Georghiou et al. 1987). Osmoconditioning of controlled deteriorated sweet pepper seeds exerted various effects on seed germination depending on deterioration rate (Lanteri et al. 1996, 1997). Lanteri et al. (1996) reported that effects of aging pepper seeds may be partly reversed by priming and part of priming effects is related to the induction of nuclear replication. In a study conducted by Siri et al. (2013), sweet pepper seeds were artificially aged by exposing to high temperature (42 °C) and high humidity (100% relative humidity) and then primed with PEG 6000 solution with the osmotic potential of −1.5 MPa for 6 days. They concluded that seed germination was improved in primed seeds due to accumulation of antioxidants and the improvement of cell membrane integrity. Ermiş et al. (2016) indicate that priming is more useful for enhancing germination of low-quality seed lots than higher-quality ones which indicates that repair of aging is one of the primary advantages of the priming treatments.

While in general it is clear that the best commercial potential for priming is the enhancement of performance of seeds of the highest quality (Perkins-Veazie and Cantliffe 1984), such repair treatments may be appropriate in the cases of valuable deteriorated stock or rare genetic material.

12.9 Storage Life of Primed Seeds

It would be highly advantageous if the promotive effects of priming on the germination of seeds were retained after drying and storage. In general, storing primed seeds reduces germination or seedling emergence, and there are conflicting results on the effect of storage life of primed seeds. As far as it concerns the storage of pepper seeds following priming treatments, Bruggink et al. (1999) reported that for pepper seeds, the desired longevity was obtained by keeping the seeds, after a priming treatment, under a mild water and/or temperature stress for a period of several hours to days. O'Sullivan and Bouw (1984) reported that a promotive effect on the germination of pepper seeds, when primed in 15% salt solution, was retained for 2 days after treatment. Perl and Feder (1981) showed a retention of seed vigor and seedling development of pepper seeds for 2 months following priming. Aljaro and Wyneken (1985) reported the effects of priming of "Y010 Y" pepper seeds to be apparent for up to 140 days prior to sowing. In pepper, osmoconditioning was found to be also an effective treatment for protecting seeds stored under high temperatures (Georghiou et al. 1987). Thanos et al. (1989) reported that primed sweet pepper seeds retained their improved vigor after a storage period of 6 months at 5–8 °C. It was determined that primed pepper seeds with PEG could be packaged in original packages and stored for 12 months under controlled conditions without any loss of effect on emergence time (Cetin and Duman 2005). They also reported that air- and moisture-proof aluminum materials were the best seed storage packaging material for the primed seeds. Korkmaz (2005) reported that 1 month storage at 4–8 °C did not reduce the performance of pepper seeds that were primed in the presence of salicylic acid. Korkmaz and Korkmaz (2009) indicate that red pepper seeds primed in 25 ppm and 50 ppm ALA (5-aminolevulinic acid) incorporated into the KNO_3 solution can be stored for 1 month at 4 °C or 25 °C and still exhibit improved germination and emergence performance at 15 °C. On the contrary, in other experiments, the promotive effect of priming was reduced rapidly after drying of the seeds (Sachs et al. 1980; O'Sullivan and Bouw 1984). The response of primed seeds to storage is species and variety dependent. Therefore, it is important that for the beneficial effects of priming to be retained, new techniques and markers have to be used to monitor the priming progress.

12.10 Effect of Priming on Physiological and Biochemical Changes in Pepper Seeds

The positive effects of priming on the germination performance of many species might be attributed to the induction of physiological and biochemical changes such as DNA replication, increased RNA and protein synthesis, greater ATP availability, increased respiratory activity, faster embryo growth, reduced leakage of metabolites, and decrease in lipid peroxidation and increased in the antioxidant activities.

12.10.1 Seed Priming and Cell Cycle Regulation

Some of the hypotheses proposing explanation for priming-induced improvement of seed performance are based on its effect on DNA in relation to activation of DNA repair mechanisms, synchronization of the cell cycle in G2, and preparation to cell division (Lutts et al. 2016). For example, Stofella et al. (1992) indicate that the increase in root/shoot ratio with hydropriming treatments may be due to the fact that priming induced nuclear replication in root tips of fresh seeds. By incorporating DNA-specific fluorescent dyes, nuclear DNA contents expressed as C values have been quantified in pepper (Lanteri et al. 1993) seeds during germination and priming. Lanteri et al. (1994) reported that priming of pepper and tomato seeds in −1.1, −1.3, and −1.5 MPa PEG solution for 14 days at 25 °C reduced the mean time to germination. Lanteri et al. (1996) reported that effects of aging pepper seeds may be partly reversed by priming and part of priming effects is related to the induction of nuclear replication. It has been reported that priming of seed promotes germination by repair of the damaged protein, RNA, and DNA (Koehler et al. 1997). An induction of 4C signals was also found after priming, indicating that during priming the cells of the embryonic root tip had replicated their DNA and arrested at the G2 phase of the cell cycle. Flow cytometric determination of nuclear DNA contents in embryos of dry, fully matured pepper seeds revealed only 2C signals. Therefore, pepper belongs to those species in which the quiescent embryo arrests nuclear division in the presynthetic G1 phase. 4C nuclei appear 1–2 days after imbibition in water, while radicle emergence starts 2 days later; at this time the proportion of the 4C nuclei is over 50% (Lanteri et al. 1992, 1993; Saracco et al. 1995). Thus, DNA replication precedes pepper seedling growth, and the 4C/2C ratio can be used to predict seedling performance of pepper (Sliwinska 2009). Using flow cytometry, Lanteri et al. (1997) observed that priming treatments in PEG solutions might induce DNA replication in the embryo root tips of pepper seeds. Lanteri et al. (1998) have also reported that the improvement of performance in pepper seeds after conditioning osmotic activity has been correlated with macromolecular repair that occur during the treatment. Changes in nuclear replication stages upon priming have been studied by flow cytometry on pepper seeds. The author observed that significant correlations were found between the frequency of priming-induced nuclear replication and the improvement of pepper seed vigor, as measured by the reduction in mean germination time (Sliwinska 2009). Flow cytometric data published by Varier et al. (2010) also reveal that the improvement of germination associated with priming is accompanied by increase in 4C nuclear DNA. This indicates that priming enhances DNA replication allowing the advancement of the cell cycle from G1 to G2 phase.

12.10.2 Effect of Priming on Seed Ultrastructure

The seeds of some species including peppers are prevented from completing germination because the embryo is constrained by seed coats and surrounding structures (Watkins and Cantliffe 1983; Bradford 1995; Bewley 1997a). After imbibition of the seeds, however, the endosperm tissue enclosing the embryo restrains the germination process acting as a physical barrier, which restricts radicle emergence (Bino et al. 1998; Lutts et al. 2016). It is necessary to reduce the resistance of these enclosing structures for germination to be completed. Therefore, weakening of the endosperm, particularly in the micropylar region adjacent to the radicle, is a prerequisite for the completion of germination (Gong et al. 2005). Priming may also contribute to rapid seed germination by reducing the mechanical restraint of endosperm on developing embryo (Mayer and Mayber 1989). In studies with pepper seeds, Watkins and Cantliffe (1983) and Watkins et al. (1985) found that the mechanical constraint by endosperm in the tip region of the radicle contributed to slow germination rate and that the weakening of the endosperm by externally applied gibberellic acid (GA_{4+7}) occurred prior to germination. The reserves in endosperm are degraded by specific enzymes after the initiation of germination to facilitate the developing seedling until photosynthesis is initiated (Bewley and Black 1994; Homrichhausen et al. 2003). In most cases, endo-beta-mannanase degrades endosperm cell walls in endospermic seeds and can be detected during germination in the seeds (Bewley 1997b; Cantliffe 2003). Andreoli and Khan (1999) indicate that the matriconditioning seed treatment provides a means to efficiently digest the endosperm cell by GA-induced enzymes and reduce the mechanical restraints of endosperm thus providing energy to start and sustain embryo growth of tomato and pepper. Lanteri et al. (2000) investigated the expression of β-tubulin in the root tips of pepper seeds as a complementary marker for priming de novo synthesis of β-tubulins in response to priming was observed prior to DNA replication after osmopriming for different durations at two water potentials. Ethylene also directly influences germination speed and percentage. Increase in ethylene production during priming may promote endo-β-mannase activity facilitating endosperm weakening and post-priming germination (Chen and Arora 2013). Priming was reported to modify the kinetics of ethylene synthesis from its precursor 1-aminocyclopropane-1-carboxylic acid (ACC) (Wu et al. 2014). It has been reported that the primed pepper seeds had higher ACC level, greater ACC-oxidase activity, and greater embryo growth potential than the nonprime seeds (Khan et al. 1995).

12.10.3 Effect of Priming on Reserve Mobilization

It is proposed that germination-related processes such as respiration, energy metabolism, and early reserve mobilization can also occur during priming (Varier et al. 2010). Higher respiratory activity is required to cover energy pool for speed up germination. Increased respiratory activity has been reportedly associated with pre-sowing treatments. For example, Halpin-Ingham and Sundstrom (1992) indicate that priming increases respiratory activity of pepper seeds. They also reported that seed priming generally reduced the oxygen-time constant and increased the standard deviation of germination responses. During seed germination, storage proteins, which provide a source of reduced nitrogen, and inorganic minerals need to be mobilized to support seedling growth (Lutts et al. 2016). Pepper seeds were examined during priming to determine if seed treatments which accelerate the rate of germination could be correlated with specific physiological changes within the seeds. Smith and Comb (1991b) found that soluble protein content increased to 109% and 120% in pepper seeds primed in −0.90 and −1.35 MPa NaCl solutions, respectively, after 12 days of priming and also revealed that there was no significant difference in the soluble protein content between two priming treatments. Khan (1992) found that two amino acids were incorporated in proteins during the first 24 h of imbibition of sweet pepper seeds in PEG solutions. QianQian et al. (2009) reported that the priming treatment with vermiculite significantly resulted in significantly higher contents of soluble protein in the pepper seedlings under NaCl stress. Hassen et al. (2017) reported that pepper seedlings developed from primed seed had improved soluble proteins at salt concentrations of 6 $g{\cdot}L^{-1}$. In contrary, in a study conducted to determine the relationship between physical and metabolic changes and germination/emergence time and rates both under optimum and stress conditions during osmotic conditioning process in pepper seeds, Ozpercin et al. (2005) reported that no significant changes were found in total protein levels in primed pepper seeds. Osmopriming induced accumulation of stress proteins, such as late embryogenesis abundant (LEA) proteins and heat shock proteins (HSP) (Gallardo et al. 2002). As LEA proteins accumulate at a high level in response to cell/tissue dehydration, they may contribute to acquisition of tolerance to drought and related stresses such as osmotic, salt, and cold stress (Lutts et al. 2016). Cortez-Baheza et al. (2007) have shown that several late embryogenesis abundant (lea) genes are strongly induced in pepper seeds when osmoprimed with PEG incorporated with GA3. A new LEA protein of 73 amino acids (Calea 73 gene) was highly induced in osmoprimed treatments in which KNO_3 was used in combination with PEG on *C. annuum* cv. Caballero seeds. Besides being induced by PEG + GA_3, the Calea 73 gene was also stimulated by PEG + KNO_3, which indicates that this gene is expressed during osmopriming regardless of the osmotic solution used (Cortez-Baheza et al. 2008).

12.10.4 Effect of Priming on Osmoprotectants

The osmoprotectants are compounds produced in plants during osmotic stress condition. Osmoprotectants are chemically small, electrically neutral molecules, which play important roles in the protection and stabilization of proteins and membranes against abiotic stresses without disrupting plant metabolism (Yancey 1994). Osmoprotectants or compatible solutes include proline, mannitol, D-ononitol, trehalose, sucrose, fructane, glycine betaine, and polyamines (Khan et al. 2015; Lutts et al. 2016). Accumulation of osmoprotectants such as proline and sugars has been reported in various plant species including peppers during a wide range of abiotic stresses. Proline is a proteogenic amino acid and accumulates both under stress and non-stress conditions as a beneficial solute in plants (Kavi Kishor et al. 2015). The main function of proline in plants is to restrain the osmotic effects by stabilizing protein structures and scavenging free radicals (Smirnof and Cumbes 1989; Biedermannova et al. 2008). Accumulation of sugars in various plant species is related to high tolerance to stress conditions (Bohnert and Jensen 1996). Sugars protect membranes and enzyme complexes from reactive oxygen species mainly by interacting with enzymes of the glutathione–ascorbate cycle (Khan et al. 2015). There are several reports showing that the osmoprotectant content increases after seed priming. For example, the priming pepper seeds with vermiculite significantly increased content of free proline and soluble sugar in the seedlings under NaCl stress (QianQian et al. (2009). Korkmaz and Şirikçi (2011) reported that pepper seed treated with glyciebetaine exhibited an enhancement in proline content under saline conditions. Hassen et al. (2014) showed that priming pepper seeds with NaCl, KCl, and $CaCl_2$ caused a marked increase in proline accumulation. Seedlings developed from primed pepper seeds had improved biomass, water content, carotenoid content, soluble sugar, and polyphenols at salt concentrations of 6 $g{\cdot}L^{-1}$ (Hassen et al. 2017).

12.10.5 Effect of Priming on Management of Oxidative Status

Management of oxidative status is also an important part of primed seed physiology (McDonald 2000; Kibinza et al. 2011). The priming could activate the response of the antioxidant system, becoming the primed seeds more prepared for possible stresses. During seed imbibition and early stages of germination, reactive oxygen species (ROS) production occurs mainly through respiratory activities of mitochondria, activities of β-oxidation pathways and enzymes such as NADPH oxidases, extracellular peroxidases, and oxalate oxidases (Wojtyla et al. 2016). Antioxidants, by scavenging the excessive ROS during early imbibition, play an essential role in ensuring successful germination, especially under stress conditions (Bailly 2004). An increased activity of antioxidant enzymes like ascorbate peroxidases (APX), catalase (CAT), peroxidase (POX/POD), glutathione reductase (GR), and superoxide dismutase (SOD) as a consequence of seed priming has been reported in various crops including peppers. For example, the priming treatment with vermiculite significantly improved the activities of superoxide dismutase (SOD) and peroxidase (POD) in the leaves of seedlings and significantly decreased the rate of producing superoxide anion (O2-) and the content of malondialdehyde (MDA), resulting in lower membrane lipid peroxidation in the seedlings under NaCl stress (QianQian et al. 2009). Kaya et al. (2010) reported that the most important effects of priming on enzymatic activities of pepper seeds were recorded in catalase which increased remarkably with priming. They further stated that, even though not the same extent, priming also increased ascorbate peroxidase and superoxide dismutase activities (Kaya et al. 2010). Korkmaz et al. (2010) reported that 5-aminolevulinic acid (ALA) pretreatment increased relative water content, stomatal conductance, and superoxide dismutase (SOD) enzyme activity and reduced membrane permeability in pepper. Pepper seeds treated with glyciebetaine exhibited a significant decrease in MDA content and enhancement in SOD enzyme activity under saline conditions (Korkmaz and Sirikci 2011). In a study focusing on the effect of priming process on quality and biochemical changes in sweet pepper seed, Siri et al. (2013) reported that osmopriming with −1.5 MPa PEG-6000 for 6 days of artificially aged seeds of sweet pepper (42 °C and 100% relative humidity) resulted in an improved germination with decreased levels of malondialdehyde (MDA) and total peroxide concentration. They further stated that accumulation of total antioxidant activity (TAA), total ascorbate, dehydroascorbate, and catalase (CAT) activity in primed seeds enhanced the defense mechanism in protecting the cell membrane damage from reactive oxygen species. Da Silva et al. (2015) showed that priming improved superoxide dismutase (SOD), catalase (CAT), and peroxidase (POX) activities.

References

Abdel-Kader MM, El-Mougy NS (2014) Controlling downy and powdery mildews of bell pepper using fungicides alternatives under plastic houses conditions. Int J Eng Innov Technol 4(2):44–48

Aljaro A, Wyneken HL (1985) Osmoconditioning of sweet pepper seed and its effect on germination and emergence. Chile Agric Techn 45(4):293–302

Aloni B, Peet MM, Pharr M, Karni L (2001) The effect of high temperature and high atmospheric CO2 on carbohydrate changes in bell pepper (*Capsicum annuum*) pollen in relation to its germination. Physiol Plant 112:505–512

Ameri A, Fatemi H, Aroiee H, Teixeira da Silva JA (2011) What's the effect of saline priming on germination factors of *Capsicum annuum* var. 'California wonder' seeds? Seed Sci Biotechnol 5(1):47–49

Amjad M, Ziaf K, Iqbal Q, Ahmad I, Riaz M, Saqib ZA (2007) Effect of seed priming on seed vigour and salt tolerance in hot pepper. Pak J Agric Sci 44(3):408–419

Ananthi M, Selvaraju P, Sundaralingam K (2014) Effect of bio-priming using bio-control agents on seed germination and seedling vigour in chilli (Capsicum annuum L.) 'PKM 1'. J Hortic Sci Biotechnol 89(5):564–568

Andreoli C, Khan A (1999) Matriconditioning integrated with gibberellic acid to hasten seed germination and improve stand establishment of pepper and tomato. Pesqui Agropecuaria Bras 34(10):1953–1958

Andrews J (1993) Chili peppers. In: Kiple KF, Ornelas KC (eds) The Cambridge world history of food. Cambridge University Press, Cambridge

Asgedom H, Becker M (2001) Effect of seed priming with nutrient solutions on germination, seedling growth and weed competitiveness of cereals in Eritrea. In: Proceedings of the Deutscher Tropentag. University of Bonn and ATSAF, Margraf Publishers Press, Weickersheim

Ashraf M, Foolad MR (2005) Presowing seed treatment a shotgun approach to ımprove germination, plant growth, and crop yield under saline and non-saline conditions. Adv Agron 88:223–271

Bailly C (2004) Active oxygen species and antioxidants in seed biology. Seed Sci Res 14:93–107

Belletti P, Quagliotti L (1988) Problems of seed production and storage of pepper. In: Proceedings of the International Symposium on integrated Management Practices, Asian vegetable Research Centre, Taipei, pp 28–41

Bennett MA, Waters L Jr (1987) Seed hydration treatments for improved sweet maize germination and stand establishment. J Am Soc Hortic Sci 112:45–49

Bennett MA, Fritz VA, Callan NW (1992) Impact of seed treatments on crop stand establishment. HortTechnology 2(3):345–349

Bewley JD (1997a) Breaking down the walls: a role for endo-β-mannanase in release from seed dormancy? Trends Plant Sci 2:464–469

Bewley JD (1997b) Seed germination and dormancy. Plant Cell 9:1055–1066

Bewley JD, Black M (1994) Seeds: physiology of development and germination, 2nd edn. Plenum Press, New York

Biedermannova L, Riley KE, Berka K, Hobza P, Vondrasek J (2008) Another role of proline: stabilization interactions in proteins and protein complexes concerning proline and tryptophan. Phys Chem Chem Phys 10:6350–6359

Bino RJ, Jalink H, Oluoch MO, Groot SPC (1998) Seed research for improved technologies. Sci Agric 55(Special Issue):19–26

Bohnert HJ, Jensen RG (1996) Strategies for engineering water-stress tolerance in plants. Trends Biotechnol 14:89–97

Bradford KJ (1986) Manipulation of seed water relations via osmotic priming to improve germination under stress conditions. Hortscience 21:1005–1112

Bradford KJ (1995) Water relations in seed germination. In: Kigel J and Galili G (eds) Seed development and germination, Marcel Dekker, Inc, New York

Bradford KJ, Steiner JJ, Trawatha SE (1990) Seed priming influence on germination and emergence of pepper seed lots. Crop Sci 30:718–721

Bray CM (1995) Biochemical processes during the osmopriming of seeds. In: Kigel J, Galili G (eds) Seed development and germination. Marcel Dekker, New York

Bruggink GT, Ooms JJJ, van der Toorn P (1999) Induction of longevity in primed seeds. Seed Sci Res 9:49–53

Callan NW, Mathre DE, Miller JB (1990) Biopriming seed treatment for biological control of *Pythium ultimum* preemergence damping-off in sh2 sweet corn. Plant Dis 74:368–372

Cantliffe DJ (1991) Benzyladenine in the priming solution reduces thermodormancy of lettuce seeds. HortTechnology 1:95–97

Cantliffe DJ (2003) Seed enhancements. Acta Hortic 607:53–59

Carter AK (1994) Effect of NaCl concentration and temperature on germination of 'Tam Veracruz' Chile seed (*Capsicum annuum* L.). J Seed Technol 18(1):16–20

Carter AK (2000) High temperature inhibits germination of Jalapeno and Cayenne Pepper. Massachusetts Agricultural Experiment Station, under Project No. MAS00786

Carter AK, Stevens R (1998) Using ethephon and GA3 to overcome thermoinhibition in 'Jalapeño M' pepper seed. Hortscience 33:1026–1027

Carter AK, Vavrina C (2001) High temperature inhibits germination of jalepeno and cayenne pepper. Hortscience 36:724–725

Cetin M, Duman I (2005) Effects of different storage conditions on germination and seedling emergence of osmotic conditioning pepper seeds. Türkiye II. Tohumculuk Kongresi 9–11 Kasım 2005 Adana, Turkey

Chartzoulakis K, Klapaki G (2000) Response of two greenhouse pepper hybrids to NaCl salinity during different growth stages. Sci Hortic 86:247–260

Chen K, Arora R (2013) Priming memory invokes seed stress-tolerance. Environ Exp Bot 94:33–45
Chen K, Fessehaie A, Arora R (2012) Dehydrin metabolism is altered during seed osmopriming and subsequent germination under chilling and desiccation in *Spinacia oleracea* L. cv. Bloomsdale: possible role in stress tolerance. Plant Sci 183:27–36
Chinnusamy V, Jagendorf A, Zhu J (2005) Understanding and improving salt tolerance in plants. Crop Sci 45:437–448
Cooksey JR, Brian A, Kahn BA, Motes JE (1994) Plant morphology and yield of paprika pepper in response to method of stand establishment. Hortscience 29(11):1282–1284
Copeland LO, McDonald MB (1995) Principles of seed science and technology, 3rd edn. Chapman & Hall, New York
Cortez-Baheza E, Peraza-Luna F, Hernadez-Allvarez MI, Aguado-Santacruz GA, Torres-Pacheco I et al (2007) Profiling the transcriptome in *Capsicum annuum* L. seeds during osmopriming. Am J Plant Physiol 2:99–106
Cortez-Baheza E, Cruz-Fernandez F, Hernandez-Alvarez MI, Peraza-Luna F, Aguado-Santacruz GA, Serratos-Arevalo JC, Ponce PP, Gonzalez-Chavira MM, Torres-Pacheco I, Guevara-Olvera L, Guevara-Gonzalez RG (2008) A New Lea Gene is induced during osmopriming of *Capsicum annuum* L. seeds. Int J Bot 4:77–84
Da Silva CD, Filho JM (2012) Priming of bell pepper seeds using a biostimulant evaluated by an automated system of seedling ımage analysis. ASA, CSSA and SSSA annual meetings (Cincinnati, OH-October 21–24, 2012). 252-2 Poster Number 616
Da Silva CD, Marcos-Filho J, Jourdan P, Bennett MA (2015) Performance of bell pepper seeds in response to drum priming with addition of 24-Epibrassinolide. Hortscience 50(6):873–878
Demir I, Okcu G (2004) Aerated hydration treatment for improved germination and seedling growth in aubergine (*Solanum melongena*) and pepper (*Capsicum annum*). Ann Appl Biol 144:121–123
Demirkaya M (2006) The effects of osmotic conditioning with polyethylene glycol and humidification applications to enhance the germination percentage and to reduce the mean germination time of pepper seeds. Erciyes Univ J Inst Sci Technol 22(1–2):223–228
Demirkaya M (2010) Effects of seaweed (*Ascophyllum nodosum*) extract application on viability and vigour of pepper and onion seeds. Erciyes Univ J Inst Sci Technol 26(3):217–224
Demirkaya M (2012) Effects of osmotic conditioning and humidification applications of pepper seeds on emergence percentage and mean emergence time. Selcuk Tarım ve Gıda Bilimleri Derg 26(4):27–32
Demirkaya M (2016) The effects of osmatic conditioning treatments with methyl Jasmonate and seaweed (*Ascophyllum nodosum*) extract on germination and average germination duration of pepper seeds at low temperatures. Selcuk Tarım Bilimleri Dergisi 3(1):42–47
Desai BB, Kotecha PM, Salunkhe DK (1997) Seeds handbook. Marcel Dekker, New York
Dutta SK, Singh AR, Boopathi T, Singh SB, Singh MC, Malsawmzuali (2015) Effects of priming on germination and seedling vigour of bird's eye chilli (*Capsicum frutescens* L.) seeds collected from eastern Himalayan region of India. The Bioscan 10(1):279–284
Ellis RH, Butcher PD (1988) The effects of priming and 'natural' differences in quality amongst onion seed lots on the response of the rate of germination to temperature and the identification of the characteristics under genotypic control. J Exp Bot 39:935–950
El-Shatoury RSA (2010) The use of seed priming to overcome stress conditions in some vegetable crops. Dissertation, Suez Canal University
Ermis S, Ozden E, Demir I (2016) Pre-treatment germination percentages affected the advantage of priming treatment in pepper seeds. Am J Exp Agric 13(1):1–7
FAO (2017) Food and Agriculture Organization of the United Nations. Food and agriculture data. http://www.fao.org/faostat/en/#data/QC. Accessed 21 May 2017
Farooq M, Basra SMA, Wahid A, Khaliq A, Kobayashi N (2009) Rice seed invigouration. In: Liehtfouse E (ed) Sustainable agriculture reviews, book series. Springer, Berlin
Flynn R, Phillips R, Ulery A, Kochevar R, Liess L, Villa M (2003) Chile seed germination as affected by temperature and salinity. New Mexico Chile Task Force, Cooperative Extension Service, Las Cruces. Report 2: 1-10

Gallardo K, Job C, Groot SPC, Puype M, Demol H, Vandekerekhove J, Job D (2002) Proteomics of Arabidopsis seed germination and priming. In: Nicholas G (ed) The biology of seeds: recent advances. CABI, Cambridge
Georghiou K, Thanos CA, Passam HC (1987) Osmoconditioning as a means of counteracting the ageing of pepper seeds during high-temperature storage. Ann Bot 60:279–285
Ghate SR, Phatak SC (1982) Preference of tomato and pepper seed germinated before planting. J Am Soc Hortic Sci 107:908–911
Gomes DP, Silva AF, Dias DCFS, Alvarenga EM, Silva LJ, Panozzo LE (2012) Priming and drying on the physiological quality of eggplant seeds. Hortic Bras 30:484–488
Gong X, Bassel GW, Wang A, Greenwood JS, Bewley JD (2005) The emergence of embryos from hard seeds is related to the structure of the cell walls of the micropylar endosperm, and not to surrounding structures. Endo-β-mannanase activity. Ann Bot 96(7):1165–1173
Grassbaugh EM, Bennett MA (1998) Factors affecting vegetable stand establishment. Sci Agric 55:116–120
Gray D (1994) Large scale seed priming techniques and their integration with crop protection treatments. In: Martin M (ed) Seed treatment: progress and prospects, Monograph no.57. British Crop Protection Council, Farnham
Hacisalihoglu G, White J (2006) Optimum matriconditioning treatments for improving pepper seed germination. Proc Fla State Hortic Soc 119:282–283
Halpin-Ingham B, Sundstrom FJ (1992) Pepper seed water content, germination response and respiration following priming treatments. Seed Sci Technol 20:589–596
Harman GE, Taylor AG, Stasz TE (1989) Combining effective strains of *Trichoderma harzianum* and solid matrix priming to improve biological seed treatments. Plant Dissease 73:631–637
Hassen A, Souguir M, Latique S, Hannachi C (2014) Germination and growth in control and primed seeds of pepper as affected by salt stress. Cercet Agronomice Moldova 47:83–95
Hassen A, Elouaer MA, Hannachi C (2017) Seed priming to improve seedling growth of pepper cultivars exposed to salt concentrations. Int J Veg Sci. https://doi.org/10.1080/19315260.2017.1326996
Heydecker W, Higgins J, Gulliver RL (1973) Accelerated germination by osmotic treatment. Nature 246:42–44
Heydecker W, Higgins J, Turner YJ (1975) Invigoration of seeds. Seed Sci Technol 3:881–888
Homrichhausen T, Hewitt J, Nongaki H (2003) Endo-beta-mannanase activity is associated with the completion of embryogenesis in imbibed carrot (*Daucus carota* L.) seeds. Seed Sci Res 13:219–227
Ibn Maaouia-Houimli S, Denden M, Dridi-Mouhandes B, Ben Mansour-Gueddes S (2011) Caractéristiques de la croissance et de la production en fruits chez trois variétés de piment (*Capsicum annuum* L.) sous stress salin. J Tropicultura 29(2):75–81
Ibrahim EA (2016) Seed priming to alleviate salinity stress in germinating seeds. J Plant Physiol 192:38–46
Ilyas S, Sutariati GA, Suwarno FC, Sudarsono S (2002) Matriconditioning improved quality and protein level of medium vigor hot pepper seed. Seed Technol 24:67–77
Ilyas S, Asie KV, Sutariati GAK (2015) Biomatriconditioning or biopriming with biofungicides or biological agents applied on hot pepper (Capsicum annuum L) seeds reduced seedborne Colletotrichum capsici and increased seed quality and yield. Acta Hortic 1105:89–96
Jisha KC, Vijayakumari K, Puthur JT (2013) Seed priming for abiotic stress tolerance: an overview. Acta Physiol Plant 35:1381–1396
Jones K, Sanders D (1987) The influence of soaking pepper seed in water or potassium salt solutions on germination at three temperatures. J Seed Technol 11:97–102
Jyoti, Malik CP (2013) Seed deterioration: a review. Int J Life Sci Bot Pharm Res 2(3):374–385
Kaewnaree P, Vichitphan S, Klarnit P, Siri B, Vichitphan K (2011) Effect of accelerated ageing process on seed quality and biochemical changes in sweet pepper (*Capsicum annum* Linn) seeds. Biotechnology 10(2):175–182

Kang JS, Choi YW, Son BG, Lee YJ, Ahn CK, Choi IS, Park HC (2003) Effect of osmotic priming and solid matrix priming to improved seed vigor and early growth of pepper and tomato seeds. J Life Sci 13(4):433–440

Kapoor R, Arya A, Siddiqui MA, Amir A, Kumar H (2010) Seed deterioration in chickpea (*Cicer arietinum* L) under accelerated ageing. Asian J Plant Sci 9(3):158–162

Kavi Kishor PB, Hima Kumari P, Sunita MSL, Sreenivasulu N (2015) Role of proline in cell wall synthesis and plant development and its implications in plant ontogeny. Front Plant Sci 6:544

Kaya G, Demir I, Tekin A, Yasar F, Demir K (2010) Effect of priming treatment on germination at stressful temperatures, fatty acid, sugar content and enzymatic activity of pepper seeds. J Agric Sci 16:9–16

Khan AA (1991) Preplant physiological seed conditioning. Hortic Rev 13:131–181

Khan AA (1992) Preplant physiological seed conditioning. Hortic Rev 13:131–181

Khan AA, Miura H, Prusinski J, Ilyas S (1990) Matriconditioning of seeds to improve emergence. In: Proceedings of the national symposium on stand establishment of horticultural crops 19–40, Minneapolis, MN

Khan AA, Maguire JD, Abawi GS, Ilyas S (1992) Matriconditioning of vegetable seeds to improve stand establishment in early field plantings. J Am Soc Hortic Sci 117:41–47

Khan AA, Ilyas S, Ptasznik W (1995) Integrating low water potential seed hydration with other treatments to improve cold tolerance. Ann Bot 75(1):13–19

Khan MA, Qayyum A, Noor E (2007) Assessment of wheat genotypes for salinity tolerance. In: Proceeding of 8th African crop science conference, October 27–31, El-Minia, Egypt pp 75–78

Khan HA, Ayub CM, Pervez MA, Bilal RM, Shahid MA, Ziaf K (2009a) Effect of seed priming with NaCl on salinity tolerance of hot pepper (*Capsicum annuum* L) at seedling stage. Soil Environ 28:81–87

Khan HA, Pervez MA, Ayub CM, Ziaf K, Bilal RM, Shahid MA, Akhtar N (2009b) Hormonal priming alleviates salt stress in hot pepper (*Capsicum annuum* L). Soil Environ 28:130–135

Khan HA, Ziaf K, Amjad M, Iqbal Q (2012) Polyamines improves germination and early seedling growth of hot pepper. Chil J Agric Res 72(3):429–433

Khan MS, Ahmad D, Khan MA (2015) Utilization of genes encoding osmoprotectants in transgenic plants for enhanced abiotic stress tolerance. Electron J Biotechnol 18:257–266

Khan FA, Maqbool R, Narayan S, Bhat SA, Narayan R et al (2016) Reversal of age-induced seed deterioration through priming in vegetable crops – a review. Adv Plants Agric Res 4(6):00159. https://doi.org/10.15406/apar20160400159

Kibinza S, Bazin J, Bailly C, Farrant JM, Corbineau F, El-Maarouf-Bouteau H (2011) Catalase is a key enzyme in seed recovery from ageing during priming. Plant Sci 181:309–315

Kikuti A, Kikuti H, Minami K (2005) Condicionamento fisiológico em sementes de pimentão. Rev Ciênc Agron 36:243–248

Koehler KH, Voigt B, Spittler H, Schelenz M (1997) Biochemical events after priming and osmoconditioning of seeds. In: Ellis RH, Black M, Murdoch AJ, Hong TD (eds) Basic and applied aspects of seed biology. Springer, New York

Korkmaz A (2005) Inclusion of acetyl salicylic acid and methyl jasmonate into the priming solution improves low-temperature germination and emergence of sweet pepper. Hortscience 40(1):197–200

Korkmaz A, Korkmaz Y (2009) Promotion by 5-amino levulenic acid of pepper seed germination and seedling emergence under low-temperature stress. Sci Hortic 119(2):98–102

Korkmaz A, Sirikci R (2011) Improving salinity tolerance of germinating seeds by exogenous application of glycinebetaine in pepper. Seed Sci Technol 39:377–388

Korkmaz A, Korkmaz Y, Demirkiran AR (2010) Enhancing chilling stress tolerance of pepper seedlings by exogenous application of 5-aminolevulinic acid. Environ Exp Bot 67:495–501

Korkmaz A, Karaca A, Kocaçinar F, Cuci Y (2017) The effects of seed treatment with melatonin on germination and emergence performance of pepper seeds under chilling stress. Tarım Bilimleri Dergisi – J Agric Sci 23:167–176

Kubik KK, Eastin JA, Eastin JD, Eskridge KM (1988) Solid matrix priming of tomato and pepper. In: Proceedings of international conference on stand establishment for horticultural crops, Lancaster

Kumar D (1995) Salt tolerance in oilseed brassicas-present status and future prospects. Plant Breed Abstr 65(10):1439–1477

Kumar S, Arya MC, Sinfh R (2010) Management of sweet pepper diseases and growth promotion by *Pseudomonas fluorescens* and *Trichoderma harzianum* in mid Hills of Central Himalayas, India. Indian Phytopathol 63(2):181–186

Lanteri S, Bino RJ, Kraak HL (1992) Flowcytometric determination of nuclear replication stages in pepper seeds during germination and after priming treatments. Capsicum Newsl 4:249–253

Lanteri S, Kraak HL, De Vos CHR, Bino RJ (1993) Effects of osmotic preconditioning on nuclear replication activity in seeds of pepper (*Capsicum annuum*). Physiol Plant 89:433–440

Lanteri S, Saracco F, Kraak HL, Bino RJ (1994) The effects of priming on nuclear replication activity and germination of pepper (*Capsicum annuum*) and tomato (*Lycopersicon esculentum*) seeds. Seed Sci Res 4:81–87

Lanteri S, Nada E, Belletti P, Quagliottil L, Binof RJ (1996) Effects of controlled deterioration and osmoconditioning on germination and nuclear replication in seeds of pepper (*Capsicum annuum* L). Ann Bot 77(6):591–597

Lanteri S, Belletti P, Marzach C, Nada E, Quagliotti L, Bino RJ (1997) Priming-induced nuclear replication activity in pepper (*Capsicum annuum* L.) seeds effect on germination and storability. In: Ellis RH, Black M, Murdoch AJ, Hong TD (eds) Basic and applied aspects of seed biology current plant science and biotechnology in agriculture, vol 30. Springer, Dordrecht

Lanteri S, Quagliotti L, Belletti P (1998) Delayed luminescence and priming-induced nuclear replication of unaged and controlled deteriorated pepper seeds (*Capsicum annuum* L). Seed Sci Technol 26(2):413–424

Lanteri S, Ports E, Bergervoet HW, Groot SPC (2000) Molecular markers for priming of pepper seeds (*Capsicum annum* L). J Hortic Sci Biotechnol 75:607–611

Lee JW, Kim KY, Chung JH (1997) Studies on priming condition PTO and priming method for bulk treatment of hot pepper seeds. Hortscience 39(1):1–8

Lee SS, Kim JH, Hong SB, Yuu SH, Park EH (1998) Priming effect of rice seeds on seedling establishment under adverse soil conditions. Korean J Crop Sci 43:194–198

Levitt J (1980) Responses of plants to environmental stresses vol II water, radiation, salt, and other stresses, 2nd edn. Academic, New York

Li X, Zou X, Liu Z (2005) On physiological and biochemical changes of artificially aged pepper seeds. J Hunan Agric Univ 31(3):265–268

Lopez-Delgado H, Dat JF, Foyer CH, Scott IM (1998) Induction of thermotolerance in tomato microplants by acetylsalicylic acid and H_2O_2. J Exp Bot 49:713–720

Lutts S, Benincasa P, Wojtyla L, Kubala S, Pace R, Lechowska K, Quinet M, Garnczarska M (2016) Seed priming: new comprehensive approaches for an old empirical technique, new challenges in seed biology. In: Susana Araújo S, Balestrazzi A (eds) Basic and translational research driving seed technology. InTech, Open, Rijeka

Maiti RK, Gupta A, Umasahankar P, Raj Kumar D, Vidyasagar P (2009) Effect of priming on seedling vigour and growth and productivity of few vegetable species: tomato, chilli, cucumber and cabbage. Int J Agric Environ Biotechnol 2(4):368–374

Maiti R, Rajkurmar D, Jagan M, Pramanik K, Vidasagar P (2013) Effect of seed priming on seedling vigour and yield of tomato and chilli. Int J Bio-Resour Stress Manag 4(2):119–125

Manonmani V, Ameer Junaithal Begum M, Jayanthi M (2014) Halo priming of seeds. Res J Seed Sci 7:1–13

Mavi K (2016) The effect of organic priming with marigold herbal tea on seeds quality in 'Aji' pepper (*Capsicum baccatum* var *pendulum* Willd). J Agric Fac Mustafa Kemal Univ 21(1):31–39

Mayer AM, Poljakoff-Mayber A (1989) The germination of seeds, 4th edn. Pergamon Press, Oxford

Maynard DN, Hochmuth GJ (1997) Knott's handbook for vegetable growers, 4th edn. Wiley, New York

McDonald MB (1999) Seed deterioration: physiology, repair and assessment. Seed Sci Technol 27:177–237

McDonald MB (2000) Seed priming. In: Black M, Bewley JD (eds) Seed technology and its biological basis. Sheffield Academic Press, Sheffield

Mehta DK, Kanwar HS, Thakur AK, Thakur KS (2010) Influence of organic seed priming on germination and seedling quality in bell pepper (*Capsicum annuum* L). J Hill Agric 1(1):85–87

Mendoza AB, Rodriguez HR, Torres VR, Davila JH, Mezquitic JGR, Tellez EB, Rangel AS, Garcia MAB (2002) Seed treatment with salicylates modifies stomatal distribution, stomatal density, and the tolerance to cold stress in pepper seedlings. In: Proceedings of international pepper conference, 10–12 November 2002, Tampico

Nascimento WM, Silva PP, Justino EV, Freitas RA (2011) Physiologıcal seed qualıty and stand establıshment of hot pepper (*Capsıcum baccatum* L var *pendulum*) under tropical conditions. Acta Hortic 898:307–310

Niu G, Rodriguez DS, Cabrera R, Jifon J, Leskovar D, Crosby K (2010) Salinity and soil type effects on emergence and growth of pepper seedlings. Hortscience 45:1265–1269

Nonnecke IL (1996) Vegetable production. Von Nostrand Reinhold Co, New York

O'sullivan J, Bouw WJ (1984) Pepper seed treatment for low-temperature germination. Can J Plant Sci 64:387–393

Ozbay N, Susuluoglu Z (2016a) Assessment of growth regulator prohexadione-calcium as priming agent for germination enhancement of pepper at low temperature. J Anim Plant Sci 26(6):1652–1658

Ozbay N, Susuluoglu Z (2016b) Influence of Prohexadione-Calcium incorporated ınto priming solutıon on germination and emergence of pepper seeds at high temperature VII International Scientific Agriculture Symposium, "Agrosym 2016" Jahorina, 6–9 October 2016, Bosnia and Herzegovina

Ozpercin I, Gursoy AB, Duman I, Eser B (2005) Determination of physical and metabolic changes in seeds undergoing osmotic conditioning process. Türkiye II Tohumculuk Kongresi 9–11 kasım 2005 Adana, Turkey

Panayotov N, Aladjadjiyan A (2014) Effect of long-term storage of pepper (*Capsicum annuum* L) seeds on their viability measured by selected thermodynamıc parameters. Acta Sci Pol Hortorum Cultus 13(2):151–162

Pandita VK, Anand A, Nagarajan S (2007) Enhancement of seed germination in hot pepper following presowing treatments. Seed Sci Technol 35:282–290

Paparella S, Araújo SS, Rossi G, Wijayasinghe M, Carboner D, Balestrazzi A (2015) Seed priming: state of the art and new perspectives. Plant Cell Rep 34:1281–1293

Parera CA, Cantliffe DJ (1994) Presowing seed priming. Hortic Rev 16:109–139

Passam HC, Karavites PI, Papandreou AA, Thanos CA, Georghiou K (1989) Osmoconditioning of seeds in relation to growth and fruit yield of aubergine, pepper, cucumber and melon in unheated greenhouse cultivation. Sci Hortic 38:207–216

Perkins-Veazie P, Cantliffe DJ (1984) Need for high quality seed for effective priming to overcome thermodormancy in lettuce. J Am Soc Hortic Sci 109:368–372

Perl M, Feder Z (1981) Improved seedling development of pepper seeds (*Capsicum annuum*) by seed treatment for pregermination activities. Seed Sci Technol 9:655–663

Pill WG (1995) Low water potential and pre-sowing germination treatments to improve seed quality. In: Basra AS (ed) Seed quality. Food Prod Press, New York

Priestly DA (1986) Seed ageing. Cornell University Press, New York

QianQian Q, Ming L, DonWei Y, Lei C, YuLin Z (2009) Effects of priming treatment with vermiculite on the seed germination and seedlings' antioxidant characteristics of hot pepper under NaCl stress. Acta Agric Shangahi 25(3):47–50

Qin C, Yu C, Shen Y, Fang X, Chen L, Min J et al (2014) Whole-genome sequencing of cultivated and wild peppers provides insights into Capsicum domestication and specialization. Proc Natl Acad Sci US Am 111:5135–5140

Rakshit A, Sunita K, Pal S, Singh A, Singh HB (2015) Bio-priming mediated nutrient use efficiency of crop species. In: Rakshit A, Singh HB, Sen A (eds) Nutrient use efficiency: from basics to advances. Springer, New Delhi

Reddy PP (2012) Bio-priming of seeds. In: Recent advances in crop protection. Springer, New Delhi

Rhoades JD, Kandiah A, Mashali AM (1992) The use of saline waters for crop production. Food and Agriculture Organization of United Nations (FAO), Irrigation and Drainage Paper 48 Fao, Rome Italy

Rivas M, Sundstrom FJ, Edwards RL (1984) Germination and crop development of hot pepper after seed priming. Hortscience 19:279–281

Rowse HR (1995) Drum priming-A non-osmotic method of priming seeds. Seed Sci Technol 24:281–294

Rowse HR (1996) Drum priming – a non-osmotic method of priming seeds. Seed Sci Technol 24:281–294

Sachs M, Cantliffe D, Watkins J (1980) Germination of pepper seed at low temperatures after various pretreatments. Proc Fla State Hortic Soc 93:258–260

Samarah NH, Wang H, Welbaum GE (2016) Pepper (*Capsicum annuum*) seed germination and vigour following nanochitin, chitosan or hydropriming treatments. Seed Sci Technol 44(3):1–15

Saracco F, Bino RJ, Bergervoet JHV, Lanteri S (1995) Influence of priming-induced nuclear replication activity on storability of pepper (*Capsicum annuum* L) seeds. Seed Sci Res 5:25–29

Sedghi M, Nemati A, Amanpour-Balaneji B, Gholipouri A (2010) Influence of different priming materials on germination and seedling establishment of milk thistle (*Silybum marianum*) under salinity stress. World Appl Sci J 11:604–609

Sharma N, Shukla YR, Metha DK (2015) Seed priming and its consequences on seedling vigour of bell pepper (*Capsicum annuum* L) under low temperature condition. Int J Bio-Resour Stress Manag 6(6):759–764

Silva PP, Freitas RA, Nascimento WM (2012) Hot pepper seed priming and germination at different temperatures. Acta Hortic 932:341–343

Singh H, Jassal RK, Kang JS, Sandhu SS, Kang H, Grewal K (2015) Seed priming techniques in field crops – a review. Agric Rev 36(4):251–264

Siri B, Vichitphan K, Kaewnaree P, Vichitphan S, Klanrit P (2013) Improvement of quality, membrane integrity and antioxidant systems in sweet pepper (*Capsicum annuum* Linn) seeds affected by osmopriming. Aust J Crop Sci 7(13):2068–2073

Sivritepe HO, Senturk B (2011) A comparison of hydro and halopriming with dehydration treatments for physiological enhancement of pepper seeds. J Agric Fac Uludag Univ 25(1):53–64

Sivritepe N, Sivritepe HO (2008) Organic priming with seaweed extract (*Ascophyllum nodosum*) affects viability of pepper seeds. Asian J Chem 20(7):5689–5694

Sivritepe HO, Sivritepe N (2016) Organic seed hydration-dehydration techniques improve seedling quality of organic tomatoes. Notulae Botanicae Hortic Agrobotanici Cluj-Napoc 44(2):399–403

Sivritepe HO, Sentürk B, Teoman S (2015) Organic priming and dehydration treatments in pepper seeds improve seedling quality and performance. J Agric Fac Uludag Univ 29(2):83–94

Sliwinska E (2009) Nuclear DNA replication and seed quality. Seed Sci Res 19:15–25

Smirnoff N, Cumbes QJ (1989) Hydroxyl radical scavenging activity of compatible solutes. Phytochemistry 28:1057–1060

Smith PT, Cobb BG (1991a) Accelerated germination of pepper seed by priming with salt solutions and water. Hortscience 26:417–419

Smith PT, Cobb BG (1991b) Physiological and enzymatic activity of pepper seeds (*Capsicum annuum*) during priming. Physiol Plant 82:433–439

Srinivasan K, Jain SK, Saxena S, Radhamani J, Uprety M (2009) Seed priming and fortification. Seed Res 37(1&2):1–13

Stofella PJ, Lipucci DP, Pardossi A, Toganoni F (1992) Seedling root morphology and shoot growth after seed priming or pre-emergence of bell pepper. J Am Soc Hortic Sci 27:214–215

Sundstrom FJ, Edwards RL (1989) Pepper seed respiration, germination, and seedling development following seed priming. Hortscience 24:343–345

Sundstrom FJ, Reader RB, Edwards RL (1987) Effect of seed treatment and planting method on tabasco pepper. J Am Soc Hortic Sci 112:641–644

Taylor AG, Harman GE (1990) Concepts and technologies of selected seed treatments. Annu Rev Phytopathol 28:321–339

Taylor AG, Klein DE, Whitlow TH (1988) SMP: solid matrix priming of seeds. Sci Hortic 37:1–11

Taylor AG, Allen PS, Bennett MA, Bradford KJ, Burris JS, Misra MK (1998) Seed enhancement. Seed Sci Res 8:245–256

Teksan BO, Kavak S (2016) Effects of marigold and rose flowers herbal teas priming treatments on germination and emergence of pepper seed. Suleyman Demirel Univ Ziraat Fak Derg 11(1):34–42

Thakur A, Thakur PS, Bharadwaj J (1997) Influence of seed osmoconditioning on germination potential and seedling performance of bell pepper. Seed Res 25(1):25–30

Thanos CA, Georghiou K, Passam HC (1989) Osmoconditioning and ageing of pepper seeds during storage. Ann Bot 63:65–69

Uche OJ, Adinde JO, Omeje TE, Agu CJ, Anieke UJ (2016) Influence of hydropriming on germination and seedling emergence of green bell pepper (*Capsicum annum* cv Goliath). Int J Sci Nat 7(1):70–75

Valdes VM, Bradford KJ (1987) Effects of seed coating and osmotic priming on the germination of lettuce seeds. J Am Soc Hortic Sci 112:153–156

Varier A, Vari AK, Dadlani M (2010) The subcellular basis of seed priming. Curr Sci, Bangalore 99(4):450–456

Vavrina CS (1994) Plastic tray inserts. Fla Veg Transplant Growers News 5:3

Venkatasubramanian A, Umarani R (2007) Evaluation of seed priming methods to improve seed performance of tomato (*Lycoperison esculentum*), eggplant (*Solanum melongena*) and chilli (*Capsicum annum*). Seed Sci Technol 35:487–493

Warren JE, Bennett MA (1997) Seed hydration using the drum priming system. Hortscience 32:1220–1221

Watkins JT, Cantliffe DJ (1983) Mechanical resistance of the seed coat and endosperm during germination of *Capsicum annuum* at low temperature. Plant Physiol 72:146–150

Watkins JT, Cantlife JD, Huber DJ, Nell TA (1985) Gibberellic acid stimulated degradation of endosperm in pepper. J Am Soc Hortic Sci 110:61–65

Welbuam GE, Shen ZX, Oluoch MO, Jet LW (1998) The evolution and effects of priming vegetable seeds. Seed Technol 20(2):209–235

Wojtyla Ł, Lechowska K, Kubala S, Garnczarska M (2016) Different modes of hydrogen peroxide action during seed germination. Front Plant Sci 7:66

Wright B, Rowse HR, Whipps JM (2003) Microbial populations on seeds during drum and steeping priming. Plant Soil B 255:631–640

Wu X, Li N, Li H, Tang H (2014) An optimized method for NMR-based plant seed metabolomics analysis with maximized polar metabolite extraction efficiency, signal-to-noise ratio, and chemical shift consistency. Analyst 139:1769–1778

Yadav PV, Kumari M, Ahmed Z (2011) Seed priming mediated germination improvement and tolerance to subsequent exposure to cold and salt stress in capsicum research. J Seed Sci 4(3):125–136

Yaklitch RW, Orzolek MD (1977) Effect of polyethylene glycol-6000 on pepper seed. Hortscience 12:263–264

Yancey PH (1994) Compatible and counteracting solutes. In: Strange SK (ed) Cellular and molecular physiology of cell volume regulation. CRC Press, Boca Raton

Yogananda DK, Vyakaranahal BS, Shekhargouda M (2004) Effect of seed invigoration with growth regulators and micronutrients on germination and seedling vigour of bell pepper Cv California wonder. Karnataka J Agric Sci 17(4):811–813

Effect of Different Seed Priming Treatments on Germination and Seedling Establishment of Two Threatened Endangered Medicinal Plants of Darjeeling Himalaya

13

Dhiman Mukherjee

Abstract

Present investigation was conducted under the aegis of Uttar Banga Krishi Viswavidyalaya, in Algarah (1900 m asl) under Darjeeling Himalaya. The aim was to study the effects of priming with different concentration of growth hormone, etc. on seed emergence characteristics and seedlings vigour of two endangered plant species of *Swertia chirayita* and *Valeriana jatamansi*. An experiment as randomized block design in 14 pretreatment seeds of *S.chirayita* and 15 pretreatment seeds of *V. jatamansi* with 3 replications was carried out in the field condition during November 2011 to November 2014. The pretreatments include osmo-priming with GA_3 50, 100, 200, 400 and 800 ppm; IAA 50, 100, 200, 400, 800 ppm and KNO_3 1, 2, 3 and 4% and hydro-priming (control with distilled water) for *S.chirayita*. In another experiment *V. jatamansi* seed osmo-priming with IBA 50, 100, 150, 200, 250 ppm; GA_3 50,100, 150, 200, 250 ppm, kinetin 50, 100, 150, 200, 250 ppm and hydro-priming (control with distilled water). The maximum seed germination was observed with GA_3 400 ppm and was at par with the GA_3 200 ppm. Onset of germination was earliest, i.e. 19.00 days after sowing as registered by GA_3 800 ppm for 12 h, which was statistically at par with IAA 200 ppm (20.33 days) and IAA 50 ppm (22 days). Mean germination time (MGT) 44.11 days in untreated (control) seeds which were significantly reduced to 27.34 days in seeds pretreated with IAA 400 ppm, which was 38.01% lower than the control and was at par with 29.63 and 30.67 days in seed pretreated with GA_3 400 ppm and KNO_3 (2%). Seeds treatment with GA_3 400 ppm for 24 h showed maximum SV-I (71.11), which was 207.71% higher than the control. The highest SV-II of 54.18 was registered by KNO_3 (2%), which was 68.16% higher than the control and was statistically at par with GA_3 200 ppm. Germination energy of 1.16 was found in seeds pretreated with

D. Mukherjee (✉)
Department of Agronomy, Bidhan Chandra Krishi Viswavidyalaya, Kalyani, Nadia, West Bengal, India

© Springer Nature Singapore Pte Ltd. 2018

A. Rakshit, H. B. Singh (eds.), *Advances in Seed Priming*,
https://doi.org/10.1007/978-981-13-0032-5_13

GA_3 400 ppm, which was statistically at par with GA_3 200 ppm, GA_3 800 ppm, KNO_3 (1%) and KNO_3 (2%). Highest germination percentage seed of *V. jatamansi* was observed with the kinetin 200 ppm treated seeds and significantly better to all other treatments measures. Days require for onset of germination was earlier recorded with the IBA 200 and kinetin 250 ppm and was at par with the kinetin 200 and IBA 150 ppm. MGT was 33.03 days in untreated (control) seeds which were significantly reduced to 22.95 days with GA_3 250 ppm, which was 31.88% lower than the control and was at par with kinetin 200, kinetin 250 and IBA 200 ppm primed seeds. Kinetin 100 ppm showed maximum SV-I (303.33), which was 484.15% higher than the control and was significantly better to all other 14 seed priming treatments. The highest SV-II of 190.06 was registered with kinetin 200 ppm, which was 362.32% higher than the control and was statistically better to other treatments. Maximum emergence index and germination energy were exhibited by kinetin 200 ppm.

Keywords
Conservation · Endangered plant · Gibberellic acid · IAA · Himalaya · *Swertia chirayita* · *Valeriana jatamansi*

Indian Himalaya has 270 endemic (62 endemic and 208 near-endemic) plant species that are used for medicinal purposes (Samant et al. 1998) most of which are harvested indiscriminately for preparation of herbal drugs and serve as important medicine to our society (Mukherjee 2009). Among these, the genus *Swertia* (family Gentianaceae) with over 32 species in IHR (15 in North East Himalaya) occupies an important place (Gupta 1987). Darjeeling-Sikkim range of Himalaya is one of the native places of different kinds of *Swertia* species and is situated between the 87°59′–88°53′E and 28°31′–27°13′N in the Eastern Himalayan region of India (Mukherjee et al. 2010). *Swertia chirayita* (Roxb. ex Fleming) Karsten, syn *S. chirata* (Wall.) Clarke, so-called authentic ‘Chirayita’, is considered a critically rare medicinal plant in the Himalayan region (Samant et al. 1998; Chakraborty et al. 2008). In general, Gentianaceae is reported to have physiological dormancy, so seed germination is under question (Kandari et al. 2007). It occurs in the temperate Himalaya at an altitude of 1500–3000 m (Kala 2006). The bitter infusion of the plant is used as a blood purifier, for skin disease and as a bitter tonic for fever and indigestion (Mukherjee 2009a). The presence of xanthones in the species is reported to remedy tuberculosis (Bhattacharjee 2001). The bitterness and anthelmintic, hypoglycaemic and antipyretic properties are attributed to amarogentin (most bitter compound till date) (Keil et al. 2000). Because of these medicinal uses, demand has increased so much in the pharmaceutical industry that annually 95,000–160,000 kg of chirayita is imported to India (Dubey et al. 2014). Germination of chirayita becomes a problem for quite some time, and its germination percentage is very low (Raina et al. 1994; Mukherjee 2008), and even some time seed germination totally failed. *Valeriana jatamansi*, which is another important endangered plant in northeastern and western Himalaya, also face problem of seed germination due to poor viability of seed and lack of dormancy breaking technology (Mukherjee et al. 2009). Seed germination, one of the most critical phases of plant growth and development, is adversely affected under hill conditions (Ghavami and Ramin 2007). Germination of seeds, one of the most critical phases of plant life (El-Keblawy and Al-Rawai 2005), is greatly influenced by environment and local condition (Mukherjee 2014). *Valeriana jatamansi* (family, Valerianaceae) is another endangered plant, which grow at an altitude of 1200–2000 m asl. It is distributed from Afghanistan to Southwest China, Burma and few part of Indian Himalayan range (Chakraborty et al. 2015a). Like many other non-timber forest products (NTFPs), this vulnerable plant (*Valeriana jatamansi* Jones; Syn. *Valeriana wallichii*) is taken as forest gift, and hence there is neither any control system in its harvest nor its domestication (Mukherjee 2009). Locally it is being used for medicinal purpose especially for headache and eye trouble. In Ayurvedic medicine, it is used as aromatic, stimulant, carminative and antispasmodic. It is also used for the treatment of epilepsy and hysteria (Polunin and Stainton 1987; Mukherjee 2016). Although the economic value of above mentioned two endangered herbs was reportedly unknown to the local people until recent past, the herb has now been widely known for its market potential. Thus, the exploitation of this plant is increasing leading to its rapid decline from its natural habitat in Himalayan range (Mukherjee 2013). Suboptimal and imbalanced disturbance of soil are the main reasons for low availability of this valuable medicinal plant species in Himalayan range and Darjeeling-Sikkim hills in particular. However, no concrete step had been taken till date regarding various possibilities of seed germination for proper conservation of this species. Low germination percentage and viability of the seeds, long gestation period and delicate field handling are some of the factors which discourage commercial cultivation of the plant (Mukherjee 2015). Similarly reckless harvesting from its natural habitat created lot of pressure on their population. In CAMP workshop chirayita and jatamansi have been listed under 'critical' category (Sharma et al. 2005). Hence according to the International Union for Conservation of Nature and Natural resources (IUCN) criteria, *Swertia chirayita* and *Valeriana jatamansi* have been categorized as critically endangered (Mukherjee et al. 2015; Chakraborty et al. 2015b). Viable seed production potential was very low in both above-mentioned plants. Seeds mature late in the season and therefore fail to produce sufficient number in both the species (Chakraborty et al. 2016). Poor germination percentage and ruthless harvesting from forest range of Himalaya, the *Swertia chirayita* and *Valeriana* plant becomes wipeout day by day, and now come in category of endangered plant. Conservation of these plants becomes a question mark to our research wings and other R&D sectors throughout the world (Chakraborty et al. 2015b). Seed priming becomes an effective tool to break through this problem up to certain extent. Seed priming is seen as a viable technology to enhance fast and uniform germination and emergence, high vigour of seedling and better yields (Basra et al. 2002). Priming can be more efficient on poorer quality seed lots, and it is an important way for enhancing of seed germination and is a helpful tool particularly for valuable medicinal plant conservation. Seed priming enhances seed performance by rapid and uniform germination, normal and vigorous seedlings, which resulted in faster and better germination in different crops (Ashraf and Foolad 2005). There are several priming techniques used for seeds. Effects of priming or pretreatment of seeds persist under suboptimal field conditions, such as acidity (Foti et al. 2008) and low soil moisture availability (Du and Tuong 2002). Seed priming mainly involves uptake of water to initiate the early events of germination but not sufficient to permit radicle protrusion, followed by drying (McDonald 1999). Priming or pretreatment of seeds persist under suboptimal field conditions, such as acidity (Foti et al. 2008) and low soil moisture availability (Du and Tuong 2002), and are very helpful for improvement of germination of valuable temperate herbs. Presowing chemical treatments can be used to enhance and improve seed germination in Himalayan medicinal plants (Mukherjee 2013a). Seed germination studies on several Himalayan medicinal plants have proved useful in developing appropriate conservation strategies (Chakraborty et al. 2015a). Gibberellins are most directly implicated in the control and promotion of germination. This chemical helps to promote growth by increasing extendibility of the cell wall followed by the hydrolysis of starch to sugars which reduces the potential in the cell, resulting in the entry of water into the cell causing elongation (Arteca 1996). Few other growth hormones such as IAA and IBA are known to stimulate germination in many species by playing a major role in cell division and differentiation (Choudhary et al. 1996). Potassium nitrate is well documented as a compound which increases the germination of photo-dormant seeds (Shanmugavalli et al. 2007). According to Bewley and Black (1994), potassium nitrate raises ambient oxygen levels by making less oxygen available for citric acid cycles. Keeping the above aspect in mind, present work was conducted to restore our valuable natural herb resource for future generation. The objective of this research was to determine the effects of best seed priming treatments on enhanced seed germination parameters and seedling vigour and other physiological parameters of two endangered medicinal plants in Darjeeling Himalaya.

13.1 Materials and Methods

Present research work is carried out in Algarah (Darjeeling hill), under the aegis of Uttar Banga Krishi Viswavidyalaya, Kalimpong (1900 m asl), during November 2011 to November 2014 (Fig. 13.1). The soil sample used for germination was analysed for physico-chemical properties as per standard procedure (Jackson 1973). Soil samples were taken before actually conducting the experiment from the depth of 0–15 cm, taking all possible precautions prescribed for soil sampling. The soil of the experimental site was sandy loam having pH 5.1, with moderate available nitrogen, phosphorus and potassium (Table 13.1). The purpose of this research was to evaluate the efficiency of employing different priming (Osmo- and hydro-priming) to mitigate germination barriers and improve seed viability and seedling establishment of two endangered plants on earth, viz. *Swertia chirayita* and

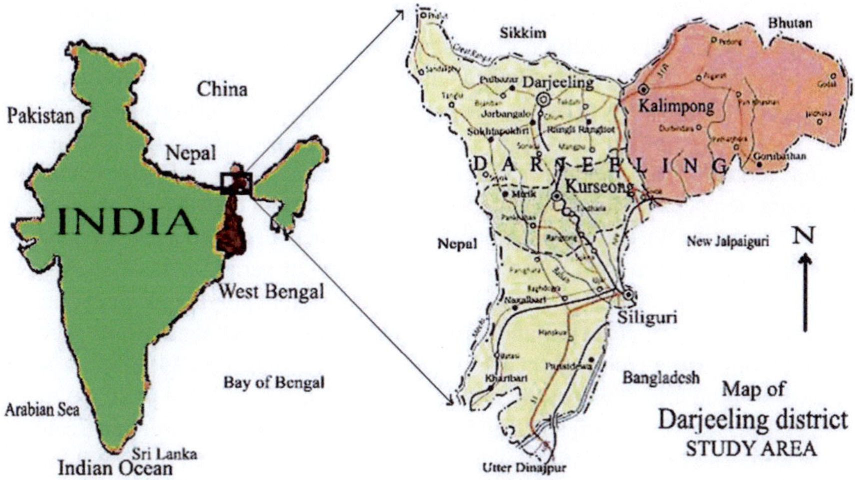

Fig. 13.1 Location of Darjeeling district (study area) of West Bengal, India

Table 13.1 Physico-chemical status of soil sample for germination study (pooled value of 2-year data)

pH	ECE (inch/cm)	Available (kg/ha) N	P_2O_5	K_2O	Total N (%)	Organic C (%)	Organic matter (%)	C/N ratio
5.1	0.17	297	26.6	226	11.16	2.71	5.08	16.01

Valeriana jatamansi. Mature seeds of *S. chirayita* and *V. jatamansi* were collected from the Sandhapukh (2200 m asl) and Lava (2000 m asl) of Darjeeling district.

The fruit of chirayita are superior bicarpellate ovoid pointed capsules with minute reticulate and irregular ovoid seeds, weighing 100 seeds to a mg (±0.96), size about 0.23–0.50 mm, length 0.15–0.50 mm, broad. Seeds of *S. chirayita* and *Valeriana* were collected during end of November 2011 and last week of February 2012, respectively, and kept in paper bags of convenient size after proper labelling. The fruit were shed dried in the open air and gently hand shaken while inside the paper bag. The minute clean seeds fall and settle in the bottom of paper bag. These fallen seeds were collected and stored in an open container. This process was repeated till the seed were almost fully collected. Finally the remaining fruits (some left over seeds) were hand crushed and passed through a fine nylon sieve and stored in a separate but similar container. This contains impurities like vegetative, clay particles and other foreign materials of almost similar size. This seed material was kept in desiccators with silica gel to avoid moisture absorption. During the investigation, seeds were surface sterilized by dipping them in 0.5% aqueous solution of $HgCl_2$ for 2 min to remove bacterial and fungal contamination and then rinsed thoroughly (four times) with distilled water and then soaked in different concentration of growth regulators for 24 h and placed in a glass plate for shed dry. Two experiments were made for seed germination investigation under field condition, with 200 seeds taken with triplicate in bed size 1.50 × 1.50 m^2 inside polyhouse under field condition. Sowing of seed in a nursery bed was done under careful observation with finally sieved and well-sterilized (in autoclave) clay soil in the 25th and 21st of April, respectively, in 2012 and 2013. Earlier work indicated that months of April and May were favourable for seed sowing in the field at high altitude regions (Anon 1997). Meteorological data indicate that during these months, day-night temperature of the experimental sites varied between 6 and 23 °C, probably meeting the optimum temperature requirement of this species and resulting in higher germination with less time for completion of germination of seeds. Two investigation programmes were conducted under different polyhouse. First one was on germination of *S. chirayita* under different hormonal treatments which includes control along with prechilled seed (−4 °C) of 15 days dipped in 14 different levels of seed priming measures mainly GA_3 (50, 100, 200, 400 and 800 ppm), IAA (50, 100, 200, 400 and 800 ppm) and KNO_3 (1, 2 , 3 and 4%) for 12 h. After this treatment seeds were washed with distilled water, shed dried and then sown under polyhouse. In the second experiment, seeds of *V. jatamansi* under different treatments include control along with different levels of IBA (50, 100, 150, 200 and 250 ppm), GA_3 (50, 100,150, 200 and 250 ppm) and kinetin (50, 100 ppm, 150, 200 and 250 ppm). Both the experiments were conducted consecutively 2 years in randomized block design with three replications in the same field.

Germination percentage was evaluated by counting the numbers of normal seedlings at the end of standard germination test. For onset of germination, seeds kept for germination were observed daily, and the day when the first seed showed germination was considered as time taken for start of germination. Completion of germination mainly includes daily count of seeds for germination and the day when the last seed showed germination. Mean germination time (MGT) was calculated according to the following formula (Sfairi et al. 2012):

$$\mathrm{MGT}\left(\mathrm{days}\right) = \Sigma\left(ni \times di\right) / N$$

where *ni* is the number of germinated seeds on day *i*, *d* is the incubation time (day) and *N* is the total number of seeds germinated.

Emergence index and germination energy calculated as per standard procedure advocated by ISTA (1965). Seedling fresh weight was measure by weighing five normal seedlings from each replicate was taken. The weight of fresh seedling samples was recorded and expressed in grams. For seed dry weight, measured seedling weight from each replicate were kept in the paper bag and dried in a hot air oven at 40 ± 10 °C temperature for 24 h (Chakraborty and Mukherjee 2010). Thereafter, seedlings were cooled for 30 min, and the weight of dried samples was recorded and expressed in grams. Seedling vigour index-I and index-II were calculated as per the following formula given by Abdul Baki and Anderson (1973). Results were analysed statistically by analysis of variance (ANOVA), using SAS software (SAS Institute Inc. 1988). Pooled analysed data of 2 years have been presented in table form. When analysis of variance showed significant treatment effects, the least significant difference test (LSD) was applied to make comparisons among the means at the 0.05 level of significance.

13.2 Results and Discussion

Seed germination is one of the most critical phases of plant life (Davies 2004) and is greatly influenced by various stresses and local governing biotic and abiotic factors (Dezfuli et al. 2008). These can affect germination and seedling growth either by creating an osmotic pressure that prevents water uptake or by toxic effects of available ions in soil (Khan et al. 2009). Epigeal type of germination of *S. chirayita* seeds was observed under field condition. The data pertaining to germination percentage of 15 invigoration treatments have revealed that all treatments under field condition showed significant variation in germination percentage (Table 13.2). The maximum seed germination percentage (61.00%) was registered in the treatment GA_3 400 ppm (Plates 13.1 and 13.2) and was at par with the GA_3 200 ppm (55.87%) and statistically better than the rest of the chemical treatment measures of seeds. This might be because GA_3 increases the synthesis of hydrolytic enzymes at aleurone layer, and by the activity of these enzymes, storage compounds convert to transferable ones (sucrose and glucose) and transfer to embryo (Sedghi et al. 2008; Mukherjee 2013a). The main factor in transferring of reservoirs is their solubility in water that helps to remobilization to embryo (Khan et al. 2009). Nearly 28–62% reduction was observed in germination when the concentration of GA_3 decreases. This corroborates with earlier finding that gibberellic acid is the most commonly used hormone for promoting seed germination and is particularly considered responsible for mobilization of nutrients (Basnet 2001). Also, gibberellic acid is used to release dormancy of seeds in many species for reducing inhibitor level or by activation GA_3 synthesis or both. On the contrary, treatment of 800 ppm GA_3 sowed inhibitory effect on seed germination and reducing germination percentage significantly over rest of the levels. Similar result also observed by earlier workers (Zheleznov et al. 1997). Lowest seed germination was observed with IAA 800 ppm; this might be due to phytotoxic effect of higher concentration of IAA on seeds. This was followed by control (Plate 13.1) and showed very poor and scanty germination of seeds. Onset of germination was earliest, i.e. 19.00 days after sowing as registered by seeds pretreated with GA_3 800 ppm for 12 h, which was statistically at par with IAA 200 ppm (20.33 days) and IAA 50 ppm (22 days) treated seeds under high altitude of Darjeeling hills. Invigoration of seeds with KNO_3 (3%) for 12 h (36.33 days) resulted in most delayed onset of germination which was significantly higher than all the other treatments except control and IAA 100 ppm treated seeds. Days required for completion of germination failed to give any statistical difference with various treated seeds; moreover maximum days required for germination was registered with the IAA 200 ppm and less time requirement for completion of germination was observed with GA_3 200 ppm. Mean germination time (MGT) of *S. chirayita* was 44.11 days in untreated (control) seeds, which was significantly reduced to 27.34 days in seeds pretreated with IAA 400 ppm. This was 38.01% lower than the control and was at

Table 13.2 Effect of seed invigoration treatments on seed germination characteristics of *Swertia chirayita* (pooled value of 2-year data)

Treatments	Germination (%)	Days require for onset of germination (days)	Days required for completion germination (days)	Mean germination time (days)	Seedling vigour		Seedling vigour index-I	Seedling vigour index-II
					Fresh biomass (g)	Dry biomass (g)		
Control	26.66 ± 2.11	36.66	44.66	44.11	1.17[a] (0.87)	0.78 (0.11)[b]	23.11	2.93
GA_3 50	36.00 ± 1.31	27.33	47.00	36.36	1.49 (1.74)	0.95 (0.40)	35.21	13.64
GA_3 100	40.44 ± 2.33	30.33	50.33	39.37	1.84 (2.89)	0.94 (0.39)	35.98	15.36
GA_3 200	55.87 ± 3.44	28.00	50.00	37.11	2.32 (4.94)	1.22 (0.98)	64.08	53.74
GA_3 400	64.00 ± 5.91	27.00	55.33	29.63	2.01 (3.54)	0.95 (0.41)	71.11	26.53
GA_3 800	33.66 ± 1.72	19.00	30.66	31.00	1.20 (0.95)	0.79 (0.12)	34.33	4.37
IAA50	30.32 ± 2.16	22.00	45.33	33.77	1.19 (0.92)	0.84 (0.21)	29.71	6.67
IAA100	44.11 ± 1.56	33.66	55.66	40.34	2.03 (3.65)	1.19 (0.91)	48.96	39.63
IAA200	34.00 ± 3.81	20.33	57.66	39.97	2.14 (4.12)	1.22 (0.98)	36.04	32.65
IAA400	30.15 ± 5.11	24.33	40.66	27.34	1.74 (2.54)	1.04 (0.59)	29.24	17.18
IAA800	2.66 ± 3.03	28.33	40.33	31.45	1.77 (2.66)	0.85 (0.23)	2.97	0.64

KNO_3 (1%)	44.11 ± 1.16	34.00	41.33	39.32	1.92 (3.21)	0.96 (0.43)	38.88	19.86
KNO_3 (2%)	53.65 ± 3.33	26.66	46.66	30.67	2.19 (4.32)	1.23 (1.01)	57.99	54.18
KNO_3 (3%)	23.54 ± 2.19	36.33	54.66	34.17	2.27 (4.66)	1.17 (0.87)	24.95	20.47
KNO_3 (4%)	32.92 ± 4.11	29.66	43.33	36.13	1.61 (2.11)	0.87 (0.25)	25.34	8.69
LSD ($P = 0.05$)	9.60	3.07	N.S	3.65	0.22	0.16	4.98	5.33

N.S Nonsignificant
[a]Values marked are square root transformed values
[b]Original values

Plate 13.1 *Swertia chirayita* under control (untreated seed) condition

Plate 13.2 *Swertia chirayita* under GA_3 400 ppm treated seeds (Best treatment)

par with 29.63 and 30.67 days in seed pretreated with GA_3 400 ppm and KNO_3 (2%). By definition, MGT is associated to the time length (day) that radicle exits. Least mean germination time was at control level because of no seed germination enhancement technique and the high with IAA 400 ppm pretreated seeds due to better mobilization of reserve food material. Seedling vigour varies significantly with various treatments, and it's showed quite promising response. Observation

Plate 13.3 Layout of the experiment

revealed that highest fresh biomass of seedling at 2-month stage registered with the GA_3 200 ppm and showed parity with the KNO_3 (3%), KNO_3 (2%) and IAA 200 ppm treated seeds. Dry biomass of seedling was highest found with the KNO_3 (3%) and was statistically similar with the KNO_3 (2%). Moreover, lowest dry biomass of seedling was observed with control and was followed by GA_3 800 ppm treated chirota seeds. The data pertaining to the effect of seed invigoration treatments on seedling vigour index-I (SV-I) in Table 13.2 reveals that IAA800 ppm and untreated (control) seeds registered minimum SV-I of 2.97 and 23.11 which was statistically poor to rest of the seed priming treatments. Seeds pretreated with GA_3 400 ppm for 12 h showed maximum seedling vigour index-I (71.11), which was 207.71% higher than the control and was significantly better results compared to the rest of the treatments. Seedling vigour index-II (SV-II) of *S. chirayita* exhibited that IAA 400 ppm treated seeds gave lowest this parameter and was followed by control. The highest SV-II of 54.18 was registered by seeds pretreated with KNO_3 (2%), which was 68.16% higher than the control and was statistically similar with GA_3 200 ppm used seeds (Plate 13.3).

Seedling length was maximum registered with the pretreatment of seeds with GA_3 200 ppm (1.16 cm) and was at par with the GA_3 400 ppm, IAA 100 ppm, KNO_3 (2%) and KNO_3 (3%) and significantly better to other treatments (Table 13.3). Lowest seedling length was observed with KNO_3 (4%) and was statistically similar with control (Fig. 13.2). Effect of seed invigoration treatments on emergence index

Table 13.3 Effect of seed invigoration treatments on seedling length, emergence index and germination energy of seeds of *Swertia chirayita* (pooled value of 2-year data)

Treatments	Seedling length (cm)	Emergence index	Germination energy
Control	0.87	0.73	0.61
GA_3 50	0.98	1.32	0.77
GA_3 100	0.89	1.33	0.81
GA_3 200	1.16	2.14	1.12
GA_3 400	1.11	2.37	1.16
GA_3 800	1.02	1.77	1.12
IAA50	0.98	1.38	0.67
IAA100	1.11	1.31	0.79
IAA200	1.06	1.67	0.59
IAA400	0.97	1.24	0.74
IAA800	0.86	0.09	0.07
KNO_3 (1%)	0.88	1.3	1.07
KNO_3 (2%)	1.08	2.01	1.15
KNO_3 (3%)	1.06	0.65	0.43
KNO_3 (4%)	0.77	1.11	0.76
LSD	0.16	0.14	0.14

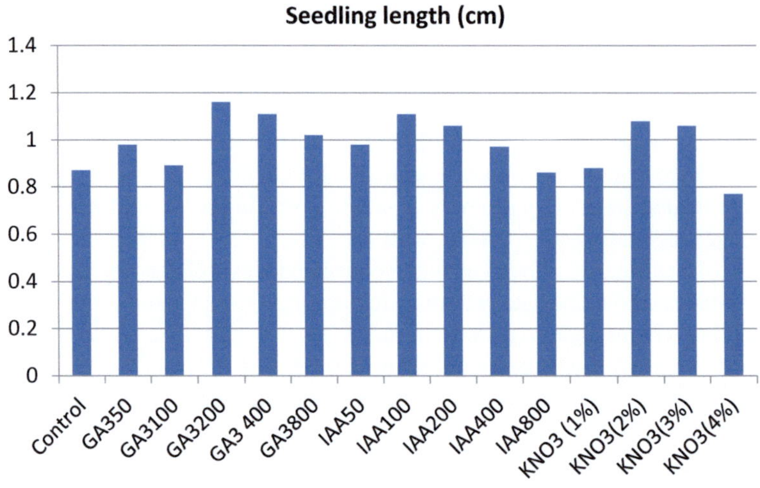

Fig. 13.2 Effect of various treatments on seedling length at 2-month stage of seedlings of *Swertia chirayita*

(EI) in chirota plant revealed that (Table 13.3) pretreated with GA_3 400 ppm gave maximum emergence index (2.37) which was significantly higher than other mentioned treatments except GA_3 200 ppm (2.14). It was followed by seeds invigored with KNO_3 (2%) and GA_3 800 ppm followed by IAA 200 ppm for 12 h (1.67) as compared to control (0.73). Further, germination energy (GE) of seeds reveals that maximum GE of 1.16 was found in seeds pretreated with GA_3 400 ppm, which was

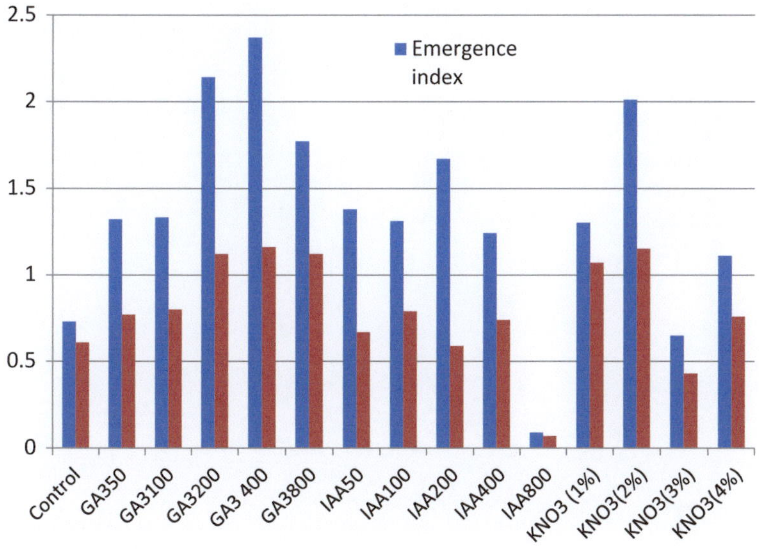

Fig. 13.3 Effect of various treatments on emergence index and germination energy of seeds of *Swertia chirayita*

statistically at par with GA_3 200 and 800 ppm, KNO_3 (1%) and KNO_3 (2%) (Fig. 13.3). EI is a parameter which reflects the seed vigour. The higher the EI the more vigorous the seed will be. The possible reasons for faster germination and higher germination energy in above-mentioned treatments in *S. chirayita* seem to be the completion of pre-germinative metabolic activities which might have resulted in attainment of the status seeds to enable them ready for radicle protrusion as compared to the control.

Valeriana jatamansi showed significant response with various seed priming measures (Plates 13.4, 13.5, 13.6 and 13.7), and highest germination percentage was observed with the kinetin 200 ppm (Plate 13.7) and significantly better to all other chemical exposed seed for priming (Table 13.4). This was followed by GA_3 200 ppm (Plate 13.6), which was at par IBA 150, GA_3 150, and 250 ppm treated seeds of *V. jatamansi*. GA_3 and kinetin were most effective chemicals that showed good response in many medicinal plant species conservation (Mukherjee 2014a). The results of our work are similar to those of Baskin and Baskin, where they found positive effect of soaking of *Osmorhiza claytonia* seeds in GA_3 solution (Baskin and Baskin 1991). Two functions for gibberellins (GA) during seed germination have been reported from various literatures. First, GA increases the growth potential of the embryo. Secondly, GA is necessary to overcome the mechanical restraint conferred by the seed covering layers, by weakening of the tissues surrounding the radical (Kucera et al. 2005). Days required for onset of germination was earlier recorded with the IBA 200 and kinetin 250 ppm and was at par with the kinetin 200 and IBA

Plate 13.4 Seeds pretreated with distill water (control)

Plate 13.5 Seeds pretreated with kinetin 100 ppm

150 ppm. More time for initiation of germination took place by control (untreated seeds). Out of 16 treatments, untreated seeds are recorded with maximum time of 29.33 days of sowing for onset of germination. Onset of germination was earliest, i.e. 20.33 days after sowing observed by seeds pretreated with IBA 200 ppm, which

Plate 13.6 Seeds pretreated with GA_3 100 ppm

Plate 13.7 Seeds pretreated with kinetin 200 ppm

was 9 days earlier as compared to control. Further Table 13.4 depicts that days required for completion of germination was least observed with the kinetin 250 ppm (23.33 days) and showed parity with the kinetin 200 (24.33 days) and GA_3 250 ppm (25.66 days) pretreated seed. In the present study, percentage of seed germination,

Table 13.4 Effect of different treatments on seed germination of *Valeriana jatamansi* (pooled value of 2-year data)

Treatments	Germination (%)	Days require for onset of germination (days)	Days required for completion of germination (days)	Mean germination time (days)	Seedling vigour		Seedling vigour index-I	Seedling vigour index-II
					Fresh biomass (g)	Dry biomass (g)		
Control	51.33 ± 6.16	29.33	36.33	33.03	2.03 (3.64)	1.02[a] (0.54)[b]	51.87	41.11
IBA 50	54.00 ± 2.12	26.00	39.00	31.59	2.22(4.43)	1.24 (1.03)	76.22	53.66
IBA 100	79.33 ± 1.33	23.33	33.33	27.83	2.37(5.12)	1.35 (1.32)	173.08	104.21
IBA 150	85.66 ± 4.13	20.66	44.66	30.56	2.85(7.65)	1.43 (1.54)	260.83	133.92
IBA 200	70.00 ± 3.11	20.00	34.33	25.15	2.13(4.02)	1.27 (1.11)	153.52	72.71
IBA 250	63.66 ± 3.92	22.66	30.66	25.66	1.81(2.79)	0.93 (0.36)	121.66	36.65
GA_3 50	61.66 ± 3.21	27.33	32.00	29.05	2.05(3.69)	1.07 (0.65)	91.19	43.07
GA_3 100	72.66 ± 3.91	27.33	30.66	27.91	2.27(4.67)	1.22 (0.98)	161.33	71.07
GA_3 150	80.15 ± 3.12	27.66	31.33	28.35	2.49(5.69)	1.26 (1.09)	156.31	86.36
GA_3 200	86.66 ± 5.60	24.00	28.33	25.16	2.61(6.32)	1.47 (1.65)	202.45	137.99
GA_3 250	80.15 ± 2.66	22.33	25.66	22.95	2.51(5.77)	1.39 (1.44)	196.66	112.42
Kinetin 50	50.21 ± 8.62	25.33	36.33	31.93	2.03(3.61)	1.13 (0.77)	74.81	33.66
Kinetin 100	58.98 ± 7.32	25.00	36.00	31.15	2.41(5.32)	1.22 (0.98)	101.83	56.81
Kinetin 150	72.36 ± 3.63	23.33	30.33	25.33	2.78(7.25)	1.48 (1.69)	220.95	121.24
Kinetin 200	94.66 ± 10.66	20.33	24.33	23.33	2.94(8.15)	1.59 (2.03)	303.33	190.06
Kinetin 250	81.33 ± 5.33	20.00	23.33	23.65	2.71(6.87)	1.32 (1.23)	201.66	100.66
LSD ($P = 0.05$)	7.21	1.98	2.05	2.11	NS	0.08	8.98	9.11

N.S Nonsignificant
[a]Values marked are square root transformed values
[b]Original values

days required for onset and final germination showed difference under different treatments. Such difference was common for many mountain species (Nautiyal et al. 2001). MGT of seeds of *V. jatamansi* which were subjected to 16 invigoration treatments showed that this was 33.03 days in untreated (control) seeds which were significantly reduced to 22.95 days in seeds soaked with GA_3 250 ppm, which was 31.88% lower than the control and was at par with 23.33, 23.66 and 25.15 days in seeds pretreated with kinetin 200, kinetin 250 and IBA 200 ppm. Fresh biomass of seedling failed to produce any statistic difference among them. However, maximum fresh weight was registered with the pretreated seeds with kinetin 200 ppm and was followed by IBA 150 ppm. Moreover, lowest fresh weight under high altitude was observed with the IBA 250 and kinetin 50 ppm treated seeds. Dry weight produced significant difference at 2-month-old seedling stage, and highest dry biomass was registered with the kinetin 200 ppm and was significantly better to all other sets of treatments. Lowest dry biomass recorded with the control followed by IBA 250 ppm. The data pertaining to the effect of seed invigoration treatments on seedling vigour index-I in *V. jatamansi* reveals that untreated (control) seeds were registered minimum SV-I of 51.87, which was statistically poor to rest of the treatments (Table 13.4). Seeds pretreated with kinetin 100 ppm showed maximum SV-I (303.33), which was 484.15% higher than the control and was significantly better to all other 14 seed priming treatments. Other treatments which gave better results on SV-I were IBA 150 ppm and kinetin 150 ppm. Further, observation reveals that kinetin 50 ppm recorded lowest SV-II and was statistically similar with IBA 250 and untreated (control) seeds registered. The highest SV-II of 190.06 was registered by seeds pretreated with kinetin 200 ppm, which was 362.32% higher than the control and was statistically better to other treatments (Fig. 13.4).

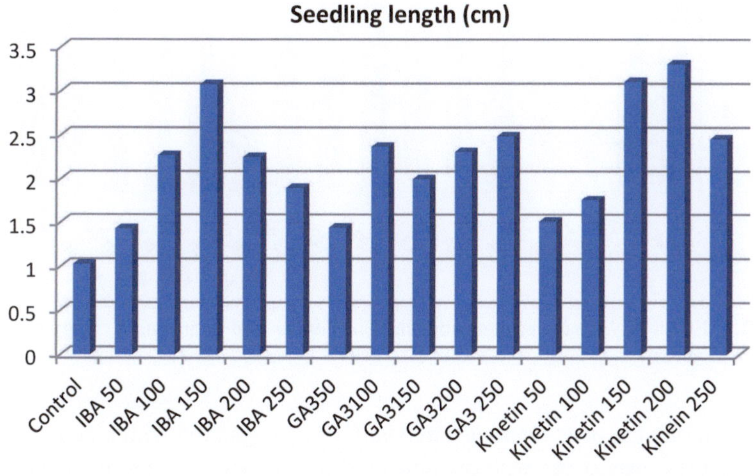

Fig. 13.4 Effect of various treatments on seedling length at 2-month stage of seedlings of *Valeriana jatamansi*

Table 13.5 Influence of different seed invigoration treatments on seedling length, emergence index and germination energy of seeds of *Valeriana jatamansi* (pooled value of 2-year data)

Treatments	Seedling length (cm)	Emergence index	Germination energy
Control	1.03	1.75	1.41
IBA 50	1.43	2.08	1.38
IBA 100	2.27	3.40	2.38
IBA 150	3.08	4.15	1.92
IBA 200	2.25	3.51	2.04
IBA 250	1.90	2.81	2.08
GA_3 50	1.44	2.26	1.93
GA_3 100	2.37	2.66	2.37
GA_3 150	2.00	2.92	2.56
GA_3 200	2.31	3.61	3.06
GA_3 250	2.49	3.59	3.12
Kinetin 50	1.51	1.98	1.38
Kinetin 100	1.76	2.36	1.64
Kinetin 150	3.11	3.10	2.38
Kinetin 200	3.31	4.65	3.89
Kinetin 250	2.46	4.07	3.49
LSD ($P = 0.05$)	0.26	0.58	0.77

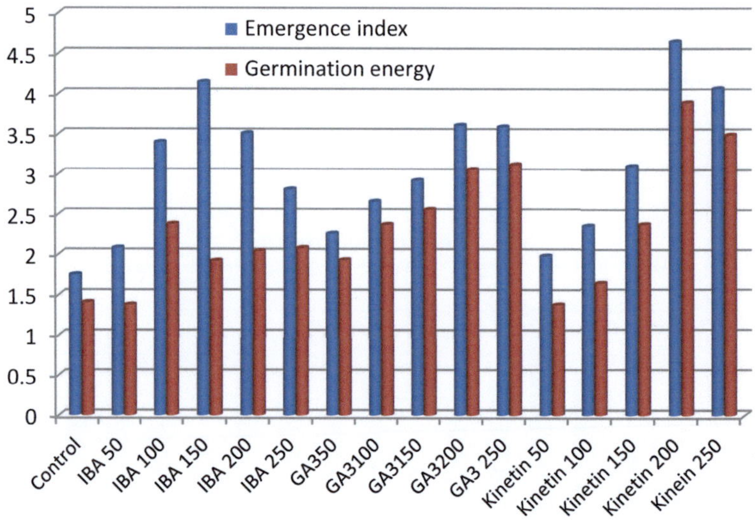

Fig. 13.5 Effect of various treatments on emergence index and germination energy of seeds of *Valerian jatamansi*

Seedling length of *V. jatamansi* maximum was registered with kinetin 200 ppm (3.11 cm) and was at par with the kinetin 200 and IBA 150 ppm pretreated seeds (Table 13.5). However, lowest seedling length was registered with the control (1.03 cm) and was significantly poor to rest all other treatments (Fig. 13.5). EI is a

Table 13.6 Effect of treatments on growth parameters of *Valeriana jatamansi* (pooled data of 2 years)

	Plant height (cm)		Fresh aerial biomass (g/plant)		Fresh underground biomass (g/plant)	
Treatments	3 months	6 months	3 months	6 months	3 months	6 months
Control	10.11	20.12	10.66	23.52	1.02	4.66
IBA 50	12.54	20.31	11.98	29.32	2.11	6.32
IBA 100	14.33	32.28	12.24	39.23	2.86	8.77
IBA 150	14.41	30.23	17.98	43.91	3.79	10.94
IBA 200	11.11	29.03	15.93	37.50	2.78	7.22
IBA 250	12.09	28.65	17.63	35.69	2.81	8.66
GA_3 50	13.65	22.11	14.83	32.65	1.88	6.11
GA_3 100	13.28	28.07	15.08	35.17	2.26	7.05
GA_3 150	15.69	30.66	13.61	40.11	2.89	8.96
GA_3 200	14.85	29.65	18.36	47.36	3.44	10.68
GA_3 250	13.36	30.01	15.25	37.39	3.03	9.11
Kinetin 50	14.23	28.02	17.06	37.12	3.29	5.91
Kinetin 100	14.69	26.09	15.89	36.01	3.20	10.81
Kinetin 150	16.82	30.78	18.53	43.95	3.87	10.99
Kinetin 200	19.35	32.66	19.11	45.32	3.98	12.01
Kinetin 250	17.32	30.40	12.76	27.33	3.11	8.24
LSD ($P = 0.05$)	NS	2.34	1.49	3.86	0.29	6.11

NS* = Nonsignificant

parameter which reflects the seed vigour and highest recorded with kinetin 200 ppm and showed parity with kinetin 250 and IBA 150 ppm pretreated seeds. Lowest value of emergence index was found with control and was followed by kinetin 50 ppm used seeds (Fig. 13.5). Effect of seed invigoration treatments on GE reveals that maximum value of 3.89 observed with kinetin 200 ppm, which was statistically at par with kinetin 250 ppm and GA_3 200 ppm (Fig. 13.5). This corroborates with the earlier finding of Sharma et al. (2005). Lowest GE was observed with the control and statistically poor to all other treatments.

Effect of various treatment measures on *V. jatamansi*, after transplanting in main field at 3-month stage, showed that plant height failed to produce any statistical difference at this stage (Table 13.6). However, at 6 months, maximum plant height was observed with the IBA 100 ppm and was at par with kinetin 200, GA_3 150 and IBA 150. GA_3 250, kinetin 150 and kinetin 250 ppm were significantly better to other treatments. Fresh plant aerial biomass also gave significant response with various seed treatment measures, and highest fresh biomass at 3-month stage was observed with the kinetin 200 ppm and was at par with the kinetin 150 and GA_3 200 ppm. At 6-month stage maximum aerial biomass was registered with the GA_3 200 ppm and was at par only with kinetin 200 ppm and statistically superior to rest of the treatments. Pretreatment with GA_3 accelerates some metabolic processes even in low potential. This causes improvement at metabolic activities in germination, especially under stress conditions, and, subsequently, the weight of radicle and plumule increases in less time, which ultimately helps to better vegetative growth of this endangered plant at 6-month stage. Fresh underground biomass, which includes root and rhizomes, was more observed with the kinetin 200 ppm and was at par with the IBA 150 and kinetin 150 ppm used seeds.

13.3 Conclusion

The present study recommends GA_3 as the best treatment for improving seed germination. Higher percentage of germination by GA_3 400 ppm indicates that it is the best source and helps to conserve this endangered plant sp. from Himalayan range. Further, pretreatment of *Valeriana jatamansi* seeds with GA_3 200 ppm and kinetin 200 ppm accelerates some metabolic processes even in low potential. This causes improvement in metabolic activities in seeds, especially under high altitude conditions and, subsequently, the weight of seedling dry and fresh weight increases in less time. Since seed priming is simple and cheap, we can propose this method to *S. chirayita* and *V. jatamansi* growers of Darjeeling-Sikkim Himalaya, so they can increase percent and homogeneity of emergence and good germination for better economic harvest of these endangered plants and help to conserve from total extinction from our earth ecosystem.

Acknowledgements Author is thankful to the Directorate of Medicinal and Aromatic Plants (ICAR), Anand, Gujarat for providing financial and technical support for this programme. The authors wish to acknowledge the critical help and support received from different forest range officers and medicinal plant growers of Darjeeling and Sikkim Himalaya for their full support throughout this programme.

References

Abdul B, Andreson A (1973) Biochemical aspects of seed vigour. Hortic Sci 15:765–771
Anon (1997) Biodiversity conservation prioritization project, conservation assessment and management plan (CAMP) for endemic medicinal plants in India. CIMP, Lucknow. 27
Arteca RN (1996) Plant growth substances principles and applications. Chapter 3: chemistry, biological effects and mechanism of action. Chapman and Hall, New York, p 66
Ashraf M, Foolad MR (2005) Pre- sowing seed treatment-a shotgun approach to improve germination growth and crop yield under saline and none-saline conditions. Adv Agron 88:223–271
Baskin CC, Baskin JM (1991) Non-deep complex morphophysiological dormancy in seeds of *Osmorhiza claytonia* (Apiaceae). Am J Bot 78:588–593
Basnet DB (2001) Evolving nursery practices and method of cultivation of high value medicinal plant *Swertia chirayita* ham. Environ Ecol 19:935–938
Basra SMA, Zia MN, Mehmood T, Afzal I, Khaliq A (2002) Comparison of different invigoration techniques in wheat seeds. Pakistan J Arid Agric 5:11–16
Bewley JD, Black M (1994) Seeds: physiology of development and germination. Plenum, New York, pp 65–87
Bhattacharjee SK (2001) Handbook of medicinal plants, vol 1. Pointer Publisher, Jaipur, pp 69–71

Chakraborty S, Mukherjee D (2010) A text book of practical approaches in seed science and technology. New Delhi Publishers, https://www.thewinkstore.com/.../Soumendra-Chakraborty-and-Dhiman-Mukherjee www.ndpublisher.in/ndp

Chakraborty S, Mukherjee D, Dasgupta T (2008) Cytological study on chromosome behaviour and new report on nature of mode of pollination of *Swertia chirayita*, a high value endangered medicinal plant of North Eastern Himalayan region. Caryologia 62(1):43–52

Chakraborty S, Mukherjee D, Baskey S (2015a) Selection of lines of *Valeriana jatamansi* Jones, a high value medicinal plant in North Eastern Himalayan region. Indian J Genet Plant Breed 75(3):1–4

Chakraborty S, Mukherjee D, Baskey S (2015b) Indian valerian, a highly endangered medicinal plant in north eastern Himalayan region. Adv Plants Agric Res 2(4):1–7

Chakraborty S, Mukherjee D, Baskey S (2016) Morphological diversity and nomenclature of *Swertia chirayita* (Gentianaceae)- recovery of endangered medicinal plant population in north eastern Himalaya. Am J Plant Sci 7:741–755

Choudhary DK, Kaul BL, Khan S (1996) Breaking seed dormancy of *Podophyllum hexandrum* Royle ex. Camb. (syn.*P. emodi* Wall.ex. Honigberger). J Non Timber For Prod 3:10–12

Davies PJ (2004) Plant hormones: biosynthesis, signal transduction, action. Kluwer Academic Press, Dordrecht, pp 89–102

Dezfuli PM, Sharif ZF, Janmohammadi M (2008) Influence of priming techniques on seed germination behavior of maize inbred lines (*Zea mays* L.). ARPN J Agric Biol Sci 3(3):22–25

Du LV, Tuong TP (2002) Enhancing the performance of dry-seeded rice: effects of seed priming, seedling rate and time of seedling. In: Pandey S, Mortimer M, Wade L, Tuong TP, Lopes K, Hardy B (eds) Direct seeding: research strategies and opportunities. International Research Institute, Manila, pp 241–256

Dubey NK, Kumar R, Tripathi P (2014) Global promotion of herbal medicine : India opportunity. Curr Sci 86:37–41

El-Keblawy A, Al-Rawai A (2005) Effects of seed maturation time and dry storage on light and temperature requirements during germination in invasive *Prosopis juliflora*. Flora 201:135–143

Foti R, Abureni K, Tigere A, Gotosa J, Gerem J (2008) The efficacy of different seed priming osmotica on the establishment of maize (*Zea mays* L.) caryopses. J Arid Environ 72:1127–1130

Ghavami N, Ramin AA (2007) Salinity and temperature effect on seed germination of milk thistle. Commenwealth Soil Sci Plant Ann 38(20):2681–2691

Gupta RK (1987) Living himalaya: aspects of environment and resource ecology. Today & Tomorrow Printer & Publishers, New Delhi, pp 66–71

International Seed Testing Association (ISTA) (1965) International rules for seed testing. Seed Sci Technol 13:300–513

Jackson ML (1973) Soil chemical analysis. Prentice Hall of India Pvt. Ltd, New Delhi, pp 183–204

Kala CP (2006) Medicinal plants of the high altitude cold desert in India, diversity, distribution and traditional use. Int J Biol Sci Manag 2:43–56

Kandari LS, Rao KS, Chauhan K, Maikhuri RK, Purohit VK, Phondani PC, Saxena KG (2007) Effect of pre-sowing treatments on the seed germination of two endangered medicinal herbs of the himalaya (*Anagelica gluca*. Edgew and *Pleurospermum aegelicoides* (Wall. Ex.DC.) Benth.Ex C.B. Clarke). Proc Indian Natl Sci Acad 73(1):11–16

Keil M, Hartle B, Guillaume A, Psiorz M (2000) Production of amarogentin in root cultures of *Swertia chirayita*. Planta Med 66:452–457

Khan HA, Ayub CM, Pervez MA, Bilal RM, Shahid MA, Ziaf K (2009) Effect of seed priming with NaCl on salinity tolerance of hot pepper (*Capsicum annuum* L.) at seedling stage. Soil Environment 28(1):81–87

Kucera B, Cohn MA, Leubner MG (2005) Plant hormone interactions during seed dormancy release and germination. Seed Sci Res 15:281–307

McDonald MB (1999) Seed priming. Seed technology and its biological basis. In: Black M, Bewley JD (eds) . CRC press, Boca Raton, pp 287–316

Mukherjee D (2008) Cultivation of *Swertia chirayita* : a high value medicinal herb in high altitude. J Med Aromat Plant Sci 30(4):350–355

Mukherjee D (2009a) Medicinal plant in Darjeeling hills. In: Rai S, Moktan MW, Ali S (eds) Krishi Sandesh. Miazik International Volunteer Center, Japan, pp 118–121

Mukherjee D (2009b) Current status, distribution and ethno-medicinal values of medicinal plant in hilly region of Darjeeling district of West Bengal. J Crop Weed 5(1):316–320

Mukherjee D (2013a) Improved agricultural practices for high value medicinal plants under mid to high altitude situation. J Crop Weed 9(1):201–206

Mukherjee D (2013b) Prospect and need of agro-forestry in the hills of Darjeeling. Indian Agric 57(1):21–27

Mukherjee D (2014a) Medicinal plant with relation to biodiversity conservation at Darjeeling hill. In: Nehra, Gothwal, Ghosh (eds) Biodiversity in India: assessment, scope and conservation. Lap Lambert Academic Publishing, Deutschland, pp 43–73

Mukherjee D (2014b) Effect of forest microhabitat on growth of high altitude plants in Darjeeling Himalaya. J Interacademicia 18(1):20–30

Mukherjee D (2015) *Swertia chirayita*: endangered medicinal plant of temperate Himalaya. In: Hemantaranjan A (ed) Advances in plant physiology, vol 16. Scientific Publishers, Jodhpur, pp 336–356

Mukherjee D (2016) Studies on ecology and outcome of various seed and organic input treatments on yield attribute and economic yield of *Valeriana jatamansi*. Himalayan Rev 1(2):193–200

Mukherjee D, Chakraborty S, Roy A, Mokthan MW (2009) Differential approach of germplasm conservation of high value of medicinal plants in north eastern himalayan region. Int J Agric Environ Biotechnol 2(4):332–340

Mukherjee D, Chakraborty S, Roy, A (2010) Innovative cultivation practices of critically endangered herb: *Swertia chirayita*, Publish from Directorate of Research, Uttar Banga Krishi Viswavidyalaya, Coochbehar. ISBN Number: 978-93-5137-910-2

Mukherjee D, Chakraborty S, Baskey S (2015) Threatened medicinal plants biodiversity of Eastern Himalaya and its conservation. J Agric Technol 2(1):101–107

Nautiyal BP, Prakash V, Chauhan RS, Purohit H, Nautiyal C (2001) An assessment of germinability, productivity and cost benefit analysis of *Picrorhiza kurrooa*, cultivated at lower altitude. Curr Sci 81:579–585

Polunin O, Stainton A (1987) Concise flowers of the Himalaya. Oxford University Press, London, p 67

Raina R, Johri AK, Srivastava LJ (1994) Seed germination studies in *Swertia chirayita* L. Seed Res 22:62–63

Samant SS, Dhar U, Palni LMS (1998) Medicinal plants of Indian himalaya: diversity, distribution and potential values. Gyanodaya Prakashan, Nainital. 3, 163

SAS Institute Inc (1988) SAS User's guide. Statistical Analysis Institute, Cary

Sedghi M, Gholipouri A, Seyedsharifi R (2008) Tocopherol accumulation and floral differentiation of medicinal pumpkin (*Cucurbita pepo* L.) in response to plant growth regulators. Not Bot Hort Agrobot Cluj 36(1):80–84

Sfairi Y, Lahcen O, Feddy M, Nana AA (2012) Dormancy breaking and salinity/water stress effects on seed germination of Atlas cypress, an endemic and threatened coniferous species in Morocco. Afr J Biotechnol 11(19):4385–4390

Shanmugavalli M, Renganayaki PR, Menaka C (2007) Seed dormancy and germination improvement treatments in fodder sorghum. Int Crops Res Inst Semi Arid Trop 3:1–3

Sharma RK, Sharma S, Shanti SS (2005) Seed germination behavior of some medicinal plants of lahaul and spiti cold desert : implications for conservation and cultivation. Curr Sci 90:1113–1118

Zheleznov AV, Solonenko LP, Zheleznova NB (1997) Seed protein of the wild and the cultivated *Amaranthus* species. Euphytica 39:77–182

Seed Priming on Germination, Growth and Flowering in Flowers and Ornamental Trees

14

Anjana Sisodia, Minakshi Padhi, A. K. Pal, Kalyan Barman, and Anil K. Singh

Abstract

Seed dormancy is an emerging problem related to germination which is common in many species of ornamental trees and flowers. Poor seed germination and subsequently poor field establishment are a common phenomenon at adverse conditions of environment. The most important problems faced are the heterogneity and lack of suitable conditions in soil that causes decrease in germination percent. Priming is a water-based technique that consents metabolic processes necessary for enhancing germination rate and seed quality by managing the temperature and seed moisture content in which the seed is taken through the first biochemical processes within the initial stages of germination but preventing the seed transition towards full germination. This is a successful way through which plants would be able to complete their growth on or before the stresses arrive (Subedi KD, Ma BL. Agron J 97(1):211–218, 2005). Seed priming technique has been practised in many countries including India, Pakistan, China and Australia, and more than thousand trials had been conducted to evaluate the performance of priming in a variety of crops. The principle of seed priming is to minimise the period of emergence and to protect seed from environmental stresses during critical phase of seedling establishment to synchronise emergence which lead to uniform establishment and improved yield. It reduces the effect of salinity on the morphological parameter of the plants. Various priming techniques, like osmopriming, biopriming, halopriming, thermopriming, hydropriming, hormonal priming and solid matrix priming, give favourable result in seeds of ornamental flowers as well as trees. This technique has been successfully carried out in flower crops like balsam, coneflower, cosmos, gladiolus, pansy, marigold,

A. Sisodia · M. Padhi · A. K. Pal · K. Barman · A. K. Singh (✉)
Department of Horticulture, Institute of Agricultural Sciences, Banaras Hindu University, Varanasi, Uttar Pradesh, India

© Springer Nature Singapore Pte Ltd. 2018

A. Rakshit, H. B. Singh (eds.), *Advances in Seed Priming*,
https://doi.org/10.1007/978-981-13-0032-5_14

periwinkle, rudbeckia, salvia, snapdragon and zinnia and trees like cassia, cypress, senegal, eucalyptus, fig, teak, pine, almond, tamarind, oak, *karanj*, *khejri*, *siris*, *subabul*, *kapok*, *gulmohar*, *kachnar*, etc.

Keywords
Priming · Gladiolus · Periwinkle · Eucalyptus · *Kachnar*

Establishment of crop is the primary importance for optimising horticultural production. Every year, mankind awaits for the miraculous transmogrification of seeds into plants and again into seeds. Poor seed germination and subsequently poor field establishment are a common phenomenon at adverse conditions of environment. It has been reported that the major difficulty to high yield and production of crop plants is due to the lack of synchronised crop establishment and adverse weather and soil conditions (Mwale et al. 2003).

Seed dormancy is another problem related to germination which is common in many species. It is an adaptation that allows a species to regulate the timing of germination for seeds in a population. Some species use environmental cues (such as drought, rainfall or temperatures) to integrate germination for most seeds at a particular time of the year. Temperature, moisture, air and light conditions are the most important factors for seed germination. Minimum temperature is the lowest temperature at which seeds can germinate effectively (Nleya et al. 2005), and the maximum is the highest temperature at which seeds can germinate. Slight change in this temperature can damage seeds or make them go into the dormancy condition. At optimal temperatures, germination is rapid and uniform. Seeds need correct moisture to initiate internal processes leading up to germination. Osmotic adjustment or priming of seeds prior to sowing is known as a potent way to increase germination and emergence rate in some species with stepwise seed development (Sivritepe 2000).

Seed priming technique has been practised in many countries including India, Pakistan, China and Australia, and more than thousand trials had been conducted to evaluate the performance of priming in a variety of crops. The need for increased seed quality has become a priority necessary to tackle the current demand for high standards of seeds in the agricultural market. Achieving rapid and uniform seedling emergence is a key point for crop performance since slow germination rates frequently expose plantlets to adverse environmental conditions and soilborne diseases (Osburn and Schroth 1989). Priming is a water-based technique that consents metabolic processes necessary for enhancing germination rate and seed quality by managing the temperature and seed moisture content in which the seed is taken through the first biochemical processes within the initial stages of germination but preventing the seed transition towards full germination. This is the way through which plants would be able to complete their growth on or before the stresses arrive (Subedi and Ma 2005).

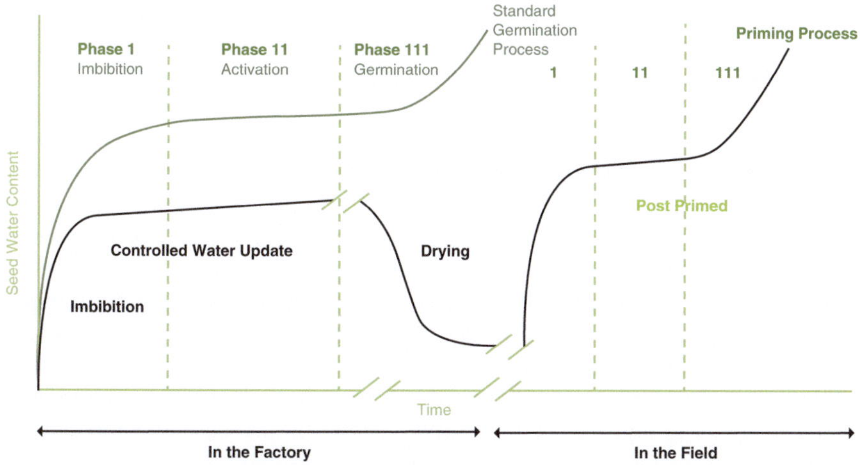

The above graph shows about the standard germination process and seed priming process in the factory and field condition. The green line corresponds to a standard germination process. During Phase I (imbibition), seeds take up water sufficiently under suitable availability of temperature and moisture. During Phase II (activation), the biochemical processes are activated and will eventually start the germination in Phase III (germination) where roots and hypocotyls emerge from the seeds. During priming, the seed invloves the activation processes followed by drying, before root can emerge from the seeds. Once conditions (temperature and moisture) are appropriate in the field, Phase III can continue, and germination occurs in a much shorter time (Hasegawa 2016).

The purpose of seed priming is to minimise the period of emergence and to protect seed from environmental stresses during critical phase of seedling establishment to synchronise emergence which lead to uniform establishment and improved yield. It reduces the effect of salinity on the morphological parameter of the plants. One of the priming techniques called osmopriming is a commercially used technique for improving seed germination and vigour. It controls imbibition of seeds to start the initial germination process followed by seed drying up to its original weight. Apart from germination process, seed priming also helps in growth and flowering of the crops. Various seed priming techniques like hormonal priming and chemopriming for enhancing flowering and growth of plant are commercially practised in- or on-farm basis. Where, on farm seed priming is energy intensive, high technology seed priming, seed hardening or seed conditioning process are available to farmers to help them in high input temperate agriculture and horticulture. (Harris et al. 2001).

The priming treatments which enhance seed germination include hydropriming (Afzal et al. 2004), biopriming, halopriming, solid matrix priming, chemopriming, thermopriming, osmopriming and hormonal priming (Afzal et al. 2006).

14.1 Physiological and Biochemical Aspects of Priming

A successful application of seed management technique depends upon the type of test, method of application, selection of crop, initial performance of the crop, selection of chemical, its concentration, duration of treatment and the purpose of implication. Priming method in seed management techniques is proven very essential factor for enhancing quality issues, germination rate, establishment, etc. Priming can inverse some of the ageing-induced deteriorative events, resulting in improved seed performance (Taylor et al. 1998). It has shown an immense effect to activate different processes related to cell cycle and to induce synthesis of nuclear DNA in radial tip cells in tomato (Liu et al. 1997).

Long duration seed storage caused a decline in the level of protein content which may cause oxidation of the amino acids, due to the increase in the respiratory activity and advance in the degradation process of the stored seeds. Seed deterioration causes loss of membrane integrity, changes in enzymatic activities and declines in protein and nucleic acid synthesis and lesions in DNA (McDonald 1999). Priming with 30% PEG for 24 h resulted in increase in the activity of superoxide dismutase (SOD) and peroxidise (POD) which enhance the intensity of respiration of plant and cause an increase in vigourity in germination (Jie et al. 2002). Priming is also thought to increase the activity of many enzymes involved in metabolism of carbohydrates (α- and β-amylases), proteins (proteases) and lipids mobilisation (isocitrate lyase) that are implicated in the stored reserves mobilisation (Varier et al. 2010; Di Girolamo and Barbanti 2012). These enzymes are vital in the breakdown of macromolecules for the development and growth of the embryo that ultimately result in early and higher seedling emergence (Farooq et al. 2006a, b; Varier et al. 2010). There are reports that priming facilitates the repair of chromosomal damage (Sivritepe and Dourado 1995), permits early DNA replication and repair, increases RNA and de novo protein synthesis and reduces the leakage of metabolites (McDonald 2000; Farooq et al. 2007a, b; Manonmani et al. 2014; Paparella et al. 2015). Thus, total seed protein, POD, PPO, RNA and de novo protein synthesis were increased significantly by seed priming. Among the various processes of priming, osmopriming may enhance rapid seed germination by reducing mechanical hindrance on the germinating embryo (Toorop et al. 1998).

14.2 Seed Priming Techniques

Seed priming includes various techniques which influences growth, establishment and germination of seeds and also influences yield of the crop. Techniques include the following.

14.2.1 Hydropriming (Drum Priming)

A major cause of poor establishment and low crop yield in agricultural areas is the lack of moisture content that reduces the ability of seedling to further emerge and growth. Hydropriming is a technique for enhancing germination without the emergence of the radicle and plumule which involves soaking of seeds in a priming agent solution followed by drying even if seeds are infected with pathogens (McDonald 1999). In hydropriming, distilled water plays a vital role for imbibition up to 10–20% (Pill 1995). It results into uncontrolled water uptake, since the process depends on seed affinity to water and the main critical point is to find and maintain optimal temperature and humidity conditions to avoid radicle protrusion (Taylor et al. 1998). Another limiting factor of hydropriming is the lack of homogeneous seed hydration which can lead to uneven germination (McDonald 2000). The main variant of hydropriming also called drum priming, patented by Rowse (1991) in which a drum that contains seeds is connected with a boiler generating vapour. The vapour condenses into liquid water inside the drum. The machine measures the increase in seed relative mass during the treatment. The time and volume of water required to complete seed rehydration are strictly controlled to reach gradual and even seed imbibition (Warren and Bennett 1997). Hydropriming is the most ancient type of priming, since the benefits of this pre-sowing treatment have been known for a long time; however, it is now applied less frequently in comparison with other methods.

14.2.2 Biopriming

Biopriming is a new technique of seed treatment that assimilates biological (inoculation of seed with beneficial organism to protect seed) and physiological aspects (seed hydration) of disease control. To respond to the negative effects of pathogens, biopriming uses beneficial microorganisms to protect against pathogens and enhance plant growth. Biological seed treatments for control of seed and seedling diseases offer the grower an alternative to chemical fungicides. Storage and application conditions are more critical than with chemical seed protectants and differential reaction to hosts, and environmental conditions may cause biological seed treatments to have a narrower spectrum of use than chemicals. Conversely, some biocontrol agents applied as seed dressers are capable of colonising the rhizosphere, potentially providing benefits to the plant beyond the seedling emergence stage (Nancy et al. 1997). Seed treatment with biocontrol agents along with priming agents may serve as an important means of managing many of the soil- and seed-borne diseases, the process often known as 'biopriming'. It involves coating seed with a bacterial biocontrol agent such as *Pseudomonas aureofaciens* AB254 and hydrating for 20 h under warm (23 °C) conditions in moist vermiculite or on moist germination blotters in a self-sealing plastic bag. The seeds are removed before radical emergence. The bacterial biocontrol agent may multiply substantially on seed during biopriming (Callan et al. 1990). Biopriming seed treatments can provide a high level of protection against root rot diseases of crop plants which was generally equal or superior to the control provided with fungicide seed treatment. So, it could be suggested that biopriming (combined treatments between seed priming and seed coating with biocontrol agents) may be safely used commercially as substitute for traditional fungicide seed treatments for controlling seed- and soilborne plant pathogens.

14.2.3 Halopriming

Halopriming is one of the methods of priming practices that includes salts like $CaCl_2$, $CaSO_4$ and NaCl in such a way that the pregermination metabolic activities start preventing radical protrusion followed by drying seeds to the original moisture level (McDonald 2000). In this method, the seeds are immersed in different salt solutions which facilitate the process of seed germination and subsequent seedling emergence even under adverse environmental conditions. Seeds treated with NaCl concentrations should be in a tolerable limit. Early initiation of metabolic activities and reserve breakdown and mobilisation might be the reason for faster germination in such type of primed seeds. Khan et al. (2009) reported salt priming induced salinity tolerance of hot pepper at seedling stage, wherein seed priming improved significantly the germination percentage and index, vigour index, plumule and radical length and dry weight of seedling as compared to the non-primed seeds (control). Improved stress tolerance of primed plants is thought to rise from the activation of cellular defence response due to halopriming (Beckers and Conrath 2007). This has been substantiated by reports on better antioxidant system in primed plants (Afzal et al. 2006) on exposure to stress. Conrath et al. (2006) proposed that halopriming could involve accumulation of signalling proteins or transcription factors. Halopriming is a simple and cheap agrotechnique and found suitable to be recommended to the farmers owing to better synchrony of emergence and crop stand under various conditions of environment (Sedghi et al. 2010).

14.2.4 Solid Matrix Priming

Solid matrix priming (SMP) is similar to osmotic priming that allows the seed to attain a threshold moisture content and pre-germinative metabolic activity but preventing radicle emergence. However, it has the advantages of allowing aeration, incorporation of biological agents to combat soilborne pathogens and improved ease of handling (Taylor et al. 1988; Wang et al. 1998). The matrix comprises of finely divided non-plant pathogenic water holding solid, which may be carbonaceous substance, preferably a lignateous solid which has a large eqilibrium water potential (ψ) and preferably has an osmotic potential component which is atleast about 90% or greater than 95% of the total water potential. Such materials include coal, especially soft coal, lignateous shale such as the leonardite shale, sold as Agro-Lig, and sphagnum moss. The matrix material when containing the water to prime the seeds must be sufficiently friable, non-clumping, etc., so that it can be mechanically separated from the treated seeds after treatment without damage to the seeds when required. The process of SMP includes the admixture of a predetermined amount of solid matrix material and a predetermined amount of water and the mixture allowed to stand preferably in a container which allows entry of air but reduces evaporative losses, resulting in sufficient amount of moisture level in seeds. Matrices can be readily used in tree seed nursery operations.

14.2.5 Thermopriming

Seed treatments carried out at various intervals of time before sowing are known as thermopriming. Information exists to specify that seeds germinate better under alternating temperature conditions compared to a constant daily temperature (Felippe 1980; Shin et al. 2006; Markovskaya et al. 2007). Changing temperature can break the dormancy of seed easily. This technique has been widely adapted to improve germination efficiency under adverse climate reducing thermo-inhibition of seed germination (Huang et al. 2002). Pre-sowing seed treatments with alternating daily temperature regimes have resulted in enhanced plant development, increased cold and/or frost resistance and higher plant productivity in cucumber and melon (Markovskaya et al. 2007). Small changes in ambient temperature can regulate flowering time via a thermosensory pathway (Franklin 2009). It was shown that cold treatment at the seedling stage can modulate the flowering of some ornamental plants (Runkle et al. 1999; Garner and Armitage 2008). Although high temperature condition has been used in some species, resulting pregermination especially for plants adapted to warm climates (Khalil and Rasmussen 1983). Seeds of white spruce (*Picea glauca* L.), lettuce, etc. that were primed with combinations with other treatments resulted in beneficial effects on germination parameters (Liu et al. 2013; Ashraf and Foolad 2005). It has been suggested that priming is responsible to repair the age-related cellular and subcellular damage of low-vigour seeds that may accumulate during seed development (Bray 1995). Wang et al. (2003) reported that both thermo- and hydro-primed seeds showed significant increase in germination performance. The resultant effect of priming depends on the method used and time of treatment.

14.2.6 Osmopriming

Osmopriming, also known as osmotic conditioning, is a widespread pre-sowing priming procedure which involves treatments of seeds with osmotic solutions at low water potential facilitating the control of water uptake into the seeds. The main goal of osmopriming is to limit the reactive oxidative species, i.e. ROS-mediated oxidative injury through insufficient water absorption. Thus, the water potential of the osmotic agent used is a crucial parameter during the priming process (Heydecker and Coolbear 1977; Taylor et al. 1998). Priming with PEG provides beneficial conditions for bacterial growth due to poor aeration (Parera and Cantliffe 1994). It shows some disadvantage when used in bulk, due to high costs and extremely high viscosity which limits oxygen transfer within the solution. Hence, research has to be done for choosing the correct chemical and its optimum dose for a crop. It is difficult to manage huge quantities of wet primed seed especially under hot tropical climate condition, while in temperate areas, maintaining the priming temperature is crucial. Some of the osmotica (osmotic compounds used for osmopriming) that can be used include potassium nitrate, potassium dihydrogen orthophosphate, dipotassium hydrogen orthophosphate, calcium chloride, zinc sulphate, borax, magnesium chloride, manganese sulphate, sodium chloride, sodium sulphate and organic compounds, viz. agrosan, cycocel, citric, furamic, succinic, malic acids, purines, pyrimidines, caffeine, uracil, xanthine and uridine diphosphate (De Chandra 1999).

14.2.7 Hormonal Priming

Hormone pretreatment is a commonly used priming approach to improve seed germination in stressful conditions (Atici et al. 2003; Gratao et al. 2005; Jisha et al. 2013; Masood et al. 2012; Hu et al. 2013). Hormonal priming in general consists of treatment of seeds with chemicals like growth regulators, sodium hypochlorite (NaOCl) or hydrochloric acid (HCl), natural substances and agrichemicals (e.g. fungicides, pesticides). It has reduced the severity of the effect of salinity, but the amelioration was found maximum due to the application of 50 ppm salicylic acid and 50 ppm ascorbic acid treatments and gives satisfied results on seedling growth, fresh and dry weights under non-saline and saline conditions, whereas hormonal priming with ABA was not effective in some of grass family crops (Afzal et al. 2006). In pepper (*Capscum annum* L.), Khan et al. (2009) showed that pretreatment with acetylsalicylic acid and salicylic acid resulted in greater uniformity of germination and establishment of seedlings under high salinity. In addition to these chemicals, ethylene was used to minimise the effect of high temperatures on seed germination of lettuce (Nascimento 2004, Nascimento et al. 2005).

14.3 Seed Priming in Ornamental Flower Crops

14.3.1 Balsam

Response of hormonal priming on flower quality, growth and germination rate of balsam was observed according to the basis of germination rate dry weight and shoot and root length. GA_3 at 10 ppm strikingly enhanced the germination percentage and speed of germination of balsam. The germination was higher in large seeds (grade A) which might be due to more supply of food material to the growing embryo. However, GA_3 at 30 ppm increased the length of shoot and root, dry weight and fresh weight of seedlings reported by Singh and Karki (2003).

14.3.2 Coneflower

Osmotic priming in polyethylene glycol (PEG) or matrix priming in expanded vermiculite had greater rate, synchrony and germination percentage at 20 °C than non-primed seeds of coneflower (*Echinacea purpurea*). Osmotic or matrix priming for 10 days at −0.4 MPa and 15 °C resulted in higher germination rate and germination percentage than short duration of exposure (5 days) or lower (−1.5 MPa) water potential. Seedling emergence rate, synchrony and percentage from osmotically or matrically primed seeds were similar in both cool (23–27 °C day) and warm (35–40 °C) glasshouse regimes. Emergence was faster in primed than from non-primed seeds in both regimes. Emergence percentage was higher (80%) from primed seeds than from non-primed seeds (50%) in the cool regime, but emergence synchrony was unaffected. Moistened vermiculite substituted for PEG solution as a priming medium for purple coneflower seeds benefits to seed germination or seedling emergence followed by priming (−0.4 MPa, 15 °C, 10 days of darkness) in these media (Pill et al. 1994).

14.3.3 Cosmos

In cosmos (*Cosmos bipinnatus*), hormonal priming plays a main role in enhancing flowering and quality of flowers. The triazoles, which include paclobutrazol (PB) and uniconazole, are more potent and persistent than most other growth retardants. PB is used most commonly in commercial practice, but non-uniform plant size can result from non-uniform spray application. Soaking of cosmos seeds to 1000 ppm PB reduced seedling shoot height but also reduced seedling emergence percentage (Pill and Gunter 2001). Seed treatment with PB also eliminates conventional fungicide seed coating treatment since the triazoles themselves are potent fungicides (Fletcher and Gilley 2000).

14.3.4 Fir

Seeds of true firs, including pacific silver fir (*Abies amabilis*), subalpine fir (*A. lasiocarpa*) and noble fir (*A. procera*), exhibit deep dormancy at maturity. To break the dormancy termination, seeds generally require prolonged moist-chilling condition (i.e. 3–4 months or longer) (Edwards 1981, 1986; Leadem 1986; Tanaka and Edwards 1986; Edwards 1996). In some cases, germination can be impaired by seed-borne pathogens where some of the seedlots have a high proportion of empty seed (Kolotelo 1998). This seed dormancy can be broken down with solid matrix priming when seeds are existed in most chilling temperature. Agro-Lig Greens Grade (humic acids with particle sizes between 0.212 and 1.29 mm), sand (particle size 1.29 mm), peat moss and sphagnum moss are used in matrix priming that help in breaking dormancy in seeds of Abies spp. and early germination in seeds with high seedling establishment (Ma et al. 2003).

14.3.5 Gladiolus

Gladiolus alatus is a prominent bulbous cut flower that sometimes gives less productivity due to low-quality seed, inadequate seedbed preparation, late sowing, poor sowing technique, inadequate soil moisture, adverse soil properties and high temperatures. Seed priming has been successfully demonstrated in this crop to improve germination and emergence in seeds. It is the enhancement of physiological and biochemical events in seeds during interruption of germination by low osmotic potential and negligible matric potential of the imbibing medium. Hydro-primed seeds compared to KNO_3-treated seeds (osmo-primed) were allowed to imbibe water for a longer time. Seeds treated with 0.25% KNO_3 attained the maximum (66.67%) level of germination followed by 60% in 0.75% KNO_3 (60%) and distilled water. After 30 days, the germination percentage was recorded maximum with treatment of 0.25% KNO_3, 0.75% KNO_3 and distilled water (83.33%) followed by non-priming (63%) and 0.5% KNO_3 (60%)-treated seeds (Mushtaq et al. 2012).

14.3.6 Meadow Fescue

Priming is highly useful for seeds under stress condition. NaCl priming (halopriming) and hydropriming on germination and early growth of *Festuca arundinacea* and *Festuca ovina* seeds were studied under salinity condition. It was observed that NaCl priming with concentrations of 15 and 45 dS/m in *Festuca arundinacea* seeds and NaCl priming with concentration 45 dS/m in *Festuca ovina* seeds had the highest performance of improved seed in both species at germination and early growth stages under salinity stress (Shakarami et al. 2011).

14.3.7 Pansy

Temperature stress is one of the most important factors that affect the growth and development of seeds of pansy (*Viola tricolor*). Thermo-inhibited seeds fail to germinate in high temperature but can germinate when temperature is reduced which may result in thermo-death of the plant which is mostly seen in pansy. So in such condition, there is a need for priming of seeds that may induce the germination rate by improving seed quality. Dorna et al. (2014) reported that hydropriming, halopriming and osmopriming gave significant effect on germination of pansy seeds. Osmopriming seeds in polyethylene glycol (PEG) solutions of −1.25 and −1.5 MPa osmotic potential at 15 °C and in PEG solution of −1.0 MPa osmotic potential at 20 °C increased the percentage of germinating seeds significantly at 30 °C. Osmopriming seeds, in all combinations used, improved the percentage of germinating seeds significantly at 35 °C. The best result was observed when seeds were primed in PEG solution of −1.0 MPa osmotic potential at 20 °C. At higher temperature hydropriming seeds in volume of water 600 μl H_2O g/seed for 3 days at 15 °C and osmopriming at 20 °C positively affected the germination rate. After osmopriming of seeds at 15 °C and after osmopriming in PEG solution of −1.0 MPa osmotic potential at 20 °C, the percentage of ungerminated seeds was lower than treatment control. Both halopriming at 20 °C and osmopriming, regardless of temperature and osmotic potential of PEG, improved significantly the uniformity of germination at 30 °C compared with untreated seeds. It was also observed that osmopriming followed by halopriming improved the speed of germination in pansy at 20 °C to the largest extent. Moisture content was also increased due to priming, but drying of seeds after priming affects the moisture content (Suleman et al. 2011).

14.3.8 Periwinkle

Seed priming (biopriming) increases antioxidant activity and seedling vigour in seeds of periwinkle (*Catharanthus roseus*). Seeds were treated with diazotrophs, i.e. *Azospirillum* and *Azotobacter* as separate treatments for 30 min. The germination percentage was calculated from 8 days after sowing (DAS) to 12 DAS. There was a significant increase in germination rate and non-significant in dry matter content. There was a significant increase in SOD, POX and CAT activities under *Azotobacter* and *Azospirillum* treatments and also an increase in the germination percentage, root length, shoot length and vigour index of the *C. roseus* (Karthikey et al. 2007).

14.3.9 Pot Marigold

Calendula officinalis, commonly known as pot marigold, was positively affected by seed priming method that influences the germination percentage and quality of seeds. Hydro- and hormonal priming give paramount results in seeds of pot marigold. Water or distilled water is used as hydrating agent in hydropriming in which seeds are needed to be soaked for few hours. As a result, sufficient water gets imbibed into the cell wall of marigold seeds that enhances the germination and further growth. In hormonal priming, GA_3 results better among all the growth regulators in relation to germination rate, flowering and total sugars (Karimi and Varyani 2016), but increase in concentration adversely affects the germination rate. However, GA_3 in addition to KNO_3 results profuse seedling establishment, shooting, rooting, maximum root length and vigour index in seeds of pot marigold. It was also noticed that the catalase activity increased significantly with GA_3 application, while enzyme activity was higher in the distilled water and KNO_3 treatments compared with the untreated seeds.

14.3.10 Rudbeckia

Rudbeckia fulgida, also known as black-eyed Susan, gloriosa daisy and orange coneflower, is a herbaceous perennial plant. Seed germination in this plant is variable with variable species. Osmopriming is a process of controlled imbibition by the seeds using different concentrations of osmotic solution containing polyethylene glycol (PEG). Accumulation of solutes and enzymatic activation during controlled seed imbibition (Bewley and Black 1978) contributed to the increased radicle emergence rate in primed seeds (Fay et al. 1994).

14.3.11 Safflower

In safflower (*Carthamus tinctorius*), hydropriming has an immense impact in increasing number of plants/m^2, capitula/plant, grains/capitulum, etc. In a field experiment, hydropriming of safflower (*Carthamus tinctorius*) seed for 12 h resulted in higher number of plants/m^2, capitula/plant, grains/capitulum, 1000 seed weight, grain yield and oil content compared to untreated seed (Bastia et al. 1999).

14.3.12 Salvia

The germination of *Salvia officinalis* L. (sage) seeds is a problem of great concern that may be overcome by employing the seed priming techniques. An effect of hydropriming positively affects the germination and seedling growth in sage (*Salvia officinalis*). Priming helps in increasing final germination percentage, improving germination rate, accelerating the synchronised seed germination, vigorous seedling establishment and stimulating vegetative growth and crop yield of many crops. Significantly, higher germination percentage and rate of germination observed in hydro-primed seeds as compared to non-primed seeds indicated a positive effect of seed priming in synchronising the seed germination process (Dastanpoor et al. 2013).

14.3.13 Snapdragon

Seed priming affects the germination growth and flowering of many flowers among which snapdragon (*Antirrhinum majus*) is one of them. There is an imperative need to work on priming techniques of these flower seeds because of the higher price and, moreover, the seeds of F_1 hybrids are difficult to germinate. The effect of biopriming influenced the germination character, shooting, rooting and flowering of snapdragon using *T. harzianum* and *B. subtilis* as bioprimers (Bhargava et al. 2015). Hormonal priming also has positive effect on biochemical changes in seeds that improves membrane integrity and metabolism of seed cell wall as well as synthesis of proteins (globulins and cruciferin) in comparison to non-primed seeds (Varier et al. 2010; Rao et al. 2009).

14.3.14 Sunflower

Biopriming in sunflower gave the utmost result for controlling fungal disease called Alternaria blight, the most important disease, and estimates yield loss up to 80%. Integration of chemicals, plant extracts and biotic agents along with priming agents for managing plant diseases has been considered as a novel approach, as it requires low amounts of chemicals, reducing the cost of control and pollution hazards while causing minimum interference with biological equilibrium (Papavizas 1973). Biopriming of sunflower seeds with *P. fluorescens* in the form of jelly can be used as an alternative method to seed treatment with chemicals (hexaconazole as foliar spray) which is eco-friendly and avoids possible residue problems.

14.3.15 Zinnia

It was found that *Alternaria zinniae* seems the most important fungal seed-borne pathogen of zinnia plants (*Zinnia elegans*), causing spotting of the petals, foliage and stems and rotting of the roots (Dimock and Osborn 1943; Richardson 1990; Łacicowa et al. 1991; Palacios et al. 1991; Wu and Yang, 1992). Łacicowa et al. (1991) reported that zinnia seeds produced in Poland were commonly infested with *A. alternata*, *A. zinniae*, *Botrytis cinerea*, *Fusarium* spp. and *Penicillium* spp. The influence of osmopriming on germination, vigour and location of pathogenic and saprotrophic fungi in zinnia seeds was studied in different varieties of zinnia (Szopinska and Tylkowska 2003) at 20 °C at 45% for 24 h that were sterilised with 1% NaOCl for 10 min at 20 °C at darkness. Disinfection of primed seeds lowered the germination capacity and increased the number of deformed seedlings in variety Jowita, Red man and Talia seeds. Seed priming considerably affected the speed of germination, regardless of sodium hypochlorite (NaOCl) treatment.

14.4 Seed Priming in Trees

Most of the tree species face the problem of extinction due to over-exploitation (Onochie 1990). Seeds of forest trees have problematic seed germination due to adverse soil and environment condition. Various salt contents in soil impaired germination of these seeds when available in excess amount, i.e. when the salt concentration increased up to 0.1%. Various salts like sodium, calcium and magnesium are the most common that contribute to salinity. High levels of fertilisation also contribute to salt accumulation and can be significant in agricultural situations (Treshow 1970). Sometimes, the lack of availability of moisture in seeds affects germination process adversely. Seed priming is an alternative way to avoid this type of problem. Some examples of seed priming of forest trees are mentioned below.

14.4.1 Baobab Tree or Senegal Tree

Baobab tree or senegal tree is scientifically known as *Adansonia digitata*, a deciduous tree. Most of the seeds of *Adansonia* fail to germinate as their propagation is adversely affected by seed coat dormancy which leads to poor growth potential. Sometimes the seeds are unable to germinate in natural condition. Osmopriming and thermopriming help in overcoming the dormancy problem in seeds of *Adansonia*. Hot water, cold water and H_2SO_4 are used for osmopriming and useful in enhancing germination percentage (70%) (Falemara et al. 2013). Different concentration of H_2SO_4 is used with variable time periods. Seeds' emergence is very fast (10 days) when treated with cold water.

14.4.2 Cassia

Cassias are ornamental plants of great beauty including more than 1000 species. Seed production is irregular with low germination percentage in natural conditions due to the integument impermeability. The vigour of stored seeds can be increased with priming, with a decrease in costs and reduction in number of collections (Vertucci 1989). Osmopriming by taking PEG with different water potentials is a helpful method for increasing germination tendency in seeds of cassia. Lowering in osmotic potential results in delaying in radicle emergence (Tarquis and Bradford 1992). Osmo-primed seeds are immediately used without drying or dehydrating as it enhances the germination ability of seeds (Heydecker and Wainwright 1976).

14.4.3 Cedar

Cedrus libani, called as Taurus cedar, is now an extinct species and difficult to grow in a large population. The temperature had the strong effect on the germination of *C. libani* seeds (Bewley and Black 1994; Schmidt 2000). The seeds of this tree exhibit seed dormancy that may be corrected by different pretreatment methods or priming methods. The seeds exhibit better germination performance at lower constant temperatures. Rise in germination temperature may develop secondary dormancy in seeds (Khan and Samimy 1982).

14.4.4 Cypress

Arizona cypress (*Cupressus arizonica*) and medite cypress (*Cupressus sempervirens*) are very important forest tree species for multiple purposes in forestry because of their ability to grow in adverse environments such as calcareous, clayish, dry and poor soils (Gallis et al. 2007). Seeds can germinate even in drought condition by using PEG-8000 as osmopriming agent. The germination percentage decreased in seeds of Arizona cypress with decreased water potential.

14.4.5 Eucalyptus

Eucalyptus has about 700 species and is commercially used for timber and pulpwood for paper and other purposes. Poor germination and seedling establishment are regarded as common problem in eucalyptus which can be corrected by pre-germination treatments. Seed size is the most important factor that affects germination in eucalyptus seeds. Spring-germinated seedlings have long growing period and attain maximum size with advanced establishment. Thus, larger seeds are more likely to survive in winter frosts as their susceptible growing tips are better adapted to cold-induced photoinhibition due to more advanced foliar pigment development (Close et al. 2000).

14.4.6 Fig

Seeds are the important planting material for propagation in *Ficus* spp. through which genetically different plants can be developed. It is also easily propagated through rooting and cuttings. Osmopriming, hydropriming and hormonal priming are useful treatment in early establishment of seedlings in *Ficus*. KNO_3, GA_3 and water are used for priming purpose. Early germination due to the priming effect of GA_3 results into longest radicle, which helps in early establishment of new seedling to produce maximum food material with the help of photosynthesis that resulted into the maximum survival of seedlings (Rawat et al. 2010)

14.4.7 Gamhar or White Teak

Gmelina arborea (gamhar) is a fast-growing deciduous tree used for furniture, carriages, sports and musical instruments as its steady timber is moderately resistant to termites. Seeds of *Gmelina arborea* (white teak or gamhar) don't have any dormancy and can easily germinate. Hydropriming in seeds may influence the germination tendency and increase the seedling emergence. Also $KMnO_4$ (0.2 M) treatment increases the germination percentage when seeds are treated with different interval of time duration.

14.4.8 Gulmohar

Delonix regia also known as gulmohar exhibits seed dormancy. For breaking the seed dormancy in seeds of *Delonix*, hormonal priming is efficiently useful by using different plant hormones, i.e. ABA, BAP, GA and IBA. Hot water treatment is also useful that tends to give the germination percentage of 95% in seeds of gulmohar. ABA does not have significant effect in breaking the dormancy. Gibberellic acid will promote the endosperm breakdown and the growth of the embryo that results in the elongation of radicle cells and cause a rupture in the micropyle, and the seed dormancy will be terminated by radicle protrusion from the seed coat. The role of temperature in breaking the dormancy to influence germination can be noticed due to the erratic nature of the germination which terms for the standardisation of both the temperature of the water and duration of cooling as suggested by Owonubi and Otegbeye (2004).

14.4.9 Henkel's Yellowwood

This ornamental tree is dioecious in nature and is used for ornamental purpose. Generally, this henkel's yellowwood tree (*Podocarpus henkelii*) is propagated through seeds. Fleshy fruit that surrounds the seed is removed as this inhibits the germination. Removal of epicuticular wax, the epidermis or the entire epimatium leads to rapid water uptake and germination (Dodd et al. 1989). Seeds stored in shade have more moisture content and reserve more starch through biochemical process. Lipids are the major reserve materials in seeds of *Podocarpus henkelii* followed by proteins, and these embryonic reserves are sufficient for early seedling establishment (Nabanyumya et al. 2015). Soaking seeds more than once a day also breaks dormancy in seeds, resulting in maximum germination percentage.

14.4.10 Whistling Pine

The whistling pine (*Casuarina equisetifolia*) is used as a windbreak in agroforestry system and also for many household uses. The wood is resistant to decomposition in soil or saltwater and is often used as roundwood for making piles, poles and fences. This is accentuated as the successful propagation of any *Casuarina spp.* by a vegetative method is very limited even with IBA hormone application (Mahmood and Possuswam 1980; Pinyopusarerk and Bolan 1990). Different concentrations of growth hormones like GA_3 and ABA, acid substances like sulphuric acid and chemical like $NaNO_3$ are helpful in promoting germination and growth in this tree. The waxy and hard seed coat in seeds of *Casuarina* prevents it to germinate which has been broken with the use of concentrated H_2SO_4 by several workers (Mayer and Poljakoff-Mayber 1963; Ballard 1973; Gill and Bamidele 1981; Etejere et al. 1982; Eze and Orole 1987).

14.4.11 Ivory Coast Almond

Terminalia ivorensis, a tropical deciduous tree, thrives in a wide range of soil and is used as multipurpose trees along with having medicinal uses. Seed germination of tropical species is influenced by several biological factors such as seed viability, seed size (Barrera and Nobel 2003) and plant growth regulators through which acts as germination stimulator (Chen et al. 2008). Seeds are very sensitive to temperature stress as it influences the germination process, and high temperature prevents the elongation of radicle and shoot by inhibiting the synthesis of protein and nucleic acid (Sivaramakrishnan et al. 1990). Soaking seeds of *T. Chebula* (Horitaki) enhances germination rate and vigour index (Hossain et al. 2005). Gibberellic acid can overcome the dormancy in seeds of *Terminalia* spp. by inducing rapid and uniform germination due to deficit in endogenous gibberellic acids (Asomaning et al. 2011). The exogenously applied gibberellic acids help in modifying the influence of cytokinins on transport across membranes and are thus able to initiate the biochemical processes necessary for germination process (Chen et al. 2008).

14.4.12 Kachnar

Bauhinia spp. (mountain ebony) are flowering plants and propagated through seeds and cuttings. This tree has been prioritised as one of the trees for conservation to enhance its contribution to health and livelihood of communities. Various methods are there to break the dormancy in the seeds of this tree that promotes its cultivation and successful regeneration. Mainly germination in kachnar depends on the size of the seeds as larger seeds germinate more percentage than smaller one. Soaking in boiled water makes the seed coats permeable to water and water gets imbibed into the cell wall which results in swelling of the seed as water cools. Recommendation of hot water must be applied judiciously without killing the seeds with excessive heating (Phartyal et al. 2005). Chemical stimulators like KNO_3 enhance seed germination by creating a balance between hormonal ratios in the seed and reducing the growth retarding substances like abscisic acid (ABA).

14.4.13 Karanj (Indian Beech)

The use of biologically active products like biomanures and biofertilisers for the production of quality planting material helps in preventing microbial inoculation of bacteria, algae and fungi in karanj (Revathi et al. 2013). Biopriming with liquid *Azospirillum* and *Phosphobacterium* in different concentrations improves germination, seedling vigour and storability of *Pongamia pinnata* seeds by reducing microbial infections. An increase in the seed germination might be due to the increased cytokinin production which actively involved in cell division (Suma et al. 2014) and production of growth-regulating substances like auxin, GA and cytokinin (Kucey 1988) when seeds are encapsulated with biofertilisers.

14.4.14 Khejri

Prosopis cineraria (khejri) is a flowering tree of which seeds sometimes show dormancy effect which can be lessened by priming methods. Fulvic acid that is extracted from compost is beneficial in enhancing germination tendency in seeds of this tree. Fulvic acids are beneficial to increase the permeability of seed coat and plant cell membranes and enhance enzymatic activity of the root system leading to increased root proliferation (Trevisan et al. 2010). Application of $CaCO_3$ can cause buffering capacity in soil that affects nutrient availability to plants, while sulfuric acid treatment has conventionally been used for breaking the dormancy and softens the seed coat through which the radicle can easily protrude.

14.4.15 Oak

Himalayan oak trees (*Quercus glauca*) have irregular fructification, and consumption of seeds by animals (Troup 1921) and loss of viability during storage for extended periods (Chalupa 1995) have overrated the problem of regeneration (artificial and natural). Propagation (clonal propagation) through stem cuttings is difficult in most of the oak species and has not been much successful in these species. Even the seeds' weight is also affected by germination percentage. Seed coat acts as a mechanical barrier in germination of oak seeds which prevents radicle emergence and is corrected by several scarification methods and priming treatments (Alderete-Chavez et al. 2011). The KNO_3 plays an osmotic role on water uptake with a nutritional effect on protein synthesis. KNO_3 is used for growth regeneration and also as germination-stimulating substance in many species (Ozturk et al. 1994).

14.4.16 Pine

It is the most widely distributed tree used for timber purposes and for making boxes, paper pulp and temporary electric poles. Seed germination in pine (*Pinus kesiya*) is a common problem which depends on the moisture level and substrate pH level in seeds (Verma and Tondon 2010). Hydropriming increases the imbibition rate and moisture level in seeds which enhances the germination rate. Available soil moisture also influences germination and early seedling survival. Influence in germination in pine is due to leaching of growth inhibitors and increase in endogenous level of growth regulators when seeds scarified or subjected to chilling treatment.

14.4.17 Siris

Albizia lebbeck (siris) has special place among all the forest crops. Despite its importance, the species is becoming scarce due to deep seed dormancy. Seed priming is a method used for producing numerous number of planting material by overcoming its seed coat-imposed dormancy (Baskin and Baskin 1998). Seeds are often used for propagation in Siris which is a cheapest method of propagation in many species of this tree. Its hard seed coat makes the germination process difficult in nursery and/or out in the fields. Sulphuric acid, hot water and gibberellic acid are used for breaking dormancy in this species (Alderete-Chavez et al. 2011; Baskin and Baskin 1974). The disintegration of the seed coat increases the imbibition and subsequent germination in seeds when treated with sulphuric acid (Egley 1989).

14.4.18 Kapok

The silk cotton tree (kapok), *Ceiba pentandra*, is a fast-growing and emergent tropical forest tree species which can grow up to a height of 60 m (Gribel et al. 1999). It is one of the extinct species among forest trees, and seeds of this tree are also limitedly available. It is pollinated by hummingbirds (Gribel et al. 1999; Toledo 1977) and non-flying primate mammals, particularly by *Saimiris ciureus* (squirrel monkey), *Cebus paella* (capuchin) and *Ateles paniscus* (spider monkey) (Janson et al. 1981). Seed soaking in seeds of silk cotton tree significantly influences the germination process. But heavy rainfall may limit the process of germination and fails to break the seed dormancy.

14.4.19 Subabul

This subabul tree (*Leucaena leucocephala*) is also called 'miracle tree' due to its multipurpose nature (Ssenku et al. 2017). Germination capacity in seeds of subabul can increase by different priming or pretreatment methods. NaCl concentration in soil shows a marked effect on the germination. With the increase in salinity, the germination initiation in this species is delayed and gets ceased (Rafiq et al. 2006). The reason behind this is that the osmotic pressure of the soil solution increases with increase in salt concentration that results in the prevention of uptake of water. It also reduces the germination energy of seeds.

14.4.20 Tamarind

Tamarind (*Tamarindus indica*) has been recognised for its potential nitrogen-fixing nature (Okoro et al. 1986). Seeds in tamarind do not have self-capacity to germinate due to the lack of the factors required for breaking dormancy. Hot water treatment and sulphuric acid affect the germination rate, and seed germination increased with increasing water temperature and soaking period.

14.4.21 Teak

Teak (*Tectona grandis*) is a tropical tree species, hardy and deciduous in nature, is valued for its durability and water resistance and is used for boat building, exterior construction, veneer, furniture, carving and other small wood decorative uses. Germination of the seeds involves pretreatment to remove dormancy arising from the thick pericarp. Pretreatment involves alternate drying and wetting of the seeds. Saline solution also affect the germination ability of seeds in teak, as increased salinity results in decreased germination ability of seeds and delayed rate of germination. This is due to the complex nature of salts that present in the solution for germination which may be sometimes having toxic effect in germination.

14.4.22 Wigandia

Wigandia caracasana, or *Caracus wigandia*, is a species of evergreen ornamental plant having purple flowers in large clusters. Sometimes, seed burial can improve the chances of establishment of seedlings through breaking the dormancy. Priming consists of a regulated hydration process that permits enhancement of some metabolic processes through the osmotic solutions or water. Burial enhanced germination process and favoured uniform and rapid germination of *W. urens* seeds (Gonzalez-Zertuche et al. 2001). This enhancement in germination is due to the increased protein synthesis in burial seeds as compared to control seeds. Burial with priming treatment in seeds increases the germination percentage, seedling growth and breaking dormancy than the priming treatment alone.

14.4.23 Willow Tree

Willow tree (*Salix babylonica*), also called sallows, is a deciduous tree as well shrub. Seeds are viable for a short period of time. Priming methods are useful in enhancing germination and growth of dormant seeds. Thermopriming gives efficient result in germination up to 100%. Seeds of willow tree results in better germination rate in lower temperature regimes (Young and Clements 2003).

14.5 Conclusion

There is a huge gap between the number of seeds sown and availability of stocky seedlings in any ornamental flower crops and several important tree species. The most important problems faced are the heterogneity and lack of suitable conditions in soil that cause decrease in germination percentage. Heterogeneous emergence, unbalanced seedling growth and competition for environmental resources such as light, nutrients and water subsequently make difference in biomass and performance of plants. Priming is helpful in reducing the risk of poor stand establishment under a wide range of environmental conditions. The purpose of these treatments is to shorten the emergence and to protect seed from biotic and abiotic factors during critical phase of seedling establishment so as to synchronise emergence, which lead to uniform stand and improved yield.

References

Afzal I, Aslam N, Mahmood F, Hameed A, Irfan S, Ahmad G (2004) Enhancement of germination and emergence of canola seeds by different priming techniques. Caderno Pesquisasérie Biol 16(1):19–34

Afzal I, Basra SM, Farooq M, Nawaz A (2006) Alleviation of salinity stress in spring wheat by hormonal priming with ABA, salicylic acid and ascorbic acid. Int J Agric Biol 8(1):23–28

Alderete-Chavez A, Cruz-Landero NDL, Guerra-Santos JJ, Guevara E, Gelabert R (2011) Promotion of germination of *Bauhinia divaricata* L. seeds by effects of chemical scarification. Res J Seed Sci 4:51–57

Ashraf M, Foolad MR (2005) Pre-sowing seed treatment-A shotgun approach to improve germination, plant growth, and crop yield under saline and non-saline conditions. Adv Agron 88:223–271

Asomaning JM, Sacandle M, Olympus NS (2011) Germination response of *Terminalia superba* Enl. And diels seed on the 2-way grant's thermogradient plate. Res J Seed Sci 4:28–39

Atici O, Agar G, Battal P (2003) Interaction between endogenous plant hormones and alphamylase in germinating chickpeas seeds under cadmium exposure. Fresenius Environ Bull 12:781–785

Ballard LA (1973) Physical barriers to germination. Seed Sci Technol 1: 285–295.

Baskin JM, Baskin CC (1974) Breaking dormancy in seeds of *Isanthus branchiathus* (Labiatae) with gibberellic acid. Phyton 32:159–165

Baskin JM, Baskin CC (1998) Ecology, biogeography and evolution of dormancy and germination. Academic, San Diego, p 666

Bastia DK, Rout AK, Mohanty SK, Prusty AM (1999) Effect of sowing date sowing methods and seed soaking on yield and oil content of rainfed safflower grown in Kalahandi. Orissa Indian J Agron 44:621–623

Beckers GJ, Conrath U (2007) Priming for stress resistance: from the lab to the field. Curr Opin Plant Biol 10(4):425–431

Bewley JD, Black M (1978) Mobilization of reserves. In: Physiology and Biochemistry of seeds in relation to germination. Springer, Berlin/Heidelberg, pp 177–244

Bewley JD, Black M (1994) Seeds: physiology of development and germination, 2nd edn. Plenum Press, New York

Bhargava B, Gupta YC, Dhiman SR, Sharma P (2015) Effect of seed priming on germination, growth and flowering of snapdragon (*Antirrhinum majus* L.). Natl Acad Sci Lett 38(1):81–85

Bray CM (1995) Biochemical processes during the osmopriming of seeds. Seed development and germination. Marcel Dekker, New York, pp 767–789

Callan NW, Mathre DE, Miller JB (1990) Bio-priming seed treatment for biological control of *Pythium ultimum* pre-emergence damping-off in *sh2* sweet corn. Plant Dis 74:368–372

Chalupa V (1995) Somatic embryogenesis in oak (Quercus spp.). In: Jain SM, Gupta PK, Newton RJ (eds) Somatic embryogenesis in woody plants, vol 2. KulwerAcademic Publishers, Dordrecht, pp 67–87

Chen SY, Kuo SR, Chien CT (2008) Roles of gibberellins and abscisic acid in dormancy and germination of red bayberry (*Myrica rubra*) seeds. Tree Physiol 28(9):1431–1439

Close DC, Beadle CL, Brown PH, Holz GK (2000) Cold-induced photoinhibition affects establishment of *Eucalyptus nitens* (Deane and Maiden) Maiden and *Eucalyptus globulus* Labill. Trees-Struct Funct 15(1):32–41

Conrath U, Beckers GJM, Flors V, Garcia-Agustin P, Jakab G (2006) Priming: getting ready for battle. Mol Plant-Microbe Interact 19:1062–1071

Dastanpoor N, Fahimi H, Shariati M, Davazdahemami S, Hashemi SMM (2013) Effects of hydropriming on seed germination and seedling growth in sage (*Salvia officinalis* L.). Afr J Biotechnol 12(11):1223–1228

De Chandra G (1999) Fundamentals of agronomy. Oxford and IBH Publishing Company, New-Delhi

De la Barrera E, Nobel PS (2003) Physiological ecology of seed germination for the columnar cactus *Stenocereus queretaroensis*. J Arid Environ 53(3):297–306
Di Girolamo G, Barbanti L (2012) Treatment conditions and biochemical processes influencing seed priming effectiveness. Ital J Agron 7(2):25
Dimock AW, Osborn JH (1943) An Alternaria disease of zinnia. Phytopathology 33:372–381
Dodd MC, Staden JV, Smith MT (1989) Seed germination in *Podocarpus henkelii*: an ultrastructural and biochemical study. Ann Bot 64(5):569–579
Dorna H, Li W, Szopinska D (2014) The effect of priming on germination and vigour of pansy (Viola× Wittrockiana Gams.) seeds. Acta Sci Pol Hortoru 13(6):15–29
Edwards DGW (1981) A new prechilling method for true fir seeds. Proceeding: Combined Meetg. Intermountain Nurseryman's Assoc. and Western For. Nurs. Assoc., Idaho. USDA 109:58–66
Edwards DGW (1996) The stratification-re-dry technique with special reference to true fir seeds. In: Landis TD, D.B. South, Technical Coordinators (eds) National proceedings, forest conservation nursery associations, General Technical Report PNW-GTR-389. U.S.D.A. Forest Service, Pacific Northwest Research Station, Portland, pp 172–182
Egley GH (1989) Water-impermeable seed coverings as barriers to germination. In: Recent advances in the development and germination of seeds. Springer, Boston, pp 207–223
Etejere EO, Fawole MO, Sani A (1982) Studies on the seed germination of *Parkia clappertoniana*. Turrialba 32(2):181–185
Eze JMO, Orole BC (1987) Germination of the seeds of *Prosopis africana*. Niger J For 17:12–16
Falemara BC, Nwadike C, Obashola EO (2013) Germination response of baobab seeds (*Adansonia digitata* L.) as influenced by three pretreatment techniques. In: Forest industry in a dynamic global environment. Proceeding: 35th annual conference of Forestry Association of Nigeria, Sokoto, Sokoto state, pp 44–55
Farooq M, Basra SMA, Hafeez K (2006a) Seed invigoration by osmohardening in coarse and fi ne rice (*Oryza sativa* L.). Seed Sci Technol 34:181–187
Farooq M, Basra SMA, Wahid A (2006b) Priming of field sown rice seed germination, seedling establishment, allometry and yield. Plant Growth Regul 49:285–294
Farooq M, Basra SMA, Khan MB (2007a) Seed priming improves growth of nursery seedlings and yield of transplanted rice. Arch Agron Soil Sci 53:1–12
Farooq M, Basra SMA, Wahid A (2007b) Improving the performance of transplanted rice by seed priming. Plant Growth Regul 51:129–137
Fay AM, Bennett MA, Still SM (1994) Osmotic seed priming of *Rudbeckia fulgida* improves germination and expands germination range. Hortscience 29(8):868–870
Felippe GM (1980) Germination of the light-sensitive seeds of and : effects of temperature. New Phytol 84(3):439–448
Fletcher RA, Gilley A (2000) Triazoles as plant growth regulators and stress protectants. Hortic Rev 24:55–138
Franklin KA (2009) Light and temperature signal crosstalk in plant development. Curr Opin Plant Biol 12(1):63–68
Gallis AT, Doulis AG, Papageorgiou AC (2007) Variability of cortex terpene composition in *Cupressus sempervirens* L. provenances grown in Crete, Greece. Silvae Genet 56(6):294–299
Garner JM, Armitage AM (2008) Cooling and long-day lighting influences growth and flowering of *Phlox paniculata* L. 'Ice Cap' used for cut flowers. Hortscience 43(3):707–709
Gill LS, Bamidele JF (1981) Seed morphology, germination and cytology of three savanna trees of Nigeria. Niger J For 2:16–23
Gonzalez-Zertuche L, Vazquez-Yanes C, Gamboa A, Sanchez-Coronado ME, Aguilera P, Orozco-Segovia A (2001) Natural priming of *Wigandia urens* seeds during burial: effects on germination, growth and protein expression. Seed Sci Res 11(1):27–34
Gratao Pl, Polle A, Lea PJ, Azevedo R (2005) Making the life of heavy metal-stressed plants a little easier. Funct Plant Biol 32:481–494
Gribel R, Gibbs PE, Queiróz AL (1999) Flowering phenology and pollination biology of Ceiba pentandra (Bombacaceae) in Central Amazonia. J Trop Ecol 15(3):247–263

Harris D, Pathan AK, Gothkar P, Joshi A, Chivasa W, Nyamudeza P (2001) On-farm seed priming: using participatory methods to revive and refine a key technology. Agric Syst 69(1):151–164
Hasegawa S (2016) What is seed priming?, Germains Seed Technology. https://germains.com/what-is-seed-priming
Heydecker W, Wainwright H (1976) More rapid and uniform germination of *Cyclamen persicum* L. Sci Hortic 5(2):183–189
Heydecker W, Coolbear P (1977) Seed treatments for improved performance survey and attempted prognosis. Seed Sci Technol 5:353–425
Hossain MA, Arefin MK, Khan BM, Rahman MA (2005) Effects of seed treatments on germination and seedling growth attributes of Horitaki (*Terminalia chebula* Retz.) in the nursery. Res J Agric Biol Sci 1(2):135–141
Hu YF, Zhou G, Na XF, Yang L, Nan WB, Zhang Y, Li J, Bi Y (2013) Cadmium interferes with maintenance of auxin homeostasis in Arabidopsis seedlings. J Plant Physiol 170:965–975
Huang YM, Wang HH, Chen KH (2002) Application of seed priming treatments in spinach (*Spinacia oleracea* L.) production. J Chin Soc Hort Sci 48:117–123
Janson CH, Terborgh J, Emmons LH (1981) Non-Flying Mammals as Pollinating Agents in the Amazonian Forest. Biotropica 13(2):1
Jie LIU, Gong she L, Dong mei Q, Fang fang L, Enhua W (2002) Effect of PEG on germination and active oxygen metabolism in wildrye (*Leymuschinensis*) seeds. Acta Prataculturae Sin 11(1):59–64
Jisha K, Vijayakumari K, Puthur J (2013) Seed priming for abiotic stress tolerance: an overview. Acta Physiol Plant 35:1381–1396
Karimi M, Varyani M (2016) Role of priming technique in germination parameters of calendula (*Calendula officinalis* L.) seeds. J Agric Sci 61(3):215–226
Karthikeyan B, Jaleel CA, Gopi R, Deiveekasundaram M (2007) Science letters: alterations in seedling vigour and antioxidant enzyme activities in *Catharanthus roseus* under seed priming with native diazotrophs. J Zhejiang Sci 8(7):453
Khalil MAK, Rasmussen RA (1983) Sources, sinks and seasonal cycles of atmospheric methane. J Geophys Res Oceans 88(9):5131–5144
Khan AA, Samimy EC (1982) Hormones in relation to primary and secondary seed dormancy. In: Khan AA (ed) Physiology and biochemistry of seed development, dormancy and germination. Elsevier, Amsterdam, pp 203–241
Khan HA, Ayub CM, Pervez MA, Bilal RM, Shahid MA, Ziaf K (2009) Effect of seed priming with NaCl on salinity tolerance of hot pepper (*Capsicum annuum* L.) at seedling stage. Soil Environ 28(1):81–87
Kolotelo D (1998) *Abies* seeds problems. Proceedings: 1995, 1996, 1997 Forest Nursery Association of British Columbia Meetings, B.C. Ministry of Forests, Surrey, B.C., Canada., pp 122–130
Kucey RMN (1988) Alteration of size of wheat root systems and nitrogen fixing bacteria measured under field conditions. Can J Microbiol 34:735–739
Lacicowa B, Kiecana I, Pięta D (1991) Mikroflora materiału siewnego roślin ozdobnych. I. Mikroflora materiału siewnego cynii (*Zinnia elegans* L.)i groszku pachnącego (*Lathyrus odoratus* L.). Inst Sadown Kwiac Ser B 16:109–116
Leadem CL (1986) Stratification of *Abies amabilis* seeds. Can J For Res 16:755–760
Liu Y, Hilhorst HW, Groot SP, Bino RJ (1997) Amounts of nuclear DNA and internal morphology of gibberellin-and abscisic acid-deficient tomato (*Lycopersicon esculentum* Mill.) seeds during maturation, imbibition and germination. Ann Bot 79(2):161–168
Liu Y, Kermode A, El-Kashaby YO (2013) The role of moist-chilling and thermo-priming on the germination characteristics of white spruce (*Picea glauca*) seed. Seed Sci Tech 41:1–15
Ma Y, Feurtado JA, Kermode AR (2003) Effect of solid matrix priming during moist chilling on dormancy breakage and germination of seeds of four fir species. New For 25(1):67–81
Mahmood AM, Possuswam PK (1980) Propagation of *Casuarina junghuhniana* by planting shoots and root suckers. Indian For 106(4):298–299
Manonmani V, Begum MAJ, Jayanthi M (2014) Halo priming of seeds. Res J Seed Sci 7(1):1–13

Markovskaya EF, Sysoeva MI, Sherudilo EG, Topchieva LV (2007) Differential gene expression in cucumber plants in response to brief daily cold treatments. Russ J Plant Physiol 54(5):607–611

Masood A, Iqbal N, Khan N (2012) Role of ethylene in alleviation of cadmium-induced capacity inhibition by sulphur in mustard. Plant Cell Environ 35:524–533

Mayer AM, Poljakoff-Mayber A (1963) The germination of seeds. Pergamon Press, London, pp 31–94

McDonald MB (1999) Seed deterioration: physiology, repair and assessment. Seed Sci Technol 27(1):177–237

McDonald MB (2000) Seed priming. In: Black M, Bewley JD (eds) Seed technology and its biological basis. Sheffield Academic Press, Sheffield, pp 287–325

Mushtaq S, Hafiz IA, Hasan SZU, Arif M, Shehzad MA, Rafique R, Iqbal MS (2012) Evaluation of seed priming on germination of *Gladiolus alatus*. Afr J Biotechnol 11(52):11520–11523

Mwale SS, Hamusimbi C, Mwansa K (2003) Germination, emergence and growth of sunflower (*Helianthus annuus* L.) in response to osmotic seed priming. Seed Sci Technol 31(1):199–206

Nabanyumya R, Obua J, Tumwebaze SB (2015) Germination of *Afrocarpus usambarensis* and *Podocarpus milanjianus* seeds in Sango Bay, Uganda. Uganda J Agric Sci 16(2):231–244

Nancy W, Don ME, James B, Charles S (1997) Biological seed treatments: factors involved in efficacy. Hortic Sci 32:179–183

Nascimento WM, Cantliffe DJ, Huber DJ (2004) Ethylene evolution and endo-beta-mannanase activity during lettuceseed germination at high temperature. Sci Agric 61(2):156–163

Nascimento AR, Mouchrek Filho JE, Mouchrek Filho VE, Martins AG, Marinho SC, Serra CLM, Alves LMC (2005) Avaliação da sensibilidade de antimicrobianos a cepas de enterobacteriaceae isoladas de amostras de alface (*Lactuca sativa*) comercializada na cidade de *São Luís-MA*. Bol Cent Pesq Processamento Aliment 23(2):265–272

Nleya T, Ball RA, Vandenberg A (2005) Germination of common bean under constant and alternating cool temperatures. Can J Plant Sci 85(3):577–585

Okoro, S.P.A., Awodola, A.M. and Itolo, G.O. (1986) The impact of selected tree species on the soil properties in a Sudan Savanna Forest. In: Proceedings of 16th Annual Conference of Forestry Association of Nigeria, pp 660–667

Onochie CFA (1990) The dying dorest, declining forest resources of Nigeria. In: Keynote address 4th annual conferences of the Botanical Society of Nigeria. University of Nigeria, Nsukka

Osburn RM, Schroth MN (1989) Effect of osmopriming sugar beet seed on germination rate and incidence of *Pythium ultimum* damping-off. Plant Dis 73(1):21–24

Owonubi JJ, Otegbeye GO (2004) Disappearing forests: a review of the challenges for conservation of genetic resources and environmental management. J Res Manag 1:1–11

Ozturk M, Gemici M, Ozdemir F, Keyikci N (1994) Tohumçimlenmesiolayındabitkiselhor mon larınveçimlenmestimülatörününtuzstresiniazaltmadakirolü. UlusalBiyolojiKongresi, Edirne, pp 44–48

Palacios MG, Smits GB, Noguera R (1991) Presencia e influencia de algunos hongos patogenos en cultivos de Zinnia elegans Jacq. en la region central de Venezuela. Agron Trop 41(5-6):237–244

Paparella S, Araujo SS, Rossi G, Wijayasinghe M, Carbonera D, Balestrazzi A (2015) Seed priming: state of the art and new perspectives. Plant Cell Rep 34(8):1281–1293

Papavizas GC (1973) Status of applied biological control of soil-borne plant pathogens. Soil Biol Biochem 5(6):709–720

Parera CA, Cantliffe DJ (1994) Pre-sowing seed priming. Hort Rev 16:109–141

Phartyal SS, Baskin JM, Baskin CC, Thapliyal RC (2005) Physical dormancy in seeds of *Dodonaea viscosa* from India. Seed Sci Res 15:59–61

Pill WG (1995) Low water potential and presowing germination treatments to improve seed quality. In: Seed quality: basic mechanisms and agricultural implications. The Haworth Press, Binghamton, pp 319–359

Pill WG, Gunter JA (2001) Emergence and shoot growth of cosmos and marigold from paclobutrazol-treated seed. J Environ Hortic 19(1):11–14

Pill WG, Crossan CK, Frett JJ, Smith WG (1994) Matric and osmotic priming of *Echinacea purpurea* (L.) Moench seeds. Sci Hortic 59(1):37–44

Pinyopusarerk K, Boland DJ (1990) *Casuarina junghuhniana*: a highly adaptable tropical casuarina. NFT Highlights 90(4):1–2

Rafiq S, Iqbal T, Hameed A, Ali RZ, Rafiq N (2006) Morphobiochemical analysis of salinity stress response of wheat. Pak J Bot 38(5):1759–1767

Rao MSL, Kulkarni S, Lingaraju S, Nadaf HL (2009) Biopriming of seeds: potential tool in the integrated management of alternaria blight of sunflower. Helia 32(50):107–114

Rawat JMS, Tomar YK, Rawat V (2010) Effect of stratification on seed germination and seedling performance of wild pomegranate. J Am Sci 6(5):97–99

Revathi R, Mohan V, Jha MN (2013) Integrated nutrient management on the growth enhancement of *Dalbergia sissoo* Roxb. seedlings. J Acad Ind Res 1(9):550–557

Richardson MJ (1990) An annotated list of seed-borne diseases. International Seed Testing Association, ZĦrich

Rowse HR (1991) Methods of priming seeds. UK, Patent No. 2192781

Runkle ES, Heins RD, Cameron AC, Carlson WH (1999) Photoperiod and cold treatment regulate flowering of Rudbeckia fulgida Goldsturm. Hortscience 34(1):55–58

Schmidt L (2000) Guide to handling of tropical and subtropical forest seed. Danida Forest Seed Centre, Denmark, p 511

Sedghi M, Nemati A, Esmaielpour B (2010) Effect of seed priming on germination and seedling growth of two medicinal plants under salinity. Emirates J Food Agric 22(2):130

Shakarami B, Dianati TG, Tabari M, Behtari B (2011) The effect of priming treatments on salinity tolerance of *Festuca arundinacea* Schreb and *Festuca ovina* L. seeds during germination and early growth. Iran J Rangelands For Plant Breed Genet Res 18(2):318–328

Shin JS, Raymer P, Kim W (2006) Environmental factors influencing germination in seeded seashore paspalum. Hortscience 41(5):1330–1331

Singh AK, Karki K (2003) Effect of grading and GA_3 on germination and seedling growth attributes in balsam. Progress Hortic 35(2):158–160

Sivaramakrishnan S, Patel VZ, Soman P (1990) Heat shock proteins of sorghum (*Sorghum bicolor* (L.) Moench) and pearl millet (*Pennisetum glaucum* (L.) R. Br.) cultivars with differing heat tolerance at seedling establishment stage. J Exp Bot 41:249–254

Sivritepe HO (2000) The effects of osmotic conditioning treatments on salt tolerance of onion seeds. In: 3rd National symposium on vegetable production, Isparta, Turkey. pp 475–481

Sivritepe HO, Dourado AM (1995) The effect of priming treatments on the viability and accumulation of chromosomal damage in aged pea seeds. Ann Bot 75(2):165–171

Ssenku JE, Ntale M, Backeus I, Oryem-Origa H (2017) Phytoremediation potential of *Leucaena leucocephala* for heavy metal-polluted and heavy metal-degraded environments. In: Phytoremediation potential of bioenergy plants. Springer, Singapore, pp 189–209

Subedi KD, Ma BL (2005) Seed priming does not improve corn yield in a humid temperate environment. Agron J 97(1):211–218

Suleman MK, Bhatt NR, Jacob S, Thomas RR (2011) Germination studies in *Ochradenus baccatus* Delile, *Peganum harmala* L. and *Gynandriris sisyrinchium* Parl. Res J Seed Sci 4:58–63

Suma N, Srimathi P, Roopa VM (2014) Influence of biofertilizer pelleting on seed and seedling quality characteristics of *Sesamum indicum*. Int J Curr Microbiol App Sci 3(6):591–594

Szopinska D, Tylkowska K (2003) Effect of osmopriming on location of seed-borne fungi in lettuce (*Lactuca sativa*) seeds. Phytopathol Pol 29:69–80

Tanaka Y, Edwards DGW (1986) An improved and more versatile method for prechilling *Abies procera* Rehd. seeds. Seed Sci Technol 14:457–464

Tarquis AM, Bradford KJ (1992) Prehydration and priming treatments that advance germination also increase the rate of deterioration of lettuce seeds. J Exp Bot 43(3):307–317

Taylor AG, Klein DE, Whitlow TH (1988) SMP: solid matrix priming of seeds. Sci Hortic 37:1–11

Taylor AG, Allen PS, Bennet MA, Bradford KJ, Burris JS, Misra MK (1998) Seed enhancements. Seed Sci Res 8:245–256

Toledo VM (1977) Pollination of Some Rain Forest Plants by Non-Hovering Birds in Veracruz, Mexico. Biotropica 9(4):262

Toorop PE, Van AC, Hilhorst HWM (1998) Endosperm cap weakening and endo-beta-mannanase activity during priming of tomato (*Lycopersicon esculentum* cv. Moneymaker) seeds are initiated upon crossing a threshold water potential. Seed Sci Res 8:483–491

Treshow M (1970) Mineral toxicity. In: Environment and plant response. McGraw-Hill, New York, pp 222–236

Trevisan S, Francioso O, Quaggiotti S, Nardi S (2010) Humic substances biological activity at the plant-soil interface: from environmental aspects to molecular factors. Plant Signal Behav 5(6):635–643

Troup RS (1921) The silviculture of Indian trees, vol 3. Clarendon Press, Oxford, pp 913–923

Varier A, Vari AK, Dadlani M (2010) Subcellular basis of seed priming. Curr Sci 99(4):450–456

Verma AN, Tandon P (2010) Seed germination and seedling growth in *Pinus kesiya* royle ex gord. I. Influence of imbibition, substrate pH and moisture level. Proc Indian Nat Sci Acad 50(3):326–331

Vertucci CW (1989) The kinetics of seed imbibition: controlling factors and relevance to seedling vigor. Seed Moisture, pp 93–115

Wang BSP, Lin TP, Chang TT (1998) Control of fungal growth with sphagnum for cold. 13(2):101–108

Wang HY, Chen CL, Sung JM (2003) Both warm water soaking and matriconditioning treatments enhance anti-oxidation of bitter gourd seeds germinated at sub-optimal temperature. Seed Sci Technol 31(1):47–56

Warren JE, Bennett MA (1997) Seed hydration using the drum priming system. Hortic Sci 32:1220–1221

Wu WS, Yang YH (1992) Alternaria blight: a seed-transmitted disease of zinnia in Taiwan. Plant Pathol Bull 1:115–123

Young JA, Clements CD (2003) Seed germination of willow species from a desert riparian ecosystem. J Range Manag 56:496–500

Role of SNP-Mediated Nitric Oxide Priming in Conferring Low Temperature Tolerance in Wheat Genotype (*Triticum aestivum* L.): A Case Study in Indian Northern Plains

15

Md. Afjal Ahmad, Pravin Prakash, and H. B. Singh

Abstract

Wheat is an economically important cereal crop grown in the vast stretch along the northern, western and part of Central Indian plains serving as livelihood tool for more than half of agrarian population of the country. The crop being grown in rabi (mid-November–mid-April) faces multifaceted abiotic stress threats among which low temperature stress being one of them. Nitric oxide has been well documented to counter many of these threats as also low temperature stress. In this context, a study was conducted to evaluate the effect of nitric oxide priming on certain morphophysiological and biochemical parameters in wheat genotype HD-2329 facing low temperature regime in the laboratory of the Department of Plant Physiology, Institute of Agricultural Sciences, Banaras Hindu University. The several parameters under observations were per cent germination, germination rate, shoot length, root length, α-amylase activity and soluble sugar content both under normal (25 °C) and low temperature (5 °C) conditions. Seed priming with sodium nitroprusside (SNP, a nitric oxide donor) at 100 μM concentration resulted in enhanced performance of these parameters as compared to non-primed seeds both under normal and low temperature, but the effect was more pronounced at low temperature regime. Hence, it was concluded that NO priming of wheat seeds had a statistical significance in conferring low temperature tolerance in the crop, thus making it a fit priming model in a climate resilient era.

Keywords

NO priming · Low temperature stress · Wheat

M. A. Ahmad (✉) · P. Prakash
Department of Plant Physiology, Institute of Agricultural Sciences, BHU, Varanasi, Uttar Pradesh, India

H. B. Singh
Department of Mycology and Plant Pathology, Institute of Agricultural Sciences, BHU, Varanasi, Uttar Pradesh, India

© Springer Nature Singapore Pte Ltd. 2018

A. Rakshit, H. B. Singh (eds.), *Advances in Seed Priming*,
https://doi.org/10.1007/978-981-13-0032-5_15

15.1 Introduction

Wheat (*Triticum aestivum* L.) is an important food crop of the world. Among cereals, it occupies top slot because of its highest consumer acceptance owing to higher nutritive value, protein and gluten content. Among major cereals, globally wheat ranks first in area and production and contributes more calories and proteins to the world's human diet than any other cereals. The total area under wheat in India is 30.47 mha with a production of 95.85 m tonnes and productivity 2.8 tonnes ha^{-1} (Anonymous 2016). India is the second largest producer of wheat in the world after China, and the crop has provided the fastest pace of agrarian growth to the Indian agriculture.

In most of wheat growing regions of northern India, there is late sowing of crop; as a result crop faces low temperature stress which significantly reduces germination and subsequent seedling emergence. Direct result of low temperature effects is on cellular macromolecules, which leads to slowing of metabolism, reduced activity of amylase enzyme, solidification of cell membranes and loss of membrane functions (Joshi et al. 2007). Low temperature stress has also been known to severely inhibit reproductive development in wheat crop.

Nitric oxide (NO), a highly reactive, membrane permeable free radical, is a widespread intercellular and intracellular messenger with a broad spectrum of regulatory functions in many physiological processes (Arasimowicz and Floryszak-Wieczorek 2007). In the past 20 years, NO has been reported to be involved in various key physiological processes of plants, including seed germination, leaf senescence, ethylene emission, stomatal closure and various plant responses to biotic and abiotic stresses (Chen et al. 2010). Exogenously applied sodium nitroprusside (SNP), as an NO donor, can enhance elongation growth in maize root segments, and application of methylene blue (MB), which inhibits NO production and/or NO action in plants, could reverse these responses (Kumar et al. 2010).

NO has emerged as a key signalling molecule in plants during the last decade, and its role has been implicated in number of physiological and developmental processes as well as response to abiotic stresses including heat and cold stress. In recent years, NO has been shown to be involved in seed germination and reduction of seed dormancy, photomorphogenesis, leaf expansion, root growth, regulation of plant maturation, senescence, suppression of floral transition (He et al. 2004) and phytoalexin production and as an intermediate downstream of ABA signalling. NO is a free radical reactive gas with many physiological functions and alleviates the deleterious effects of ROS and establishes a stress resistance response thus conferring low temperature tolerance (Amooaghaie 2011). It has got many important functions to play in plant system like ubiquitous signal involved in diverse physiological processes that include germination, root growth, stomatal closing and adaptive response to biotic and abiotic stresses. Nitric oxide production in plants is involved in acquiring cold acclimation or cold tolerance (Chen et al. 2010). NO is produced immediately as a plant response to cold stress, and it participates in the regulation of cold-responsive gene expression (Dahal et al. 1990). The involvement of NO in imparting cold tolerance has been indicated by its exogenous application in certain cold-sensitive plant species such as maize, tomato and wheat. Its mechanism in providing protection against cold has been attributed to its antioxidative feature and suppression of peroxidative metabolism caused by stress (Bailly 2004).

Seed priming with SNP has been proved to be a promising technology to enhance rapid germination, better seedling emergence, high vigour and better yields in rice crops (Farooq et al. 2007). This technology is commonly employed to decrease the time between seed sowing and seedling emergence and for synchronization of seedling emergence. Priming induces some biochemical and physiological changes to help better and faster germination of seeds (Chung et al. 2007).

The present investigation was carried out in the laboratory of the Department of Plant Physiology, Institute of Agricultural Sciences, Banaras Hindu University, Varanasi. A detailed account of the materials and methods used and results obtained thereafter are discussed below.

15.2 Experimental Details

15.2.1 Source of Experimental Material

Disease-free, healthy seeds of wheat variety HD-2329 were procured from Division of Genetics and Plant Breeding, IARI, New Delhi. Bold and uniform seeds were selected, whereas small, discoloured, infected seeds were discarded. The seeds were then brought to the laboratory and washed with distilled water before giving priming treatments.

15.2.2 Treatment Imposition

Healthy seeds were selected for imposing the priming treatment. There were six priming treatments, viz. hydropriming (control), osmopriming (PEG), SNP priming (NO donor), methylene blue (MB) priming and MB+PEG and MB+SNP priming. Each priming treatments was performed in three replications, and each replication contained 20 seeds in a large-sized petriplates (150 × 25 mm). The seeds of respective treatments were put in a clean beaker, and priming solutions were put in almost equal volume as of seeds. The beaker was shaken to mix the priming solution with that of seeds and kept at room temperature for 16 h of priming period. After priming period was over, seeds were shade dried for another 6 h. Twenty seeds each were placed in the petriplates containing germination sheet. Distilled water was applied to all the petriplates irrespective of treatments in such a way that half of the seeds immersed in water. As per research objectives, two BOD incubators set at 25 °C (normal temperature) and 5 °C (low temperature) were kept ready. Eighteen petriplates (each treatment in three triplicates) were put in each incubator for studying the temperature effects on germination and subsequent growth. Observation for germination and other growth parameters were taken beginning from 1 day after treatment imposition. Distilled water was periodically applied in each petriplates. One hundred milligrams of seeds from each patriplates were taken for biochemical estimations.

15.2.3 Treatment Details

There were six priming treatments (T1, control, distilled water; T2, osmopriming PEG 6000 150 g L^{-1}; T3, sodium nitroprusside, SNP 100 μM; T4, methylene blue, MB 100 μM; T5,MB 100 μM + PEG 6000 150 g L^{-1}; and T6, MB 100 μM + SNP 100 μM) given to the cultivar HD-2329 in three replications each in optimum and low temperature regime.

15.2.4 Sampling Procedures and Observations

Observations were recorded on changes in germination percentage, germination rate, shoot length, root length, α-amylase activity and soluble sugar content in all the priming treatments under the normal and low temperature regime. Samples for α-amylase activity were taken after 18 h of inducing temperature treatment and that of soluble sugar were taken 48 h after the treatment.

15.2.5 Statistical Analysis

Factorial completely randomized design (FCRD) was followed, and analysis of variance was performed on the data as described by Panse and Sukhatme (1967). Critical difference values were calculated at 5% level of significance in order to compare the treatment means.

15.2.6 Results and Discussion

The effect of different priming treatments on per cent germination especially with respect to nitric oxide priming and osmopriming coupled with use of methylene blue (NO inhibitor) under low temperature is shown in Fig. 15.1. The maximum per cent germination was observed in SNP (100 μM) treatment at 25 °C, whereas minimum germination percentage was recorded in methylene blue (MB 100 μM) primed seeds at 5 °C. Among the priming treatments both at optimal (25 °C) and suboptimal temperature (5 °C), the maximum and minimum per cent germination was reported in SNP (100 μM) and MB (100 μM), respectively. Significant differences were observed in germination rate (Fig. 15.1) both under temperature regimes and among the priming treatments. The effects pertaining to shoot length (cm) showed favourable effect of NO priming post germination both at normal and low temperature treatments (Fig. 15.2). The effect on root length with respect to nitric oxide priming and osmopriming showed significant differences (Fig. 15.2). The

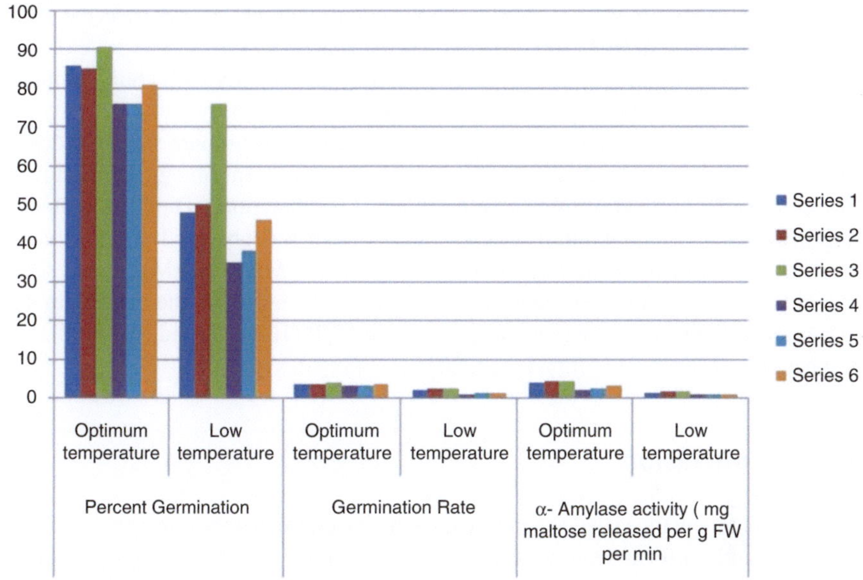

Fig. 15.1 Effect of NO priming on per cent germination, germination rate and α-amylase activity in wheat genotype both under optimum and low temperature regime. (Note: Series 1–6 refers to six priming treatments discussed in the text)

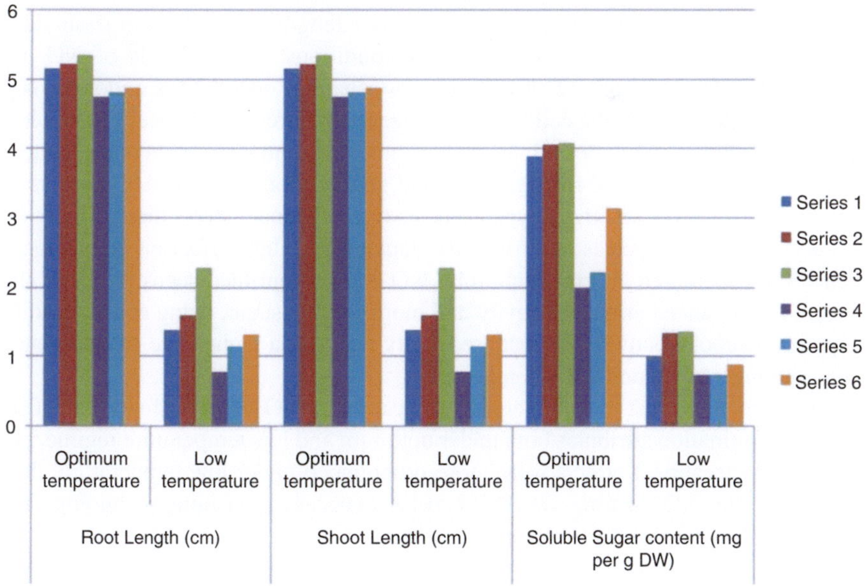

Fig. 15.2 Effect of NO priming on root length, shoot length and soluble sugar content in wheat genotype both under optimum and low temperature regime. (Note: Series 1–6 refers to six priming treatments discussed in the text)

maximum root length was observed in SNP-treated seeds at 25 °, whereas the minimum was recorded in MB (100 μM)-treated seeds at 5 °C.

Significant differences were observed in α-amylase activity both under temperature regimes and among the priming treatments (Fig. 15.1). The maximum α-amylase was observed in SNP-treated seeds at 25 °C, whereas the minimum was recorded in MB (100 μM)-treated seeds at 5 °C. The soluble sugar content (mg g^{-1} DW) showed favourable effect of NO priming post germination both at normal and low temperature treatments. The maximum soluble sugar content was observed in SNP-treated seeds at 25 °C, whereas the minimum was recorded in MB (100 μM)-treated seeds at 5 °C. Among the priming treatments both at 25 °C and 5 °C, the maximum and minimum soluble sugar content was reported in SNP (100 μM) and MB (100 μM), respectively (Fig. 15.2) (Ahmad et al. 2015).

15.3 Conclusion

Treatments to enhance seed vigour have been proven to be very effective in achieving rapid and uniform seed germination. The priming treatment may constitute a useful tool in overcoming the barriers created by abiotic stresses and, thus, assuring a high probability of successful establishment for each seed planted. Since some wheat genotypes are more sensitive to cold than others, it is possible that priming could make these genotypes germinate as fast as the cold-tolerant ones. As temperature decreased from optimum, there was a concomitant reduction in per cent germination, germination rate, shoot length and root length, as well as in α-amylase activities and sugar content. Under stress conditions, higher levels of NO are required for the maintenance of cell homeostasis, and once NO is endogenously generated or gets inside the cell from an exogenous source, it enhances stress tolerance. The tolerant genotypes might produce more NO under stress. In control, there is sufficient NO produced for signalling and germination to proceed at faster rate, and there is lesser of reactive oxygen species formed. Under stress there is increase in reactive oxygen species and oxidative damage and their effect on metabolism; henceforth, increased NO production by NO and osmoprimed seeds might be the reason for enhanced amylase activity and more soluble sugar, being able to readily support metabolic activities and consequently result in a higher rate of emergence and seedling growth under low temperature.

Our results suggest that exogenous application of NO has a profound effect on several germination attributes both under optimum and low temperature regime, but the priming treatment effects were more pronounced under low temperature, thus indicating that NO priming via SNP can be a successful priming technology for conferring low temperature tolerance in wheat.

Acknowledgement The authors would like to acknowledge the cooperation obtained from the technical staff of the Department of Plant Physiology, I. Ag. Sc. BHU for smooth conduct of the research work.

References

Ahmad MA, Sushmita, Prakash P (2015) Role of nitric oxide priming on germination attributes in wheat (*Triticum aestivum* L.) under low temperature. Environ Ecol 33:1245–1248

Amooaghaie R (2011) The effect of hydro- and osmopriming on alfalfa seed germination and antioxidant defense under high salt concentration stress. Afr J Biotechnol 33:6269–6275

Anonymous. (2016) State of Indian Agriculture (2015–16) Report, Ministry of Agriculture & Farmer's Welfare, Goverment of India

Arasimowicz M, Floryszak-Wieczorek J (2007) Nitric oxide as a bioactive signalling molecule in plant stress responses. Plant Sci 172:876–887

Bailly C (2004) Active oxygen species and antioxidants in seed biology. Seed Sci Res 14:93–107

Chen F, Wang F, Sun H, Cai Y, Mao W, Zhang G, Vincze E, Wu F (2010) Genotype-dependent effect of exogenous nitric oxide on Cd-induced changes in antioxidative metabolism, ultrastructure, and photosynthetic performance in barley seedlings (*Hordeum vulgare*). J Plant Growth Regul 29:394–408

Chung Y, Still DW, Jones RL (2007) The *Arabidopsis thaliana* aleurone layer responds to nitric oxide, gibberellin, and abscisic acid and is sufficient and necessary for seed dormancy. Plant Physiol 143:1173–1188

Dahal P, Bradford KJ, Jones RA (1990) Effects of priming and endosperm integrity on seed germination rates of tomato genotypes. I. germination at suboptimal temperatures. J Exp Bot 41:1431–1439

Farooq M, Basra SMA, Khan MB (2007) Seed priming improves growth of nursery seedlings and yield of transplanted rice. Arch Agron Soil Sci 53:311–322

He Y, Tang RH, Hao Y, Stevens RD, Cook CW et al (2004) Nitric oxide represses the Arabidopsis floral transition. Science 305:1968–1971

Joshi AK, Mishra B, Chatttarth R, Ortiz FG, Singh RP (2007) Wheat improvement in India, preset status, emerging challenges and future prospects. Euphytica 157:431–446

Kumar P, Tewari RK, Sharma PN (2010) Sodium nitroprusside-mediated alleviation of iron deficiency and modulation of antioxidant responses in maize plants. AoB Plants:plq002. https://doi.org/10.1093/aobpla/plq002.173–183

Panse VG, Sukhatme PV (1967) Statistical methods for agricultural workers. ICAR Publication New Delhi pp 167–174

Seedling Bio-priming with *Trichoderma* spp. Enhances Nitrogen Use Efficiency in Rice

16

Preeti Priya, Kartikay Bisen, Amitava Rakshit, and H. B. Singh

Abstract

Present study was undertaken to investigate the effect of seedling bio-priming with *Trichoderma harzianum* (ATCC No. PTA-3701), varied N fertilizer doses (100%, 75%, 50%, and 25% RDN), and different soil type on rice plant under greenhouse condition. Results showed that the plant growth, chlorophyll content, nutrient use efficiency, and grain N content were significantly increased in various treatments. Significantly higher growth promotion, chlorophyll content, nutrient use efficiency, and grain N content were found with recommended dose of fertilizer (RDF) NPK @ 120-60-60 kg ha^{-1} followed by seedling treated with *T. harzianum* + 3/4th N and RDF of PK. Seedling bio-priming with *T. harzianum* also enhanced the plant growth and nutrient use efficiency in rice plant. Agronomic use efficiency and physiological use efficiency were maximum in alluvial soil as compared to black soil and red soil. Results indicated the suitability of seedling bio-priming with *T. harzianum* for better plant growth and nutrient use efficiency in rice plants.

Keywords

Bio-priming · *Trichoderma* · Nitrogen use efficiency (NUE)

P. Priya · A. Rakshit
Department of Soil Science and Agricultural Chemistry, Institute of Agricultural Sciences, BHU, Varanasi, Uttar Pradesh, India

K. Bisen (✉) · H. B. Singh
Department of Mycology and Plant Pathology, Institute of Agricultural Sciences, BHU, Varanasi, Uttar Pradesh, India

© Springer Nature Singapore Pte Ltd. 2018

A. Rakshit, H. B. Singh (eds.), *Advances in Seed Priming*,
https://doi.org/10.1007/978-981-13-0032-5_16

16.1 Introduction

Nitrogen is often considered as most essential limiting nutrient for the crop growth and yield in many of world's agriculture systems. Thus, N fertilizers are one of the most important inputs for cereal production systems worldwide, and a sharp boost has occurred in consumption of N fertilizers from 11.6 million tonnes in 1961 to 104 million tonnes in 2006 (Hoang and Alauddin 2010; Mulvaney et al. 2009). Over the past five decades, consumption of mineral N fertilizers has increased about 7.4-fold, whereas crop production has only increased by only 2.4-fold (Tilman et al. 2002) indicating a sharp decline in N use efficiency. Unfortunately, the dynamic nature of N and its susceptibility to loss from plant-soil systems creates a challenge to its efficient management. Large amounts of N fertilizers were lost due to the combined effect of surface runoff, leaching, volatilization,denitrification and gaseous plant emission. Low efficient use of N fertilizers is not only responsible for higher cost of production but also caused serious environmental problems such as soil acidification, groundwater contamination, and N_2O emission. Nitrogen emissions to the air – notably those of nitrous oxide (N_2O) – are contributing to climate change. Nitrogen use efficiency (NUE) for production of cereal crop including rice, wheat, corn, barley, sorghum, oat, and rye is approximately 33%, and the rest unaccounted 67% represents a US$ 15.9 billion annual loss of N fertilizer worldwide (Raun and Johnson 1999). Therefore, maintaining agricultural productivity in such a way that minimizes the harmful effects of fertilizers on environmental is need of the hour.

Various techniques have been advocated to increase the NUE. Incorporation of various sustainable agricultural practices, including crop rotation, fertilizer use rationalization, establishment of ground cover, green manuring, and burial of crop residues, is widely recommended. Research for new improved crop varieties, through genetic modification and breeding that take up more organic or inorganic N from the soil N and utilize the absorbed N more efficiently have been carried out since decades. Legislation aimed at protecting the environment from nutrient runoff has been enacted by some governments, and policies based on this legislation are being implemented. However, a little success has been achieved in improving nitrogen use efficiency in plants. Microbial inoculants offer promising integrated solutions to agro-environmental problems through plant growth promotion, disease protection, and enhancing nutrient availability and uptake in plants. *Trichoderma* spp. are most common filamentous free-living saprophytic fungi better known for plant disease suppression and dominate the global biopesticide market (Singh 2016; Keswani et al. 2014). However, intense investigation since last decades acknowledged various traits in *Trichoderma* that promote its applications beyond crop protection. Various species of *Trichoderma* are identified which are capable of plant growth promotion, managing abiotic stresses, and improving nutrient use efficiency and uptake (Bisen et al. 2015; Mehetre and Mukherjee 2015). Beneficial microbes mediated improvement of nutrient use efficiency in plant is gaining a lot of attention worldwide in the context of soil fertility and productivity deterioration (Rakshit et al. 2015a, b). *Trichoderma* spp. can enhance crop productivity by virtue of both enhanced decomposition of biomass and improving uptake of inorganic fertilizers. The main objective of present work is to evaluate the response of seedling bio-priming with *Trichoderma* on nitrogen use efficiency in rice.

16.2 Materials and Methods

16.2.1 Experimental Soils

The soils of three orders, entisol, inceptisol, and alfisol, were collected from different locations. Alluvial soil sample was collected from the Agricultural Research Farm, Institute of Agricultural Sciences, Banaras Hindu University, Varanasi. The red soil was collected from the Semara village, Mirzapur, and black soil was collected from Arzilime block, Varanasi. Physiochemical properties of experimental soils were analyzed (Table 16.1).

16.2.2 Plant Material

Seeds of rice variety PHB-71 were obtained from the market. Seeds were surface sterilized with 0.1% $HgCl_2$ for 30 s, then washed with sterile distilled water thrice, and sown in nursery.

16.2.3 Trichoderma Strain

Trichoderma harzianum (ATCC No. PTA-3701) were obtained from the culture repository of Plant Health Clinic Laboratory of the Department of Mycology and Plant Pathology, Institute of Agricultural Sciences, Banaras Hindu University, Varanasi, India.

Table 16.1 Physiochemical properties of experimental soils

Physiochemical parameters	Alluvial soil	Black soil	Red soil
pH	7.2	7.4	6.4
Bulk density (g/cm^3)	1.39	1.52	1.29
Particle density	2.56	2.61	2.49
EC (dSm^{-1}) 0.39	0.44	00.61	0.31
CEC (Cmol$_{(p+)}$ kg^{-1})	28.98	30.92	18.0
Organic carbon (%)	0.37	0.39	0.31
Available N (kg ha^{-1})	229	236	173
Available P (kg ha^{-1})	17	13	8
Available K (kg ha^{-1})	230	235	109

16.2.4 Seedling Bio-priming with Trichoderma Formulation

Liquid formulation of *T. harzianum* containing approximately 10^8 spores ml^{-1} was used for seedling treatment. Roots of 40-day-old rice seedlings prepared for transplanting were dipped in liquid formulation for 30 min, and then the seedlings were kept in shade for half an hour before transplanting. Control seedlings were only treated with water.

16.2.5 Pot Experiments

The experiments have been conducted on three different soil types at the Institute of Agricultural Sciences, BHU. Seedlings of rice cultivar treated with *Trichoderma* were grown with different level of N fertilizer dose. Five treatments were applied:

T_1- Control N: P: K @ 0: 0: 0 Kg h^{-1}
T_2- Recommended dose of N: P: K @ 120: 60: 60
T_3- Seedling treated with *T. harzianum* + 3/4th N: P: K @ 90: 60: 60
T_4- Seedling treated with *T. harzianum* + 1/2 N: P: K @ 60: 60: 60
T_5- treated with *T. harzianum* + 1/4th N: P: K @ 30: 60: 60

16.2.6 Study of Effect of Seedling Bio-priming on Growth Parameter and Yield Attributes of Plants

Growth parameters including height of plants, leaf area, and dry weight were measured at 30, 60, and 90 days after transplanting. Yield attributes including number of tiller pot^{-1} and length of panicle of randomly selected plants from each treatment were recorded. Test weight and yield pot^{-1} were recorded after harvesting.

16.2.7 Estimation of Nitrogen in Plant and Grains Sample

N content in plant and grain sample was determined by modified Kjeldahl method as outlined by Peng et al. (1996). N content was calculated by the following formula:

$$\text{Percent N in plant material} = \frac{0.02 \times T \times 0.014 \times 50 \times 50}{5 \times 50}$$

T = sample reading – blank reading

16.2.8 Chlorophyll Content

The chlorophyll content of leaves was calculated as described by Arnon (1949). Leaf samples were obtained from 60-day-old rice crop. 0.5 g of leaf sample was crushed with 80% acetone in mortar and pestle and filtered through Whatman No. 1 filter. Supernatant was collected, and volume raised up to 50 ml. Absorbance was recorded at 645 nm using spectrophotometer (RayLEIGH UV-2601). Total chlorophyll content was determined as follows:

$$\text{Total chlorophyll}\left(\text{mg g}-1\,\text{fresh wt.}\right)=\left(20.2\times \text{A}645\right)+\left(8.02\times A663\right)\times\frac{\text{V}}{\text{W}}\times 1/1000$$

16.2.9 Estimation of N Use Efficiency

Nitrogen use efficiency (NUE) formulas

$$\text{Physiological use efficiency}\left(\text{PUE}\right)=\frac{\text{Yield F kg}-\text{Yield C kg}}{\text{Nutrient uptake F kg}-\text{Nutrient uptake C kg}}$$

$$\text{Agronomic use efficiency}\left(\text{AUE}\right)=\frac{\text{Yield F kg}-\text{Yield C kg}}{\text{Quantity of nutrient applied}\left(\text{kg}\right)}$$

where F is plant receiving fertilizers and C is plants receiving no fertilizer

16.3 Results (Tables 16.2, 16.3 and 16.4)

16.4 Discussion

16.4.1 Chlorophyll Content

Chlorophyll content in rice leaves was significantly affected by combination of N-fertilizer doses with seedling bio-priming by *T. harzianum*. Maximum chlorophyll content was recorded in plants shown in alluvial soil (2.64 mg g^{-1}) followed by black (2.23 mg g^{-1}) and red soil (1.5 mg g^{-1}). Treatment T2 showed significantly higher chlorophyll content in all soil type in with maximum in alluvial soil (3.46 mg g^{-1}). Singh and Singh (2011) has reported that the alluvial soil showed better

Table 16.2 Effect of soil types, seedling biopriming, graded N dose application on plant height

Treatments	30 days			60 days			90 days		
	Alluvial soil	Red soil	Black soil	Alluvial soil	Red soil	Black soil	Alluvial soil	Red soil	Black soil
T1	55.92±2.17^{hi}	50.05±1.65^{j}	52.0±1.95j	82.86±1.5^{ef}	70.86±1.5^{j}	73.81±1.0^{hi}	89.36±2.13d	80.83±1.29h	85.85±2.0^{ef}
T2	80.55±2.15^{a}	74.22±1.45^{bc}	77.42±1.3^{ab}	98.15±1.1^{a}	90.18±1.0^{cd}	93.28±0.94^{b}	103.44±2.35a	94.36±0.8c	97.83±1.9^{b}
T3	74.51±2.14^{bc}	70.30±0.79^{d}	71.24±1.8^{cd}	93.66±1.5^{b}	81.20±1.9^{f}	87.83±2.5^{cd}	100.85±1.85a	90.18±1.7d	95.47±1.5^{bc}
T4	69.17±2.0^{de}	63.0±1.9^{g}	66.72±1.98^{ef}	87.67±2.5^{cd}	76.47±1.5^{gh}	80.28±1.0^{f}	95.19±1.4bc	83.98±0.4fg	94.31±0.7^{c}
T5	64.96±2.90^{fg}	52.81±1.7^{ij}	56.33±0.9^{h}	85.19±1.9^{de}	73.57±1.6^{i}	77.52±1.0^{g}	91.13±1.9d	81.65±0.8gh	88.37±1.9^{de}

Table 16.3 Effect of soil types, seedling bio-priming, and graded N dose application on chlorophyll content

Treatments	Alluvial soil	Red soil	Black soil
T1	1.43 ± 0.1^{a}	0.66 ± 0.1^{a}	1.06 ± 0.1^{a}
T2	3.46 ± 0.2^{d}	2.83 ± 0.1^{d}	3.03 ± 0.1^{d}
T3	3.26 ± 1.0^{d}	2.43 ± 0.8^{c}	2.83 ± 0.9^{c}
T4	2.83 ± 0.1^{c}	2.16 ± 0.1^{c}	2.53 ± 0.1^{c}
T5	2.23 ± 0.1^{b}	1.5 ± 0.8^{b}	1.7 ± 0.1^{b}

$P \leq 0.05$

potential with graded N fertilizer application and seed bio-priming by *T. harzianum*. In present study, 100% RDF-treated plants showed maximum chlorophyll content followed by seedling treated with *T. harzianum* + 3/4th N and RDF of P and K fertilizers in comparison to control.

16.5 Nitrogen Uptake

Application of 100% RDF in pots resulted in significantly higher N uptake in plants at all growth stages in rice crop as compared to control. The N uptake in rice plants was gradually increased and reaches maximum at 90 DAS in all soil type under T2 followed by T3>T4>T5>T1. Meena et al. (2016) has reported that the N accumulation in wheat plants was increased with N fertilizer doses and was recorded maximum at 100% RDF. Significantly higher N uptake was noticed at 60 DAS under following treatment T2 >T3> T4 >T5 >T1.

At 90 DAS increased N uptake was recorded and ranged from 6.4 to 61.86 and 3.79–24.93 and 4.47–50.78 mg plant^{-1} under alluvial, red, and black soil, respectively. Higher N uptake by plants may be attributed to the ability of *Trichoderma* to facilitate the plant growth, nutrient availability, and plant health (Meena et al. 2016; Adesemoye et al. 2009; Singh and Singh 2011).

16.5.1 Grain Nitrogen Content

Significantly higher grain nitrogen content was observed in treatment T2 under alluvial, black, and red soil, respectively, in comparison to control (Fig. 16.1). Nitrogen content in grain of rice was also influenced by different soil types. Meena et al. (2016) have reported maximum grain N content in wheat plant under alluvial soil treated with 100% RDF. In present study T3 seedling treated with *T. harzianum* + 3/4th N:P:K @ 90:60:60 showed significant increase in grain N content in comparison to control.

Table 16.4 Effect of soil types, seedling bio-priming, and graded N dose application on nitrogen uptake by plant (mg $plant^{-1}$) at different growth stages of rice

	30 days			60 days			90 days		
	Alluvial soil	Red soil	Black soil	Alluvial soil	Red soil	Black soil	Alluvial soil	Red soil	Black soil
T1	14.0 ± 1.2[a]	8.6 ± 0.6[a]	10.9 ± 0.06[a]	39.2 ± 1.0[a]	25.30 ± 0.7[a]	34.58 ± 1.1[a]	6.47 ± 0.5[a]	3.79 ± 0.8[a]	4.47 ± 0.6[a]
T2	100.4 ± 1.7[e]	68.4 ± 0.7[e]	85.4 ± 1.1[e]	150.3 ± 1.3[e]	95.32 ± 0.7[e]	136.61 ± 1.1[e]	61.86 ± 1.2[e]	24.93 ± 1.3[d]	37.38 ± 0.8[c]
T3	90.7 ± 1.0[d]	61.2 ± 1.3[d]	70.18 ± 1.3[d]	119.5 ± 0.9[d]	75.65 ± 1.1[d]	102.05 ± 1.2[d]	27.43 ± 0.6[d]	21.78 ± 1.3[c]	26.59 ± 1.1[b]
T4	66.3 ± 0.7[c]	49.8 ± 0.8[c]	56.1 ± 1.2[c]	70.8 ± 0.9[c]	52.23 ± 0.8[c]	82.05 ± 1.3[c]	20.73 ± 1.3[c]	11.6 ± 0.7[b]	50.78 ± 0.9[d]
T5	37.5 ± 1.7[b]	26.4 ± 0.9[b]	33.9 ± 1.1[b]	58.1 ± 0.8[b]	46.33 ± 0.8[b]	50.45 ± 0.9[b]	11.59 ± 1.2[b]	9.49 ± 0.7[b]	27.93 ± 1.3[b]

$P \leq 0.05$

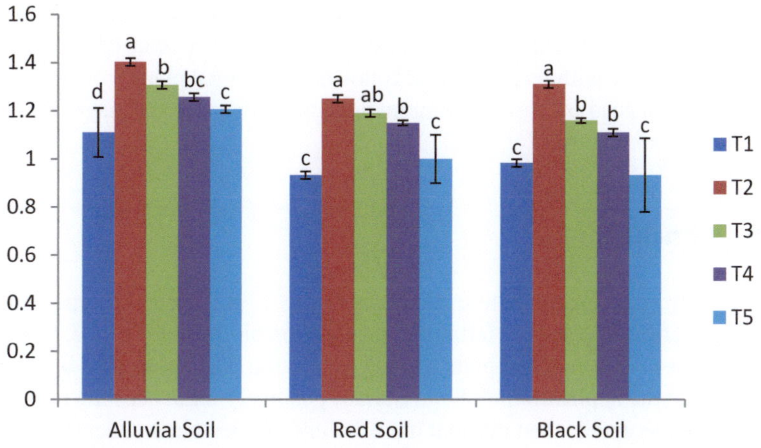

Fig. 16.1 Effect of soil types, seedling bio-priming, and graded N dose application on grain N content

Table 16.5 Effect of soil types, seedling bio-priming, and graded N dose application on agronomic use efficiency and physiological use efficiency

	Agronomic use efficiency			Physiological use efficiency		
	Alluvial soil	Red soil	Black soil	Alluvial soil	Red soil	Black soil
T2	10.61±0.5[a]	5.57±0.6[b]	8.54±1.0[a]	0.37±0.012[a]	0.25±0.01[a]	0.33±0.07[a]
T3	12.73±0.7[b]	5.62±0.4[b]	9.87±0.5[a]	0.56±0.012[b]	0.31±0.03[a]	0.45±0.01[b]
T4	15.20±0.8[c]	5.45±1.1[b]	10.48±0.9[a]	0.65±0.012[c]	0.42±0.03[b]	0.50±0.01[b]
T5	17.99±1.1[d]	3.75±0.3[a]	8.62±0.6[a]	0.71±0.012[d]	0.41±0.05[b]	0.51±0.05[b]

16.6 Agronomic Use Efficiency (AUE)

Present study clearly showed that AUE was significantly influenced by soil type and different level of N fertilizer. Higher AUE was observed in plants grown in alluvial soil followed by black soil and red soil (Table 16.5). Highest AUE in alluvial soil was recorded under T5 (17.99) followed by black soil (15.20) and red soil (12.73), respectively. It may be due to the combined application of graded N fertilizer and *T. harzianum* (Meena et al. 2016; Martinez-Medina et al. 2011; Tripathi et al. 2013).

16.7 Physiological Use Efficiency (PUE)

Pot with alluvial soil showed maximum PUE (0.57 mg g^{-1}) followed by black soil (0.44 mg g^{-1}) and red soil (0.34 mg g^{-1}) of rice plant in comparison to control. PUE of rice plants ranged from 0.37 to 0.71, 0.25–0.41, and 0.33–0.51 in alluvial, red, and black soil, respectively. Data showed in Table 16.5 indicated that in alluvial soil plant under T5 showed maximum PUE (0.71) followed by T4 (0.65) and T5 (0.56). Same pattern was followed in different soil type. Higher PUE under alluvial soil may be due to the better growth condition provided by soil and proliferation of microbes with different doses of fertilizers (Meena et al. 2016). Higher value of PUE indicates nutrient sufficiency with reference to P followed by optimal plant growth and stress mitigation influenced by various combinations of reccomended dosage of fertilizer and *T. harzianum* (Murphy et al. 2013).

16.8 Conclusions

Present study was conducted to evaluate the effect of different soil type and combined application of graded N fertilizer and seedling bio-priming with *T. harzianum* on nutrient use efficiency and plant growth promotion. Data presented in this study indicated that the application of *T. harzianum* positively influences the growth promotion, nutrient uptake, and nutrient use efficiency in rice plant under different soil conditions. One hundred percent RDF of NPK was the most effective treatment for rice plants in all soil types. Seedling bio-priming with *T. harzianum* + 3/4th N and RDF of P&K treatment was found to be most effective in comparison to control and other combined graded N and *T. harzianum* application.

References

Adesemoye AO, Torbert HA, Kloepper JW (2009) Plant growth-promoting rhizobacteria allow reduced application rates of chemical fertilizers. Microb Ecol 58:921–929

Arnon DI (1949) Copper enzymes in isolated chloroplasts. Polyphenoloxidase in *Beta vulgaris*. Plant Physiol 24:1

Bisen K, Keswani C, Mishra S, Saxena A, Rakshit A, Singh HB (2015) Unrealized potential of seed biopriming for versatile agriculture. In: Rakshit A, Singh HB, Sen A (eds) Nutrient use efficiency: from basics to advances. Springer, New Delhi, pp 193–206

Hoang VN, Alauddin M (2010) Assessing the eco-environmental performance of agricultural production in OECD countries: the use of nitrogen flows and balance. Nutr Cycl Agroecosyst 87:353–336

Keswani C, Mishra S, Sarma BK, Singh SP, Singh HB (2014) Unraveling the efficient application of secondary metabolites of various *Trichoderma*. Appl Microbiol Biotechnol 98:533–544

Martínez-Medina A, Martınez-Medina RA, Jose AP (2011) Performance of a *Trichoderma harzianum* bentonite–vermiculite formulation against fusarium wilt in seedling nursery melon plants. Hortscience 44:2025–2027

Meena SK, Rakshit A, Meena VS (2016) Effect of seed bio-priming and N doses under varied soil type on nitrogen use efficiency (NUE) of wheat (*Triticum aestivum* L.) under greenhouse conditions. Biocatalyst Agric Biotechnol 6:668–675

Mehetre ST, Mukherjee PK (2015) *Trichoderma* improves nutrient use efficiency in crop plants. In: Rakshit A, Singh HB, Sen A (eds) Nutrient use efficiency: from basics to advances. Springer, India, pp 173–180

Mulvaney RL, Khan SA, Ellsworth TR (2009) Synthetic nitrogen depleted soil nitrogen: a global dilemma for sustainable cereal production. J Environ Qual 38:2295–2314

Murphy PNC, O'Connell K, Watson S, Watson CJ, Humphreys J (2013) Seasonality of nitrogen uptake, apparent recovery of fertilizer nitrogen and background nitrogen supply in two Irish grassland soils. Ir J Agric Food Res 52:17–38

Peng S, Garcia FV, Laza RC, Sanico AL, Visperas RM, Cassman KG (1996) Increased N- se efficiency using a chlorophyll meter on high-yielding irrigated rice. Field Crop Res 47:243–252

Rakshit A, Singh HB, Sen A (2015a) Nutrient use efficiency: from basics to advances. Springer, New Delhi

Rakshit A, Sunita K, Pal S, Singh A, Singh HB (2015b) Bio-priming mediated nutrient use efficiency of crop species. In: Rakshit A, Singh HB, Sen A (eds) Nutrient use efficiency: from basics to advances. Springer, New Delhi, pp 181–191

Raun WR, Johnson GV (1999) Improving nitrogen use efficiency for cereal production. Agron J 91:357–363

Singh HB (2016) Seed biopriming: a comprehensive approach towards agricultural sustainability. Indian Phytopathol 69:203–209

Singh SP, Singh HB (2011) Effect of consortium of *Trichoderma harzianum* isolates on growth attributes and *Sclerotinia sclerotiorum* rot of brinjal. Veg Sci 39:144–148

Tilman D, Cassman KG, Matson PA, Naylor R, Polasky S (2002) Agricultural sustainability and intensive production practices. Nature 418:671–677

Tripathi P, Singh PC, Mishra A, Chaudhry V, Mishra S, Tripathi RD, Nautiyal CS (2013) *Trichoderma* inoculation ameliorates arsenic induced phytotoxic changes in gene expression and stem anatomy of chickpea (*Cicer arietinum*). Ecotoxicol Environ Saf 89:8–14

MIX
Papier aus verantwortungsvollen Quellen
Paper from responsible sources
FSC® C105338

If you have any concerns about our products,
you can contact us on
ProductSafety@springernature.com

In case Publisher is established outside the EU,
the EU authorized representative is:
Springer Nature Customer Service Center GmbH
Europaplatz 3, 69115 Heidelberg, Germany

Printed by Libri Plureos GmbH
in Hamburg, Germany